Preparation of Radiopharmaceuticals and Their Use in Drug Development

Preparation of Radiopharmaceuticals and Their Use in Drug Development

Editor

Svend Borup Jensen

Basel • Beijing • Wuhan • Barcelona • Belgrade • Novi Sad • Cluj • Manchester

Editor
Svend Borup Jensen
Department of Nuclear Medicine
University Hospital Aalborg
Aalborg C
Denmark

Editorial Office
MDPI
St. Alban-Anlage 66
4052 Basel, Switzerland

This is a reprint of articles from the Special Issue published online in the open access journal *Molecules* (ISSN 1420-3049) (available at: https://www.mdpi.com/journal/molecules/special_issues/preparation-radiopharmaceuticals).

For citation purposes, cite each article independently as indicated on the article page online and as indicated below:

Lastname, A.A.; Lastname, B.B. Article Title. *Journal Name* **Year**, *Volume Number*, Page Range.

ISBN 978-3-03928-627-0 (Hbk)
ISBN 978-3-7258-0000-1 (PDF)
doi.org/10.3390/books978-3-7258-0000-1

© 2024 by the authors. Articles in this book are Open Access and distributed under the Creative Commons Attribution (CC BY) license. The book as a whole is distributed by MDPI under the terms and conditions of the Creative Commons Attribution-NonCommercial-NoDerivs (CC BY-NC-ND) license.

Contents

About the Editor vii

Preface ix

Rodrigo Teodoro, Matthias Scheunemann, Winnie Deuther-Conrad, Barbara Wenzel, Francesca Maria Fasoli, Cecilia Gotti, et al.
A Promising PET Tracer for Imaging of α_7 Nicotinic Acetylcholine Receptors in the Brain: Design, Synthesis, and *in Vivo* Evaluation of a Dibenzothiophene-Based Radioligand
Reprinted from: *Molecules* **2015**, *20*, 18387–18421, doi:10.3390/molecules201018387 1

Thomas Ebenhan, Mariza Vorster, Biljana Marjanovic-Painter, Judith Wagener, Janine Suthiram, Moshe Modiselle, et al.
Development of a Single Vial Kit Solution for Radiolabeling of ^{68}Ga-DKFZ-PSMA-11 and Its Performance in Prostate Cancer Patients
Reprinted from: *Molecules* **2015**, *20*, 14860–14878, doi:10.3390/molecules200814860 36

Gábor Máté, Jakub Šimeček, Miroslav Pniok, István Kertész, Johannes Notni, Hans-Jürgen Wester, et al.
The Influence of the Combination of Carboxylate and Phosphinate Pendant Arms in 1,4,7-Triazacyclononane-Based Chelators on Their ^{68}Ga Labelling Properties
Reprinted from: *Molecules* **2015**, *20*, 13112–13126, doi:10.3390/molecules200713112 55

Clinton Rambanapasi, Nicola Barnard, Anne Grobler, Hylton Buntting, Molahlehi Sonopo, David Jansen, et al.
Dual Radiolabeling as a Technique to Track Nanocarriers: The Case of Gold Nanoparticles
Reprinted from: *Molecules* **2015**, *20*, 12863–12879, doi:10.3390/molecules200712863 70

Susann Schröder, Barbara Wenzel, Winnie Deuther-Conrad, Rodrigo Teodoro, Ute Egerland, Mathias Kranz, et al.
Synthesis, ^{18}F-Radiolabelling and Biological Characterization of Novel Fluoroalkylated Triazine Derivatives for *in Vivo* Imaging of Phosphodiesterase 2A in Brain via Positron Emission Tomography
Reprinted from: *Molecules* **2015**, *20*, 9591–9615, doi:10.3390/molecules20069591 87

Benjamin H. Rotstein, Steven H. Liang, Vasily V. Belov, Eli Livni, Dylan B. Levine, Ali A. Bonab, et al.
Practical Radiosynthesis and Preclinical Neuroimaging of [^{11}C]isradipine, a Calcium Channel Antagonist
Reprinted from: *Molecules* **2015**, *20*, 9550–9559, doi:10.3390/molecules20069550 112

Chunxiong Lu, Quanfu Jiang, Minjin Hu, Cheng Tan, Huixin Yu and Zichun Hua
Preliminary Biological Evaluation of ^{18}F-FBEM-Cys-Annexin V a Novel Apoptosis Imaging Agent
Reprinted from: *Molecules* **2015**, *20*, 4902–4914, doi:10.3390/molecules20034902 122

Catharina Neudorfer, Amir Seddik, Karem Shanab, Andreas Jurik, Christina Rami-Mark, Wolfgang Holzer, et al.
Synthesis and *in Silico* Evaluation of Novel Compounds for PET-Based Investigations of the Norepinephrine Transporter
Reprinted from: *Molecules* **2015**, *20*, 1712–1730, doi:10.3390/molecules20011712 135

Manuela Kuchar and Constantin Mamat
Methods to Increase the Metabolic Stability of ^{18}F-Radiotracers
Reprinted from: *Molecules* **2015**, *20*, 16186–16220, doi:10.3390/molecules200916186 154

Irina Velikyan
^{68}Ga-Based Radiopharmaceuticals: Production and Application Relationship
Reprinted from: *Molecules* **2015**, *20*, 12913–12943, doi:10.3390/molecules200712913 . . . . **189**

About the Editor

Svend Borup Jensen

Following graduation, Svend worked as an organic chemist and as an analysis chemist for private companies. However, his main career has been in radiochemistry, first at Aarhus University Hospital as head of production, and since 2009 as the head of radiochemistry/QP (Qualified Person) at Aalborg University Hospital. Svend's responsibility as QP is for the production and release of radiopharmaceutical drug used in human for diagnosis. He has also been strongly involved in designing and qualifying the new and heavily enlarged cleanroom facilities for the Dept. of Nuclear Medicine at "Nyt Aalborg Universitetshospital" (NAU) to ensure that the facilities are in compliance with the legislation for production of sterile pharmaceuticals and contains the suitable apparatus for radioactive drug production.

Research in the field of radioactive drugs is very versatile. Most of Svend research has the synthesis of new radiopharmaceutical drug or the optimization of known syntheses as the focal point, but the analysis and identification of the radioactive drugs and the by-products has also received attention.

As a researcher Svend has authored or co-authored 49 research papers in peer-reviewed journals, 3 book chapters, and 11 articles in Danish professional journals with a more educational angle.

Preface

The purpose of this Special Issue was to bring together research and review papers on the use of radiopharmaceuticals in drug development.

Labeling a potential drug with a radionuclide for imaging and injecting radiopharmaceuticals into animals or humans can immediately provide information about the in vivo destiny of the compound through the application of PET or SPECT scanners. Does a radioactive drug accumulate in the site where the drug is supposed to have its effect or does it not reach the target organ or tumor at all? If radionuclides are employed throughout the whole process of developing new drugs, especially if the results are negative and the potential drug can be discarded early on in the process, considerable effort and money can be saved. An alternative way of obtaining information about the in vivo destiny/effect of a potential drug is to imagine, for example, a target organ or a tumor with radiopharmaceuticals before and after treatment with a drug. This technique will most likely become widespread when it comes to personalized medicine when a scan with a radiopharmaceutical can be used to determine whether a certain drug has its desired effect or if the physician has to employ another method of treatment.

Determining how a drug is metabolized is an integrated part of drug development. Labeling a drug with a radionuclide and providing it with a radioactive tag can help in subsequently determining its location and quantifying metabolites. If the metabolites can be extracted, the solutions can be examined by means of common analytical measurements.

Using ionizing radiation in the treatment of tumors is also an application for radiopharmaceuticals which is expected to expand greatly in the years to come. By labeling a specific compound with a radionuclide used for diagnostics, we can determine whether it accumulates in the tumor. The same compound is thereafter combined with a radionuclide for treatment.

This Special Issue contains seven original research papers, one communication, and two reviews, all within the area of radiopharmaceuticals in drug development.

Svend Borup Jensen
Editor

Article

A Promising PET Tracer for Imaging of α₇ Nicotinic Acetylcholine Receptors in the Brain: Design, Synthesis, and *in Vivo* Evaluation of a Dibenzothiophene-Based Radioligand

Rodrigo Teodoro [1,†], Matthias Scheunemann [1,†], Winnie Deuther-Conrad [1,†,*], Barbara Wenzel [1], Francesca Maria Fasoli [2], Cecilia Gotti [2], Mathias Kranz [1], Cornelius K. Donat [1], Marianne Patt [3], Ansel Hillmer [4], Ming-Qiang Zheng [4], Dan Peters [5], Jörg Steinbach [1], Osama Sabri [3], Yiyun Huang [4] and Peter Brust [1]

1. Helmholtz-Zentrum Dresden-Rossendorf, Institute of Radiopharmaceutical Cancer Research, Permoserstraße 15, Leipzig 04318, Germany; E-Mails: r.teodoro@hzdr.de (R.T.); m.scheunemann@hzdr.de (M.S.); b.wenzel@hzdr.de (B.W.); m.kranz@hzdr.de (M.K.); cdonat@nru.dk (C.K.D.); j.steinbach@hzdr.de (J.S.); p.brust@hzdr.de (P.B.)
2. Consiglio Nazionale delle Ricerche, Institute of Neuroscience, Biometra-Institute University of Milan, Via Luigi Vanvitelli 32, Milano 20129, Italy; E-Mails: f.fasoli@in.cnr.it (F.M.F.); c.gotti@in.cnr.it (C.G.)
3. Department of Nuclear Medicine, University Hospital Leipzig, Liebigstraße 18, Leipzig 04103, Germany; E-Mails: marianne.patt@medizin.uni-leipzig.de (M.P.); osama.sabri@medizin.uni-leipzig.de (O.S.)
4. PET Center, Yale University, P.O. Box 208048, 801 Howard Avenue, New Haven, CT 06520-8048, USA; E-Mails: ansel.hillmer@yale.edu (A.H.); ming-qiang.zheng@yale.edu (M.-Q.Z.); henry.huang@yale.edu (Y.H.)
5. Dan PET AB, Rosenstigen 7, Malmö SE-21619, Sweden; E-Mail: info@danpet.eu

† These authors contributed equally to this work.

* Author to whom correspondence should be addressed; E-Mail: w.deuther-conrad@hzdr.de; Tel.: +49-341-234-179-4613; Fax: +49-341-234-179-4699.

Academic Editor: Svend Borup Jensen

Received: 15 July 2015 / Accepted: 28 September 2015 / Published: 9 October 2015

Abstract: Changes in the expression of α₇ nicotinic acetylcholine receptors (α₇ nAChRs) in the human brain are widely assumed to be associated with neurological and neurooncological processes. Investigation of these receptors *in vivo* depends on the availability of imaging agents such as radioactively labelled ligands applicable in positron emission tomography (PET). We report on a series of new ligands for α_7 nAChRs designed by the combination of dibenzothiophene dioxide as a novel hydrogen bond acceptor functionality with diazabicyclononane as an established cationic center. To assess the structure-activity relationship (SAR) of this new basic structure, we further modified the cationic center systematically by introduction of three different piperazine-based scaffolds. Based on *in vitro* binding affinity and selectivity, assessed by radioligand displacement studies at different rat and human nAChR subtypes and at the structurally related human 5-HT$_3$ receptor, we selected the compound 7-(1,4-diazabicyclo[3.2.2]nonan-4-yl)-2-fluorodibenzo-[*b,d*]thiophene 5,5-dioxide (**10a**) for radiolabeling and further evaluation *in vivo*. Radiosynthesis of [^{18}F]**10a** was optimized and transferred to an automated module. Dynamic PET imaging studies with [^{18}F]**10a** in piglets and a monkey demonstrated high uptake of radioactivity in the brain, followed by washout and target-region specific accumulation under baseline conditions. Kinetic analysis of [^{18}F]**10a** in pig was performed using a two-tissue compartment model with arterial-derived input function. Our initial evaluation revealed that the dibenzothiophene-based PET radioligand [^{18}F]**10a** ([^{18}F]DBT-10) has high potential to provide clinically relevant information about the expression and availability of α_7 nAChR in the brain.

Keywords: α_7 nAChR; pharmacophore; positron emission tomography; neuroimaging; fluorine-18

1. Introduction

The long-standing interest in molecular imaging of nicotinic acetylcholine receptors (nAChRs) is driven by findings from preclinical and clinical studies, which have demonstrated that dysfunction of neuronal nAChRs is involved in the pathophysiology of many disorders [1]. The nAChRs are a superfamily of ligand-gated ion channels that consist of a pentamer of protein subunits. In the mammalian brain, different subunits assemble with much diversity; however, the α_7 and $\alpha_4\beta_2$ nAChR subtypes predominate [2]. The α_7 subunit is highly expressed in the hippocampus and hypothalamus, as well as in cell types where receptor-mediated ion currents have not been reported. Therefore, α_7 nAChR-mediated effects such as cognitive enhancement [3] may depend on the ionotropic signaling while other effects may be independent of ion-channel currents [4]. Altered functional availability of α_7 nAChR has been implicated in a number of diseases of the human central nervous system (CNS), including Alzheimer's and Parkinson's disease, schizophrenia and autism, as well as in lung cancer and heart disease [5]. With highly selective ligands, α_7 nAChRs are approachable targets not only for therapeutic interventions, but also for non-invasive imaging such as Positron Emission Tomography (PET), which can be used to investigate disease pathophysiology *in vivo* and support drug development.

While the cationic pharmacophore sufficient for full activation of all neuronal nAChRs is the tetramethylammonium cation, several α_7 nAChR selective motifs have been identified based on the observation that an additional hydroxyl group present in choline or quinuclidinol activates largely the homomeric α7 but not the heteromeric nAChRs [6]. So far, at least nine structurally related families of high-affinity small-molecule ligands of the α7 nAChR have been characterized [6,7], including anabaseine-, pyrrolidine-, and diazabicyclononane-based PET radiotracers such as [^{11}C]GTS-21 [8], [^{11}C]A-844606 and [^{11}C]A-582941 [9], [^{11}C]CHIBA-1001 [10], [^{11}C]NS14492 [11], [^{18}F]NS10743 [12] and [^{18}F]NS14490 [13]. Recent progress in α7 nAChR PET imaging comes from the discovery of the α7 nAChR selective binding of the antiviral interferon inducer tilorone [14], containing a tricyclic fluorenone nucleus. Further structural modification by replacing this fluorenone moiety with dibenzothiophene sulfone as alternative hydrogen bond acceptor (HBA) functionality and introduction of 1,4-diazabicyclo[3.2.2]nonane as cationic center resulted in a novel compound with markedly increased binding affinity for α7 nAChR (K_i = 56 nM for tilorone and 0.023 nM for compound **48**, Figure 1, A1) [15]. This discovery by Schrimpf *et al.*, in 2012 prompted us to develop a series of novel derivatives as references for ^{18}F-labeled ligands of α7 nAChRs and to further investigate the SAR of this new pharmacophore. At the time of working on the radiolabeling of the most promising ligand for PET imaging studies [16], researchers from the Johns Hopkins University had successfully completed the radiosynthetic work on respective ligands [17]. While the evaluation of the resulting [^{18}F]ASEM in baboons and humans was published [18–20], we felt encouraged by the inherent potential of the Abbott lead structure to continue our research on ^{18}F-labelled diazabicyclononane-containing α7 nAChR ligands [16,21] by modifying both the HBA functionality and the cationic center.

Besides the elaboration of a more effective approach towards compound [^{18}F]**10a** ([^{18}F]DBT-10), which we have selected for further development [16] and is identical to compound [^{18}F]**7c** reported by Gao *et al.* [17], we have synthesized a series of derivatives of this lead structure to further analyze structural elements that determine the affinity of potential α7 nAChR ligands to other pentameric ligand-gated ion channels such as heteromeric nicotinic or 5-HT3 receptors. It has been shown previously that the affinity for α4β2 nAChR of dibenzothiophene-diazabicyclononanes (e.g., [^{18}F]ASEM, K_i = 562 nM [17]) is notably higher than that of oxadiazolyl-substituted diazabicyclononanes (e.g., [^{18}F]NS10743 K_i > 10 μM [12]). Therefore, we replaced the cationic center with alternative tertiary amine motifs. Insertion of three different diazabicycloalkanes resulted in *N*-methyl-substituted derivatives of the ethylene- and propylene-bridged piperazines diazabicyclo[3.2.1]octane and diazabicyclo[3.3.1]nonane, respectively [22,23].

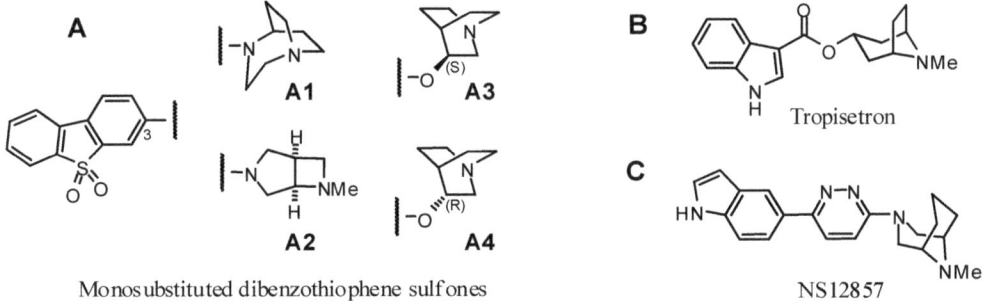

Figure 1. Structures of recent α7 nAChR ligands. Four examples of bicyclic amines coupled to dibenzothiophene sulfone (**A**) [15], a tropane derived compound (**B**) [24] and a propylene-bridged *N*-methylpiperazine (**C**) [25] are shown.

Previous reports have also shown that structural changes on the amino side chain (cationic center) of a given aromatic unit may affect the binding profile with respect to $\alpha_7/\alpha_4\beta_2$ nAChR affinity [26,27]. Furthermore, a small series of bicyclic amines coupled to dibenzothiophene sulfone, including chiral quinuclidines (Figure 1, A3 and A4) and a fused azetidine (3,6-diazabicyclo[3.2.0]heptane) (Figure 1,A2) have been investigated by Schrimpf *et al.* [15]. In addition to nitrogen bridgeheaded aza- or diazabicylic molecules derived from quinuclidine or diazabicyclo[3.2.2]nonane, several alkylene bridged piperidines (e.g., tropane derivatives such as tropisetron [24]) or piperazines (e.g., *N*-methyl-3-azagranatanine derivative such as NS12857 [25]) have been identified as potent α_7 nAChR selective ligands (Figure 1B,C). Considering this, we were inspired to prepare a small series of fluorinated dibenzothiophene sulfones with a set of (three) differently arranged piperazine-based diazabicyclic scaffolds.

To test for the effect of the designed structural modifications and to evaluate the target selectivity of potential α_7 nAChR imaging probes, the binding affinities of all fluoro-substituted derivatives for the human α_7 nAChR, as well as the three most important subtypes of heteromeric nAChRs (human $\alpha_4\beta_2$, human $\alpha_3\beta_4$, and rat $\alpha_6\beta_2^*$) and the structurally similar 5-HT$_3$ receptor, were determined. For the most suitable ligand **10a**, a radiolabeling procedure was developed and successfully implemented, then subsequently optimized and transferred to a fully automated setup. Finally, the pharmacokinetics of [^{18}F]**10a** was evaluated in both piglets and monkey by dynamic PET scans under baseline and blocking conditions.

2. Results and Discussion

2.1. Chemistry

2.1.1. Rationale

1,4-Diazabicyclo[3.2.2]nonane-substituted dibenzothiophene sulfone **10** was found to have exceptional high affinity for α_7 nAChR and effective brain uptake [15,17]. This prompted us to develop novel derivatives as candidates for ^{18}F-labeled ligands and to further investigate the SAR of this new pharmacophore (Table 1).

Initially, the three fluoro-substituted 1,4-diazabicyclo[3.2.2]nonane-containing derivatives **10a**–**c** were designed. While **10a** and **10b** are identical to compounds **7c** and **7a**, respectively, recently published by Gao *et al.* [17], the isomer **10c** has not been described so far. To gain further insight into the SAR of dibenzothiophene-based scaffolds we embarked on the synthesis of a second series of compounds. Previous results on the systematic side chain rigidification have highlighted the importance of conformational restriction of the aminocycle for high α_7 selectivity [15]. Since the tropane core has been identified as a structural motif that could provide this selectivity [4,6] it was of interest to include a set of three tropane-like diamines into the synthetic plan. With the carbon atom replaced at position 3 both in tropane and in *N*-methylgranatanine [28] by an nitrogen atom, the resulting secondary amino group of each *N*-methyl piperazine building block could serve as position of attachment to the tricyclic arene system.

Table 1. Structures of fluoro-substituted dibenzothiophene-based ligands for α7 nAChR.

Compound	R	X^1	X^2	X^3
10	(diazabicyclo)	H	H	H
10a	(diazabicyclo)	H	F	H
10b	(diazabicyclo)	F	H	H
10c	(diazabicyclo)	H	H	F
12a	(diazabicyclo)	H	F	H
12b	(diazabicyclo)	F	H	H
13b	(diazabicyclo)	F	H	H
14b	(diazabicyclo)	F	H	H

Four novel fluorine-substituted derivatives were envisioned, in which the homopiperazine-based 1,4-diazabicyclo[3.2.2]nonane was replaced by conformationally restricted piperazine-based substituents, bridged by either ethylene (8-methyl-3,8-diazabicyclo[3.2.1]octane = azatropane (**13b**), 3-methyl-3,8-diazabicyclo[3.2.1]octane (**14b**)) or propylene units (9-methyl-3,9-diazabicyclo[3.3.1]nonane = *N*-methyl-3-azagranatanine (**12a,b**)) [22].

2.1.2. Synthesis of α7 nAChR Ligands

Dibenzo[*b,d*]thiophene (**1a**) and dibenzo[*b,d*]thiophene 4-boronic acid (**1b**) served as starting materials for the synthesis of dibenzo[*b,d*]thiophene-5,5-dioxides functionalized in the *para*- (Scheme 1) or *ortho*- position (Scheme 2) to the sulfone group. The final coupling to the *N*-atom of a diazabicyclic moiety was performed at position 7 or 3 of the respective tricyclic scaffold to afford the potential α7 nAChR ligands **10**, **10a–c**, **12a,b**, **13b**, and **14b**, as shown in Table 1.

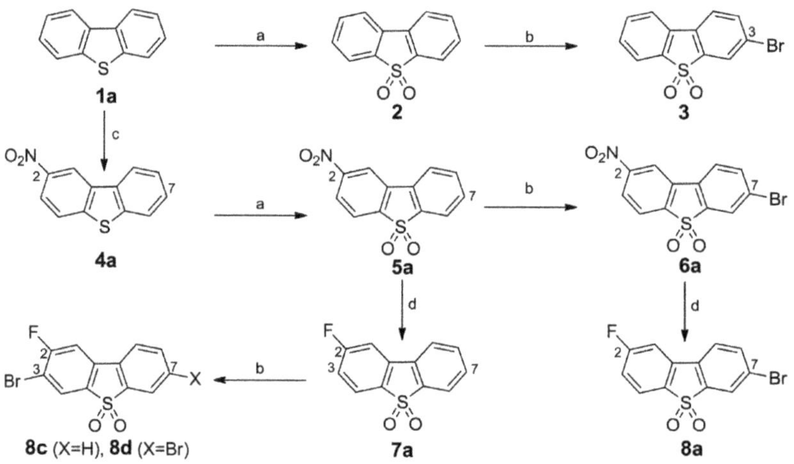

Scheme 1. Synthesis of 3-bromo dibenzothiophene intermediates **3**, **6a**, **8a** and **8c**. Reagents and conditions: (a) H$_2$O$_2$ (50%), HOAc, 120 °C, 5 h, 97% for **2** and 96% for **5a** (b) NBS, conc. H$_2$SO$_4$, r.t., 24 h, 44% for **3**, 64% for **6a**, 46% for **8c**, and 18% for **8d**; (c) HNO$_3$ (65%), HOAc, 20–40 °C, 24 h, 26% for **4a**; (d) TMAF·4 H$_2$O, cyclohexane/DMSO, ↑↓, 6 h (azeotropic drying with separation of H$_2$O), compound **5a** or compound **6a**, 95 °C, 5 h, 75% for **7a** and 74% for **8a**.

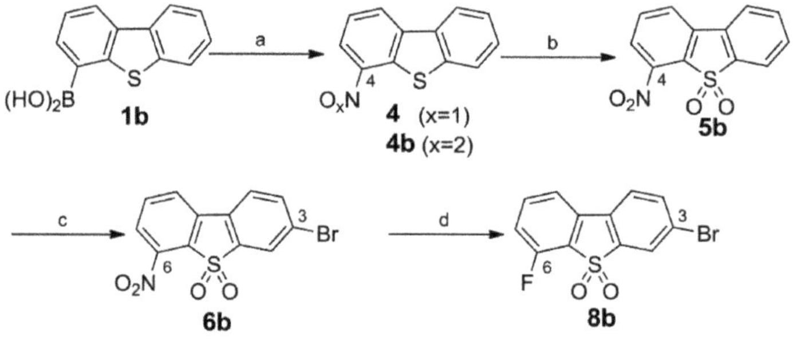

Scheme 2. Synthesis of 3-bromo dibenzothiophene intermediates **6b** and **8b**. Reagents and conditions: (a) *tert*-butyl nitrite, MeCN, 55 °C, 24 h, mixture of **4**/**4b** (~88:12); (b) H$_2$O$_2$ (50%), HOAc, 120 °C, 5 h, 72% (two steps); (c) NBS, conc. H$_2$SO$_4$, r.t., 24 h, 78%; (d) TMAF·4H$_2$O, cyclohexane/DMSO, 6 h (azeotropic drying with separation of H$_2$O), compound **6b**, 95 °C, 5 h, 73%.

Synthesis of Bromo Dibenzothiophene Intermediates **3**, **6a** and **8a**–**8c**

The brominated intermediates were obtained from dibenzo[*b,d*]thiophene (**1a**) by three different routes. The 3-bromo-derivative **3** was prepared via sulfide to sulfone oxidation of **1a** with H$_2$O$_2$ in acetic acid followed by bromination with an excess of **2** in the NBS/H$_2$SO$_4$ system in 44% yield [29]. Isolation of compound **3** from the 3,7-dibrominated by-product (not shown) along with unreacted starting material was obtained by fractional crystallization of the dibromo compound, followed by a chromatographic separation of pure **3** from the filtrate. The 2-nitrodibenzothiophene sulfone **5a** was prepared by nitration of **1a** using HNO$_3$ (65%) in acetic acid [30,31] affording the 2-nitro compound **4a** in moderate yield (26%), followed by oxidation with H$_2$O$_2$ in acetic acid to give **5a** [32]. The 2-nitro-7-bromo-dibenzothiophene sulfone **6a** was obtained by bromination of **5a** according to the corresponding conversion of **2** to **3**. Because the one benzo ring in **5a** is strongly deactivated by the NO$_2$ and SO$_2$R groups, no over-brominated by-products were detected and the 2-nitro-7-bromo-dibenzothiophene sulfone **6a** was afforded in 64% yield. To obtain different mono-fluorinated building blocks in a short synthesis, the fluorine substitution either at positions *para* or *ortho* to the sulfone were intended to be carried out with a simple uniform synthetic pathway. In addition to the known three-step conversion of **6a** into the 2-fluoro-7-bromo derivative **8a** based on the Balz-Schiemann reaction with 45% overall yield [17], we deemed a one-step conversion from **6a** to **8a** via fluorodenitration as preferable [33,34]. Tetramethylammonium fluoride (TMAF) proved to be a good source for active fluoride after azeotropic drying of commercial TMAF tetrahydrate in refluxing cyclohexane/DMSO for 6 h. A smooth conversion of **6a** occurred at 95 °C within 5 h to give **8a** in 75%yield.Likewise, the nitro group in the 2-nitrodibenzothiophene 5,5-dioxide **5a** was displaced by fluorine to give **7a** in 74% yield. The vicinal 3-bromination of **7a** was accomplished by applying NBS in H$_2$SO$_4$ (98%) at room temperature for 24 h to give compound **8b**. In addition to **8c**, the 3,7-dibromo derivative **8d** was detected. Chromatographic separation gave **8d** in 18% yield along with 2-fluoro-3-bromo compound **8c** in 46% yield.

Synthesis of the 3-Bromo Dibenzothiophene Intermediates **6b** and **8b**

As result of the two-step synthesis of **5a** from **1a** (Scheme 1), the *ortho*-derivative **4b** was concomitantly formed as a minor nitration product. However, because it was difficult to separate from the excess of **4a**, this approach was considered inappropriate to gain an adequate amount of the 4-nitro dibenzothiophene sulfone **5b**. The intermediate **5b** was recently reported [17]. The described procedure for **4b** is a nitration of commercially available arylboronic acid **1b** with two equivalents of Bi(NO$_3$)$_3$ as described by Maiti *et al.* [35]. In contrast, we successfully applied a metal-free approach by performing an *ipso* nitrosation of **1b** with *tert*-butyl nitrite in acetonitrile [36,37]. As shown in Scheme 2, the nitroso derivative **4** was formed as the main product along with a minor amount of **4b**. Without further purification, the **4**/**4b** mixture was readily oxidized in a one-pot reaction with H$_2$O$_2$/acetic acid to give **5b** in 72% yield over two steps from **1b**. After bromination, the resulting **6b** was converted directly into **8b** in 73% yield via our fluorodenitration protocol, in contrast to the earlier described independent four-step synthesis applying a Pschorr reaction [17].

Synthesis of **10**, Fluorinated Reference Compounds **10a–c**, and Nitro Precursor **11**

The 3-bromodibenzo[*b,d*]thiophene sulfones **3**, **6a** and **8a–c** were reacted with 1,4-diazabicyclo [3.2.2]nonane (**9a**) under Pd-catalyzed Buchwald-Hartwig conditions to provide the previously reported compounds **10a** and **10b** [17] and the novel fluorinated isomer **10c** in 52%, 48% and 44% yields, respectively. The non-fluorinated compound **10** [15] and the nitro derivative **11a** [17] were also obtained under Buchwald-Hartwig conditions in 73% and 53% yields, respectively, as depicted in Scheme 3.

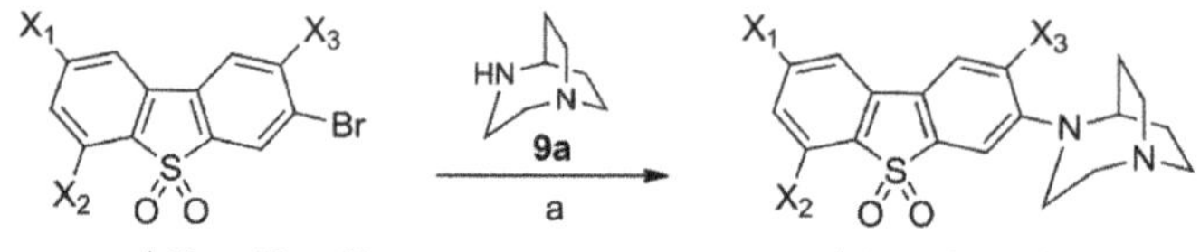

	X_1	X_2	X_3
3	H	H	H
8a	F	H	H
8b	H	F	H
8c	H	H	F
6a	NO_2	H	H

	X_1	X_2	X_3
10	H	H	H
10a	F	H	H
10b	H	F	H
10c	H	H	F
11a	NO_2	H	H

Scheme 3. Synthesis of **10**, fluorinated reference compounds **10a–c**, and nitro precursor **11**. Reagents and conditions: (a) $Pd_2(dba)_3$, BINAP, Cs_2CO_3, toluene, 90 °C, 24–36 h, 52% for **10a**, 48% for **10b**, 44% for **10c**, 73% for **10**, and 53% for **11a**.

Synthesis of the Fluorinated Reference Compounds **12a,b**, **13b**, and **14b**

In order to expand the structural diversity of compounds derived from the novel pharmacophore dibenzothiophene for SAR characterization, in a second series of compounds the homopiperazine-based 1,4-diazabicyclo[3.2.2]nonane moiety was replaced by piperazine-based substituents, bridged by either ethylene or propylene units. The bromine substituent in the *meta* position to the sulfone group was used to incorporate these more rigid diamines. The Buchwald-Hartwig coupling of 3-bromodibenzo[*b*,*d*]thiophene sulfones **8a** and **8b** with 9-methyl-3,9-diazabicyclo[3.3.1]nonane (**9b**), 8-methyl-3,8-diazabicyclo[3.2.1]octane (**9c**) and 3-methyl-3,8-diazabicyclo[3.2.1]octane (**9d**) gave compounds **12a,b**, **13b**, and **14b** in moderate yields (Scheme 4).

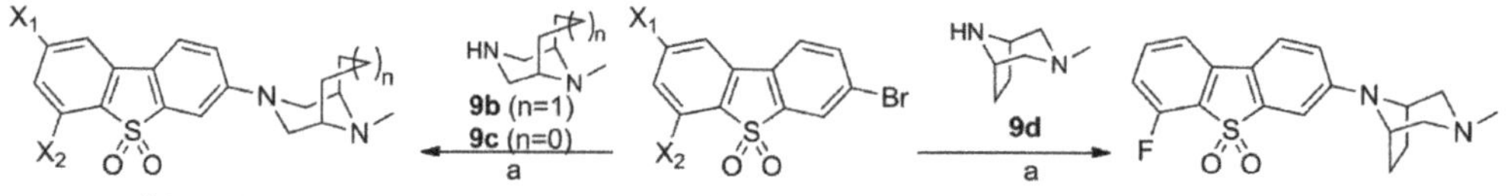

	X_1	X_2	n
12a	F	H	1
12b	H	F	1
13b	H	F	0

	X_1	X_2
8a	F	H
8b	H	F

14b

Scheme 4. Synthesis of the fluorinated reference compounds **12a,b**, **13b**, and **14b**. Reagents and conditions: (a) $Pd_2(dba)_3$, BINAP, Cs_2CO_3, toluene, 90 °C, 24–36 h, 38% for **12a**, 42% for **12b**, 47% for **13b**, and 67% for **14b**.

All final compounds (**10**–**14**) are crystalline solids and were fully characterized by NMR spectroscopy (^1H, ^{13}C, ^{19}F, COSY, HSQC) and high resolution mass spectrometry.

2.2. *In Vitro Affinity Assays*

All compounds were evaluated *in vitro* to measure their affinity and selectivity for the target receptor α_7 nAChR in relation to the heteromeric receptor subtypes $\alpha_4\beta_2$, $\alpha_6\beta_2$, and $\alpha_3\beta_4$. The results of the respective binding assays are summarized in Table 2.

Table 2. *In vitro* binding affinities towards human homomeric α_7, heteromeric $\alpha_4\beta_2$ and $\alpha_3\beta_4$ nAChR, rat $\alpha_6\beta_2$* nAChR subtypes, and human 5-HT$_3$ receptor.

| Compound | Affinity (K_i in nM) | | | | 5-HT$_3$ [d,e] | Selectivity (K_i ratio) | |
| | nAChR Subtype | | | | | | |
	hα_7 [a]	h$\alpha_4\beta_2$ [b]	h$\alpha_3\beta_4$ [b]	r$\alpha_6\beta_2$* [c]		$\alpha_7/\alpha_4\beta_2$	$\alpha_7/\alpha_3\beta_4$
10	0.51 ± 0.32	318 ± 43.3	49.6 ± 14.7	517 ± 186	(35%)	623	97
10a	0.60 ± 0.44	517 ± 375	119 ± 29.0	589 ± 217	440 (2%)	862	198
10b	0.84 ± 0.16	211 ± 108	42.3 ± 4.73	435 ± 152	(42%)	251	50
10c	8.53 ± 1.74	507 ± 212	279 ± 24.4	1390 ± 340	(26%)	59	33
12a	30.9 ± 8.72	141 ± 11.7	96.0 ± 1.48	1180 ± 360	(10%)	5	3
12b	105 ± 23.9	301 ± 148	94.8 ± 6.48	1190 ± 230	(11%)	3	1
13b	40.9 ± 7.77	426 ± 197	224 ± 51.7	2260 ± 540	(10%)	10	5
14b	9.26 ± 2.23	>4000	>5000	1450 ± 370	(4%)	>400	>500

[a] Human α_7 nAChR in stably transfected SH-SY5Y cells, with radiotracer [^{3}H]methyllycaconitine (0.5–1 nM), K_D = 2.0 nM. [b] Human $\alpha_4\beta_2$ and $\alpha_3\beta_4$ nAChR in stably transfected HEK-293 cells, with radiotracer [^{3}H]epibatidine (0.5–1 nM), K_D = 0.025 nM for h$\alpha_4\beta_2$ nAChR, K_D = 0.117 nM for h$\alpha_3\beta_4$ nAChR. [c] Rat $\alpha_6\beta_2$* obtained from rat striatum by immunoimmobilization using anti-rα_6 nAChR antibody, with radiotracer [^{3}H]epibatidine (0.1 nM), K_D = 0.025 nM. [d] K_i value in nM; human 5-HT$_3$ receptor recombinant-HEK293 cells, with radiotracer [^{3}H]GR65630 (working concentration n = 0.69 nM; K_D = 0.2 nM). [e] percentage of inhibition at 0.1 µM concentration of test compound.

2.2.1. Affinity for α_7 nAChR

The affinity for α_7 nAChR was assessed for the human receptor protein expressed in a stably transfected cell line and labeled with the selective α_7 nAChR antagonist [^{3}H]methyllycaconitine. The affinity of the lead of the series (compound **10**), K_i = 0.51 nM, is almost identical to the value reported by Gao *et al.*, obtained on [^{125}I]α-bungarotoxin labelled rat cortical membranes. The binding affinities of the fluoro-substituted compounds **10a** (K_i = 0.6 nM *vs.* 1.4 nM of compound **7c** [17]) and **10b** (K_i = 0.84 nM *vs.* 0.4 nM of compound **7a** [17]) are also comparable with previously published results. All novel fluoro-containing compounds **10c**, **12a,b**, **13b**, and **14b** bind with remarkably lower affinity to α_7 nAChR. Interestingly, the vicinal substitution of the dibenzothiophene with both the fluorine and the diazabicycle is inappropriate in terms of binding as reflected by the 2-fluoro-3-amino derivative **10c**, which possesses an about 10-fold lower affinity at the α_7 subtype relative to the 2-fluoro-7-amino derivative **10a** and the 6-fluoro-3-amino derivative **10b**.

In particular, we assume that steric effects impair the interactions between the cationic center of compound **10c** and the binding site of α_7 nAChR [38]. Furthermore, as noticed for the new series of dibenzothiophene derivatives (**12a,b**, **13b** and **14b**) an increase in the flexibility of the tertiary amine in the cationic center is also not tolerated. Replacement of the NC-bridged homopiperazine moiety, containing a bicyclic tertiary amine as a basic structural element of the reference compound **10** by three different CC-bridged piperazine ring systems and carrying a methyl-substituted tertiary amino group, has negative effects. The 9-methyl-3,9-diazabicyclo[3.3.1]nonane substituted compounds **12a** and **12b** have significantly diminished affinity for α_7 nAChR relative to the matched pairs of 1,4-diazabicyclo[3.2.2]nonane-substituted **10a** and **10b**. A similar effect is observed by substitution with an azatropane in compound **13b** (8-methyl-3,8-diazabicyclo[3.2.1]octane) and its isomer **14b** (3-methyl-3,8-diazabicyclo[3.2.1]octane).

Thus, in terms of affinity for the α_7 nAChR we considered compound **10a** as the most suitable compound reported herein. The structural modifications within the cationic center of the molecules, which we performed to increase selectivity by reducing off-target binding, unexpectedly impaired the binding to α_7 nAChR to such an extent that the resulting compounds were not included in the development of radioligands within this study.

2.2.2. Affinities for $\alpha_4\beta_2$, $\alpha_3\beta_4$, and $\alpha_6\beta_2$ nAChR

To evaluate off-target binding of the compounds towards heteromeric nAChR subtypes, respective binding assays were performed with [^{3}H]epibatidine-labeled human $\alpha_4\beta_2$ or human $\alpha_3\beta_4$ nAChR, both stably expressed on human HEK cells, as well as native rat $\alpha_6\beta_2$* nAChR immobilized by a subunit-specific antibody.

Besides the $\alpha_4\beta_2$* nAChR, the $\alpha_6\beta_2$* nAChR belongs to the high-affinity family of β_2-containing nAChRs [39]. While functionally similar to $\alpha_4\beta_2$* nAChR, the expression of the $\alpha_6\beta_2$* subtype is rather selective, primarily in dopamine neurons in the brain [40–43]. Overlapping expression of α_7 nAChR with $\alpha_4\beta_2$* nAChR and also $\alpha_6\beta_2$* nAChR within the mesolimbic axis [44,45] requires careful analysis of subtype affinity of radioligands for nAChR. In heteromeric nAChR subtypes, the orthosteric ligand binding site is formed at the interface of an α subunit (principal component) and an adjacent non-α subunit (complementary component), equivalent to the (+) surface and the (−) surface of an α subunit in homomeric nAChRs [46,47].

Consistent with the observation that in heteromeric neuronal nAChRs the non-α subunit is of particular importance [38], as it makes the affinity for nicotine dependent on the presence of a β_2 subunit regardless of the α subunit [47], the herein investigated compounds bind with generally moderate (K_i values > 100 nM) and almost equal affinities towards both human $\alpha_4\beta_2$ and rat $\alpha_6\beta_2$* nAChRs. It is worth noting that the affinities for the $\alpha_4\beta_2$ subtype of the α_7 nAChR ligands investigated herein, with the exception of **14b**, are significantly higher than those of the compounds in the NeuroSearch (NS) series [12]. Because these compounds contain identical cationic centers, differences in the receptor subtype binding affinities are probably related to differences in the HBA and hydrophobic functionalities. In the NeuroSearch compound series, an oxadiazole moiety appears as HBA which is spatially more separated from and not forced in plane with the hydrophobic fluorophenyl moiety, while compounds **10a**–**10c** possess fused functionalities with both the HBA and the hydrophobic moiety represented by the fluorine-substituted dibenzothiophene dioxide ring system. We hypothesize that in particular the sulfonyl moiety of these novel compounds promotes binding to the $\alpha_4\beta_2$ subtype, comparable to the carbonyl moiety acting as a hydrogen acceptor along with a cationic pharmacophore element in the off-target binding of α_7 nAChR ligand CHIBA-1001 [48,49]. However, for the most selective fluorine-containing ligand of the current study, compound **10a** (selectivity > 800), interfering effects on α_7 nAChR PET imaging due to its binding to β_2-containing receptor subtypes are highly unlikely. The conformational changes due to the shift of the bridgehead carbons in the diazabicyclo[3.2.1]octane that we assume to contribute to the significantly improved selectivity of compound **14b** in comparison to **13b** will be analyzed in future studies. It appears that the basic *N*-methyl group is sterically more deshielded in contrast to the azatropane **13b**, resulting in a detrimental effect both on $\alpha_4\beta_2$ and $\alpha_3\beta_4$ nAChR binding with only weakly attenuated affinity to the α_7 subtype (Table 2).

An overlap in the receptor expression between α_7 and $\alpha_3\beta_4$* nAChRs, in particular within autonomic neurons [50,51], necessitates the investigation of the selectivity of potential ligands. By comparing data obtained for compounds **10a** and **10b** (K_i: 119 nM and 43 nM) with the affinity values of the corresponding compounds **7c** and **7a** reported by Gao et al., (K_i: 5000 nM and 709 nM) [17], species-specific differences become obvious. Although the most selective compound **10a** (selectivity ~200) binds with sufficiently low affinity to human $\alpha_3\beta_4$ nAChRs to afford imaging of the α_7 subtype in human brain, more than 10-fold higher affinities at the human than the rat receptor subtype illustrate the significance of species-specific targeting at the early stage of the development of PET ligands.

The *in vitro* binding affinity of all compounds from this series at the 5-HT$_3$ receptor, as determined by percentage of inhibition of the binding of [^{3}H]GR65630 at 10 μM, indicates high α_7 nAChR over 5-HT$_3$ selectivity (Table 2). From this direct comparison a much higher affinity of ASEM (**10b**) than that of **10a** for 5-HT$_3$ may be expected considering the inhibition data at 100 nM (2% for **10a**, 42% for ASEM). Based on this value, compound **10a** is assumed to possess an at least fivefold higher selectivity than all other compounds, except **14b**, reported herein. The K_i of **10a** was estimated to be 440 nM for 5-HT$_3$.

In addition, investigation of interaction of **10a** at 1 μM with further off-targets (α_1 nAChR, SERT, DAT, NET, VMAT, and choline transporter) by radioligand binding assays revealed no significant binding. Altogether, the target affinity and nAChR subtype specificity of **10a**, the most suitable fluorine-substituted α_7 nAChR ligand reported herein, is certainly sufficient to ensure that assessment of α_7 nAChR in humans will not be confounded by binding of the respective PET radiotracer to relevant off-target sites. Therefore, compound **10a** was selected for radiolabeling and further evaluation *in vivo*.

2.3. Optimization of Manual Radiosynthesis of [^{18}F]**10a** for Translation to an Automated Module

Manual radiosynthesis of [^{18}F]**10a** was performed starting from the nitro precursor **11a** and systematically optimized by varying the base, solvent, reaction time and heating system (Scheme 5). The amount of nitro precursor **11a** was kept constant at 1 mg according to our previous practice [52]. A nitro-to-fluoro substitution of activated aryl moieties in general requires moderately high temperatures (e.g., 130–160 °C) [53], reflected by low labeling efficiencies of <1% at 90 °C using MeCN. With increasing temperature (150 °C) and the use of DMSO, labeling efficiency of up to 30% were achieved.

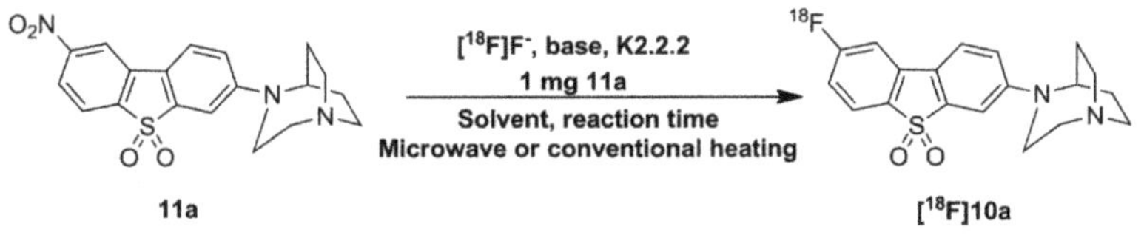

Scheme 5. Radiosynthesis of [^{18}F]**10a**. Solvents tested: MeCN, DMSO, and DMF at different temperatures with microwave irradiation or conventional heating.

In contrast to decomposition of the corresponding [^{18}F]**7c** reported by Gao et al. [17], we did not observe decomposition of [^{18}F]**10a** in the presence of the strong base potassium carbonate in DMSO at this temperature. In an attempt to improve labelling efficiencies prior to high-scale production of [^{18}F]**10a** in a commercially available automated device, we investigated the aromatic radiofluorination of **11a** using DMF as solvent based on our recent experience [52] under different conditions (thermal *vs.* microwave heating). Influence of the basicity of the metal salt on the labelling yield (potassium carbonate *vs.* potassium oxalate) was also investigated.

Under thermal heating and the use of K_2CO_3, we determined a time- and temperature-dependent increase in the nitro-to-fluoro conversion with maximum labelling efficiencies of about 90% after 10 min of reaction time at 120 °C. Under these conditions, K_2CO_3 is preferred, since we observed a significantly lower labelling efficiency of ~30% using the weaker base $K_2C_2O_4$. Our attempt to combine microwave dielectric heating with DMF as solvent with a mean-to-high dielectric constant rendered comparably high labeling efficiencies (≈94%) under microwave pulse mode (power cycling mode, 150 °C, 75 W, 5 min).

A semi-automated radiosynthesis coupled with microwave was recently reported for [^{18}F]**10b** ([^{18}F]ASEM) using the DMSO/K_2CO_3 system at 160 °C, resulting in a labeling efficiency of 45%–50% [54]. In the attempt to gain insight into the influence of the solvent on labeling efficiencies, we radiolabeled the nitro precursor **11b** to obtain [^{18}F]**10b** using the DMF/K_2CO_3 system under conventional and microwave-assisted heating at 120 °C. With only 1 mg of the nitro precursor **11b**, labeling efficiencies of 60%–70% were achieved with DMF as solvent in both heating modes. These findings reinforce the superiority of DMF as a solvent of choice for this compound class, and allows the fully automated radiosynthesis of [^{18}F]**10a** and [^{18}F]**10b** for transfer to clinical radiopharmacies.

Overall, the radiolabelling reactions proceeded cleanly. No significant amount of ^{18}F-labeled by-products was detected according to radio-TLC analyses (data not shown). The crude reaction mixture was applied directly onto a semi-preparative HPLC column (Figure 2a) and [^{18}F]**10a** was successfully isolated in high radiochemical purities (≥98%) with a retention time of about 17 min. Analytical radio-HPLC analysis of the final product spiked with the reference compound confirmed the identity of [^{18}F]**10a** (Figure 2b). [^{18}F]**10a** proved to be stable in physiological solutions and organic solvents at 40 °C for up to 90 min.

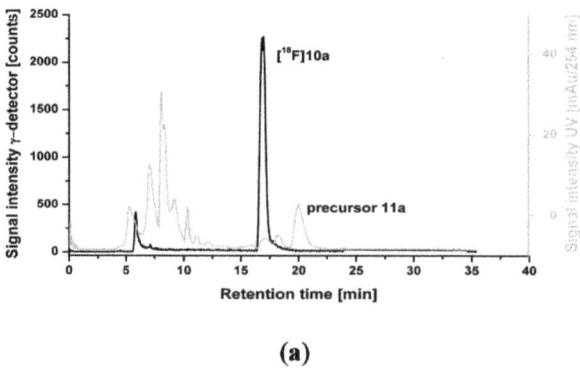

(a)

Figure 2. *Cont.*

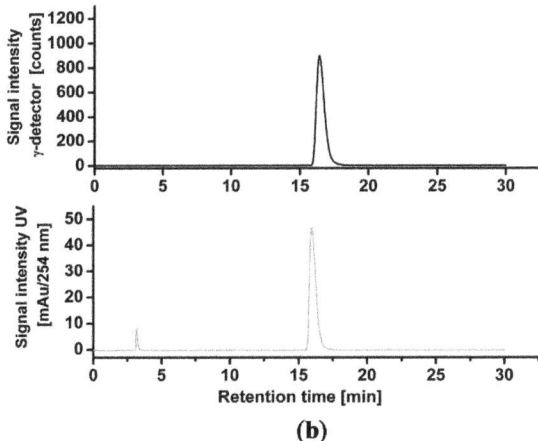

Figure 2. (a) Representative radio- and UV-chromatograms obtained for isolation of [^{18}F]**10a** from crude reaction mixture by semi-preparative HPLC (Reprosil-Pur C18-AQ column, 35% MeCN/H$_2$O/0.05% TFA, Flow rate: 10 mL·min^{-1}); (b) Analytical radio-chromatogram (top) and UV-chromatogram (bottom) of purified [^{18}F]**10a** spiked with the reference compound **10a**.

We determined the distribution coefficient of [^{18}F]**10a** in the *n*-octanol-PBS system experimentally by the shake-flask method with a log $D_{7.2}$ value of 1.3 ± 0.1 (*n* = 3), making sufficient blood-brain barrier permeability of [^{18}F]**10a** likely.

The translation to remotely controlled radiosynthesis of [^{18}F]**10a** was performed using the TRACERLAB™ FX$_{FN}$ module. The reaction was performed using the Kryptofix®222/K$_2$CO$_3$ system at 120 °C in 10 min with 1 mg of precursor **11a** in DMF. Also in the automated process [^{18}F]**10a** was obtained with extremely high labeling efficiencies of 86% ± 3%. After isolation via semi-preparative HPLC, [^{18}F]**10a** was trapped on a pre-conditioned Sep-Pak® (Waters, Milford, MA, USA) C18 light cartridge, and formulated in isotonic saline containing 10% of EtOH (*v*/*v*). The average decay-corrected radiochemical yield was 35% ± 9% (*n* = 6) calculated at the end of the synthesis (EOS). Radiochemical purity of >99% and high specific activities of 855 ± 302 GBq·μmol^{-1} (*n* = 6) were obtained in a total synthesis time of 70 min. This rapid and versatile automated radiosynthesis will enhance the accessibility of [^{18}F]**10a** for widespread production in future clinical studies.

2.4. Dynamic PET Studies of [^{18}F]10a in Piglets

2.4.1. Baseline Studies

After intravenous injection of 425 ± 78 MBq [^{18}F]**10a** (*n* = 3), we observed high uptake of radioactivity in the brain with peak concentrations at 8–10 min p.i., followed by washout. The highest accumulation of radioactivity (SUV$_{max}$ > 2.2) occurred in the thalamus, colliculi, and midbrain. Somewhat lower accumulation (SUV$_{max}$ ~1.9–2.1) was observed in other brain regions (Figure 3). The summed PET image (inset of Figure 3) shows the corresponding distribution pattern in the brain of a female piglet from 0 to 20 min p.i.

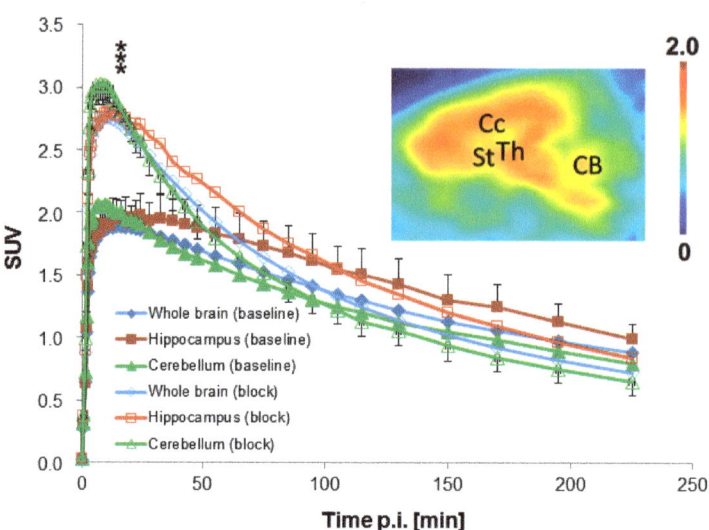

Figure 3. Time-activity curves in different brain regions obtained by dynamic PET scans of [^{18}F]**10a** in piglets under baseline and blocking conditions (mean values; $n = 3$). Standard deviations are shown for two examples (hippocampus baseline and cerebellum block). *** $p < 0.001$ vs. baseline. Inset: Brain image in sagittal plane acquired from 0 to 20 min p.i. after injection of [^{18}F]**10a** in a female piglet. Data are expressed as standardized uptake value (SUV). Regions of interest (ROI) were drawn based on an overlay with T1-weighted MR images of a pig brain. Abbreviations: CB = Cerebellum, CC = Corpus callosum, St = Striatum, Th = Thalamus.

2.4.2. Metabolite Analysis

Radioactive metabolites of [^{18}F]**10a** were assessed in plasma of pigs for up to 120 min p.i. (Figure 4a). For preparation of RP-HPLC samples, the proteins were precipitated and extracted two times with MeCN with a reproducible recovery of >90% of the starting radioactivity in the supernatant. The patterns of radioactive metabolites obtained by radio-TLC and radio-HPLC correlated well with each other. Parent fraction accounted for 88% ± 6%, 24% ± 4%, and 19% ± 4% of total radioactivity at 2, 60, and 120 min p.i., respectively.

One major radioactive metabolite was detected, which was more hydrophilic than [^{18}F]**10a** and is therefore assumed not to pass the blood-brain barrier (Figure 4b). Based on previous experience from our group with the diazabicyclononane-containing α$_7$ nAChR ligand [^{18}F]NS10743 [21], we believe that this metabolite is formed by enzymatic oxidation of the nitrogen at position 1 in the identical motif of **10a**.

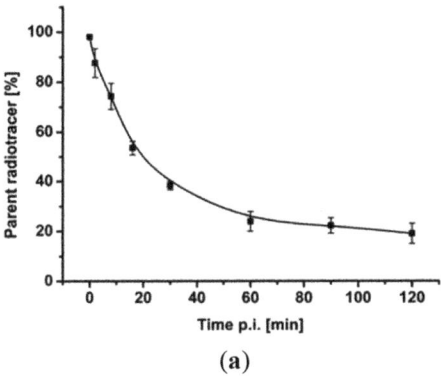 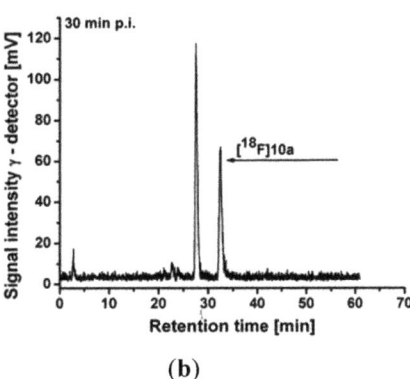

Figure 4. (**a**) Tri-exponential function fitting of the percentage of [^{18}F]**10a** under baseline conditions ($n = 3$) which was used to obtain metabolite-corrected arterial plasma input functions for modelling; (**b**) Representative radio-chromatogram of a pig plasma sample obtained at 30 min after intravenous injection of [^{18}F]**10a** under baseline conditions.

2.4.3. Blocking Studies

To evaluate the binding specificity of [^{18}F]**10a** *in vivo*, we performed blocking studies ($n = 3$). Animals were pretreated with 3 mg·kg^{-1} of the α7 nAChR ligand NS6740 [3] at 10 min prior to injection of 429 ± 45 MBq [^{18}F]**10a**, followed by a continuous infusion of NS6740 at 1 mg·kg^{-1}·h^{-1} during the course of the PET scan [21].

Despite a significant increase of about 40%–50%, in the brain uptake, radioactivity uptake under blocking conditions was about 15%–20% lower than that under baseline conditions in all brain regions evaluated from about 90 min p.i. until the end of the study (Figure 3). The early increase of brain uptake is accompanied by a significant increase of the influx rate constant K_1 (Table 4). The most likely explanation is a blood flow driven effect. There is strong evidence that α7 nAChRs are expressed on vascular endothelial and smooth muscle cells [55–57], which we assume were activated by NS6740 despite its classification as a "weak agonist" [3].

2.4.4. Modeling

One-tissue and two-tissue compartment models (1TCM and 2TCM) were evaluated for mathematical description of [^{18}F]**10a** time-activity curves (TACs) in pig brain regions (Figure 3) using metabolite-corrected arterial plasma input functions. The 2TCM was identified as the more appropriate model based on its smaller Akaike information criterion (AIC) value. The four rate constants K_1, k_2, k_3, and k_4 were fitted and the non-displaceable volume of distribution ($V_{ND} = K_1/k_2$) and the binding potential ($BP_{ND} = k_3/k_4$) as defined by Innis *et al.* [58] were calculated for each region of interest (Tables 3 and 4).

The values of the influx rate constant K_1 and the clearance rate constant k_2 were comparable between the different brain regions with a mean of 0.362 mL·cm^{-3}·min^{-1} and 0.131 mL·cm^{-3}·min^{-1}, respectively, in the whole brain. The total volume of distribution V_T calculated from $K_1/k_2(1 + k_3/k_4)$ [58] was 17 mL·cm^{-3} in thalamus and 13 mL·cm^{-3} in cerebellum, *i.e.*, slightly lower than the comparable values for [^{18}F]ASEM in baboons [18].

Table 3. Kinetic rate constants, non-displaceable volume of distribution (V_{ND}), and non-displaceable binding potential (BP_{ND}) of [^{18}F]**10a** in different regions of piglet brain using 240 min of scan data under baseline conditions ($n = 3$).

Brain Region	Rate Constant				V_{ND}	BP_{ND}
	K_1 (mL·cm^{-3}·min^{-1})	k_2 (min^{-1})	k_3 (min^{-1})	k_4 (min^{-1})		
Whole	0.362 ± 0.038	0.131 ± 0.062	0.093 ± 0.074	0.027 ± 0.018	3.32 ± 1.74	3.31 ± 1.02
Frontal Cortex	0.387 ± 0.059	0.182 ± 0.153	0.142 ± 0.165	0.027 ± 0.015	3.32 ± 2.06	4.22 ± 2.93
Parietal Cortex	0.399 ± 0.038	0.140 ± 0.089	0.126 ± 0.135	0.031 ± 0.021	3.88 ± 2.49	3.47 ± 1.65
Occipital Cortex	0.395 ± 0.053	0.123 ± 0.063	0.099 ± 0.087	0.030 ± 0.023	4.07 ± 2.23	3.16 ± 1.11
Hippocampus	0.400 ± 0.023	0.188 ± 0.083	0.145 ± 0.101	0.025 ± 0.013	2.43 ± 0.96	5.50 ± 1.01
Striatum	0.382 ± 0.028	0.144 ± 0.079	0.114 ± 0.093	0.027 ± 0.014	3.35 ± 1.81	3.98 ± 1.40
Thalamus	0.448 ± 0.027	0.150 ± 0.071	0.111 ± 0.085	0.030 ± 0.018	3.67 ± 2.10	3.51 ± 1.13
Colliculi	0.450 ± 0.072	0.134 ± 0.074	0.070 ± 0.042	0.027 ± 0.018	4.19 ± 2.37	2.68 ± 1.42
Midbrain	0.436 ± 0.045	0.162 ± 0.117	0.121 ± 0.132	0.029 ± 0.022	4.04 ± 3.06	3.40 ± 1.73
Pons	0.440 ± 0.045	0.149 ± 0.059	0.068 ± 0.042	0.026 ± 0.019	3.32 ± 1.40	2.76 ± 0.81
Cerebellum	0.421 ± 0.048	0.131 ± 0.052	0.067 ± 0.051	0.026 ± 0.020	3.62 ± 1.47	2.53 ± 0.57

Table 4. Kinetic rate constants, non-displaceable volume of distribution (V_{ND}), and non-displaceable binding potential (BP_{ND}) of [^{18}F]**10a** in different regions of piglet brain using 240 min of scan data under blocking conditions ($n = 3$).

Brain Region	Rate Constant				V_{ND}	BP_{ND}
	K_1 (mL·cm^{-3}·min^{-1})	k_2 (min^{-1})	k_3 (min^{-1})	k_4 (min^{-1})		
Whole	0.598 ± 0.108 *	0.062 ± 0.015	0.018 ± 0.022	0.011 ± 0.018	10.27 ± 3.94 *	0.84 ± 1.07 *
Frontal Cortex	0.647 ± 0.117 *	0.055 ± 0.011	0.014 ± 0.019	0.010 ± 0.020	12.34 ± 4.27 *	0.86 ± 0.92
Parietal Cortex	0.659 ± 0.120 *	0.055 ± 0.013	0.015 ± 0.020	0.011 ± 0.021	12.58 ± 4.52 *	0.82 ± 0.88 *
Occipital Cortex	0.668 ± 0.103 *	0.064 ± 0.019	0.022 ± 0.032	0.013 ± 0.023	11.36 ± 4.41 *	0.95 ± 1.03 *
Hippocampus	0.603 ± 0.118 *	0.063 ± 0.023 *	0.024 ± 0.030	0.011 ± 0.015	10.81 ± 5.02 *	0.84 ± 1.70 *
Striatum	0.624 ± 0.143 *	0.057 ± 0.015	0.020 ± 0.025	0.011 ± 0.017	11.78 ± 4.91 *	0.75 ± 1.29 *
Thalamus	0.762 ± 0.129 *	0.070 ± 0.025	0.023 ± 0.031	0.012 ± 0.021	12.09 ± 5.31 *	0.88 ± 1.00 *
Colliculi	0.735 ± 0.182 *	0.066 ± 0.007	0.011 ± 0.013	0.007 ± 0.015	11.41 ± 3.90 *	0.73 ± 0.74 *
Midbrain	0.685 ± 0.131 *	0.063 ± 0.010	0.013 ± 0.014	0.012 ± 0.018	11.17 ± 3.64 *	0.61 ± 0.77 *
Pons	0.670 ± 0.092 *	0.084 ± 0.030	0.019 ± 0.023	0.012 ± 0.016	8.93 ± 3.96 *	0.76 ± 1.16 *
Cerebellum	0.711 ± 0.148 *	0.076 ± 0.014	0.016 ± 0.019	0.011 ± 0.016	9.75 ± 3.57 *	0.71 ± 1.10 *

* $p < 0.05$ vs. baseline (Table 3).

Under blocking conditions K_1 was significantly increased by 50% to 70% in all brain regions while k_2 was decreased between 42% and 70% resulting in a significant (up to 4-fold) increase of V_{ND} (Table 4) and preventing the use of the occupancy plot [59] for calculation of BP_{ND}. Therefore BP_{ND}, as the ratio (at equilibrium) of specifically bound radioligand to that of nondisplaceable radioligand in tissue [58], was directly calculated form the rate constants k_3, which is proportional to the association rate constant, and k_4, which is proportional to the dissociation rate constant from the specific compartment. The baseline study revealed that the rate constant k_3 was highest in the hippocampus and lowest in the cerebellum and the values of k_4 were low, rather uniform and ranged from 0.025 to 0.031 min^{-1}. BP_{ND} values between 5.5 (hippocampus) and 2.5 (cerebellum) were reliably calculated from these data ($SD_{whole\ brain}$ ~30%). A similarly high BP_{ND} has recently been reported in baboons for the structurally related [^{18}F]ASEM [18] but not for any previous α7 nAChR PET radioligands [60]. Under blocking conditions k_3 was decreased by 70% to 90% in various brain regions while k_4 was decreased between 60% and 70%, resulting in a significant decrease of BP_{ND} by 72% to 85% in all regions but the frontal cortex (Table 4), suggesting specific binding of [^{18}F]**10a** to α7 nAChR in a similar range as the recently reported [^{18}F]ASEM [18].

2.5. Comparative PET Study of [^{18}F]10a ([^{18}F]DBT-10) and [^{18}F]10b ([^{18}F]ASEM) in a Rhesus Monkey

A PET study in a single rhesus monkey was performed for direct comparison of [^{18}F]DBT-10 (injected dose: 167 MBq) and [^{18}F]ASEM (injected dose: 185 MBq). Regional time-activity curves for both radioligands in the monkey brain are presented in Figure 5. Initial uptake levels were lower, and kinetics slower for [^{18}F]DBT-10 than [^{18}F]ASEM. For both radioligands the uptake levels follow the order of thalamus > frontal cortex = putamen > caudate > hippocampus > occipital cortex > cerebellum.

The total volume of distribution V_T for both radioligands was similar with the highest values in the thalamus ([^{18}F]DBT-10: 48.1 mL·cm^{-3}; [^{18}F]ASEM: 46.7 mL·cm^{-3}) and the lowest in the cerebellum ([^{18}F]DBT-10: 28.5 mL·cm^{-3}; [^{18}F]ASEM: 26.8 mL·cm^{-3}). These values are almost two-fold higher than those reported for [^{18}F]ASEM in baboons [18], supporting the species differences in α7 nAChR binding and distribution.

Altogether these preliminary data support the equal potency of both [^{18}F]DBT-10 and [^{18}F]ASEM for PET imaging of α7 nAChRs, with [^{18}F]ASEM having slightly faster kinetics.

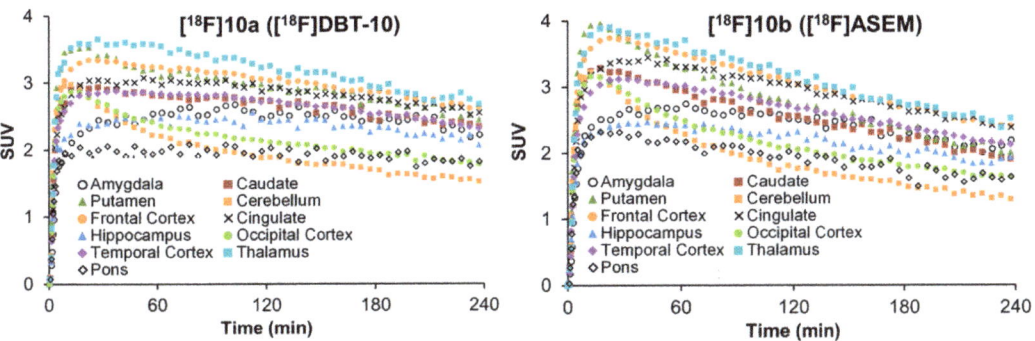

Figure 5. Time-activity curves of [^{18}F]DBT-10 and [^{18}F]ASEM in selected brain regions of a rhesus monkey.

2.6. Toxicity Studies in Rats

To prepare [^{18}F]**10a** for use in humans, toxicity studies were performed in rats according to EU cGLP. Compound **10a** was administered by a single intravenous injection to rats followed by an observation period of 2 or 15 day. The study was performed with 4 test groups, including 1 control and 3 dose groups (6.2, 62 and 620 μg·kg^{-1}), with 60 male and 60 female Wistar rats divided into two experiments: Day 2 (40 males and 40 females) and Day 15 (20 males and 20 females). During clinical observation the animals displayed no notable clinical effects. No statistically significant differences in body weights between control and treated groups in either gender were detected. Food consumption corresponded with body weight development.

All haematology parameters on Day 2 and Day 15 were within physiological range for this species. Individual divergences of some haematology parameters were small and not correlated with treatment. No test item effect on haematology parameters was observed.

There were no findings in clinical chemistry parameters which could be definitively considered as adverse. The average values of all test groups were within the historical control ranges. Occasional changes had no dose relationship, and they were therefore considered as a result of intra- and inter-individual variability for this species. The results of pathology examination indicated that **10a** after single intravenous administration did not result in changes in pathological and histopathological parameters on either Day 2 or Day 15. The no observed effect level (NOEL) of **10a** was determined to be 620 µg·kg^{-1}.

3. Experimental Section

3.1. General Information

Analytical thin-layer chromatography (TLC) was performed with Macherey-Nagel precoated plastic sheets with fluorescent indicator UV$_{254}$ (Polygram® SIL G/UV$_{254}$, Düren, Germany). Visualization of the spots was effected by illumination with an UV lamp (254 nm and 366 nm). Dry-column flash chromatography (DCFC) [61] was performed with vacuum on silica gel 60 (particle size 15–40 µm, Ref. 815650) from Macherey-Nagel (Macherey-Nagel GmbH & Co. KG, Düren, Germany). NMR spectra (^{1}H, ^{13}C, ^{13}C-APT, ^{19}F, COSY, HSQC, HMBC) were recorded with Varian spectrometer (Varian Mercury-300BB and Mercury-400BB; Agilent Technologies, Santa Clara, CA, USA). Chemical shifts are reported as δ (δ_H, δ_C, δ_F) values. Coupling constants are reported in Hz. Multiplicity is defined by s (singlet), d (doublet), t (triplet), and combinations thereof; br (broad) and m (multiplet). ESI/Ion trap mass spectra (LRMS) were recorded with a Bruker Esquire 3000 plus instrument (Billerica, MA, USA). High resolution mass spectra were recorded on an FT-ICR APEX II spectrometer (Bruker Daltonics; Bruker Corporation, Billerica, MA, USA) using electrospray ionization (ESI) in positive ion mode. Melting points were determined on a Linström capillary apparatus (Wagner & Munz GmbH, Vienna, Austria) in open capillary tubes and are uncorrected. Reagents and solvents were purchased from commercial sources and used without further purification unless otherwise noted. Compounds **2**, **4a**, and **5a** were prepared starting from **1a** (dibenzo[*b,d*]thiophene; Acros Organics, Geel, Belgium) according to literature methods [30–32]. The preparation of intermediates **3**, **5b**, **8a**, **8b**, and the final compound **10** have been described previously, but with different synthetic routes or from starting materials not used in our protocols [17,31,62]. Intermediate **6a** (two steps from **4a**), **6b** (one step from **5b**) and final compounds **10a**, **10b**, **11a** and **11b** (one step each, from amine **9a**), which have been prepared according to the literature [62], were also reported by Gao *et al.* [17]. To the best of our knowledge, intermediates **7a**, **8c**, **8d** and final compounds **10c**, **12a,b**, **13b**, and **14b** have not been reported so far.

3.2. Chemistry

4-Nitrosodibenzo[b,d]thiophene (**4**). *tert*-Butyl nitrite (2.14 mL, 1.86 g, 18 mmol) was added in one portion to a stirred suspension of dibenzo[*b,d*]thiophen-4-yl boronic acid (**1b**, 1.37 g, 6 mmol; Frontier

Scientific, Logan, UT, USA) in MeCN (24 mL) under argon at 22 °C. The mixture was stirred at 50–55 °C while its colour turned from yellow to dark brown in the course of 24 h. The solvent was evaporated and the residual solid was dissolved in CH_2Cl_2 (80 mL), washed with H_2O (25 mL), dried ($MgSO_4$) and evaporated to leave a dark greenish brown residue (1.38 g), which was used for the oxidation step. A sample (40 mg) was dissolved in cyclohexane/CH_2Cl_2 (2:3, v/v, 3 mL) and filtered through a short plug of silica gel (15–40 μm, 2 g). Subsequent elution with cyclohexane/CH_2Cl_2 (2:3, v/v, 50 mL) gave a green eluate which was evaporated to yield 30 mg of the title compound **4** (R_f = 0.56, heptane/EtOAc, 3:1) as a green powder, m.p. 113–115 °C (lit. m.p. 115–117 °C [36]). ^{1}H-NMR (300 MHz, $CDCl_3$): δ 7.48–7.63 (m, 2H, 7-H, 8-H), 7.88–7.94 (m, 1H, 6-H), 7.94 (t, J = 7.7 Hz, 1H, 2-H), 8.13–8.22 (m, 1H, 9-H), 8.49 (dd, J = 7.7, 1.1 Hz, 1H, 3-H), 9.58 (dd, J = 7.6, 1.1 Hz, 1H, 1-H). ^{13}C-NMR (75 MHz, $CDCl_3$): δ 120.10 (C), 121.68 (CH), 123.67 (CH), 125.44 (CH), 125.84 (CH), 127.82 (CH), 128.21 (CH), 132.38 (C), 137.94 (C), 138.11 (C), 142.04 (C), 161.91 (C), 165.95 (C). The product is 88% pure, containing 12% of 4-nitrodibenzo[b,d]thiophene (**4b**), as determined by comparison with ^{1}H-NMR of a pure sample.

2-Nitrodibenzo[b,d]thiophene 5,5-dioxide (**5a**) [32]. A mixture of 2-nitrodibenzo[b,d]thiophene (**4a**, 2.3 g, 10 mmol) in acetic acid (64 mL) was stirred at 80 °C while H_2O_2 (50%, 6.8 mL, 0.12 mol) was added. The temperature was raised to 120 °C and after 1 h a second portion H_2O_2 (50%, 4.5 mL, 0.08 mol) was added. Stirring was continued at 120 °C for 1 h and at 80 °C for 3 h. After cooling, the mixture was poured into water (180 mL) to form a precipitate which was filtered, washed and dried to yield the pure title compound **5a** (2.52 g, 9.64 mmol, 96% yield, R_f = 0.22, cyclohexane/$CHCl_3$, 1:3) as a pale yellow powder, m.p. 256.5–257.5 °C (lit. m.p. 258 °C [32]); ^{1}H-NMR (400 MHz, DMSO-d_6) δ$_H$ 7.74 (td, J = 7.6, 0.9 Hz, 1H$_{Ar}$, 7-H), 7.87 (td, J = 7.6, 1.0 Hz, 1H$_{Ar}$, 8-H), 8.07 (d, J = 7.7 Hz, 1H$_{Ar}$, 6-H), 8.29 (m, 1H$_{Ar}$, 9-H), 8.42 (dd, J = 8.4, 2.0 Hz, 1H$_{Ar}$, 3-H), 8.48 (d, J = 7.8 Hz, 1H$_{Ar}$, 4-H), 9.04 (d, J = 2.0 Hz, 1H$_{Ar}$, 1-H). ^{13}C-NMR (100 MHz, DMSO-d_6): δ$_C$ 118.26 (CH), 122.28 (CH), 123.60 (CH), 123.95 (CH), 126.10 (CH), 129.10 (C), 132.09 (CH), 132.98 (C), 135.04 (CH), 137.08 (C), 141.44 (C), 151.72 (C). LRMS: m/z (ESI) = 284.0 (M + Na)$^+$.

4-Nitrodibenzo[b,d]thiophene 5,5-dioxide (**5b**). In a procedure similar to the preparation of **5a**, compound **4** (1.33 g, 5.8 mmol) was reacted with H_2O_2 (6.6 mL, 0.11 g, 116 mmol) in acetic acid to give 1.33 g of a dark yellow solid. The product was further purified by column chromatogaphy (silica gel 15–40 μm, 24 g) with cyclohexane/CH_2Cl_2 (1:1→1:2) to yield the pure title compound **5b** (1.1 g, 4.2 mmol, 72% yield, R_f = 0.2, $CHCl_3$) as pale yellow crystals, m.p. 245.5–251 °C. ^{1}H-NMR (400 MHz, DMSO-d_6): δ$_H$ 7.75 (td, J = 7.6, 0.8 Hz, 1H$_{Ar}$, 7-H), 7.86 (td, J = 7.6, 1.0 Hz, 1H$_{Ar}$, 8-H), 8.04 (d, J = 7.7 Hz, 1H$_{Ar}$), 8.08 (t, J = 8.0 Hz, 1H$_{Ar}$), 8.29 (d, J = 7.7 Hz, 1H$_{Ar}$), 8.39 (dd, J = 8.2, 0.7 Hz, 1H$_{Ar}$), 8.65 (dd, J = 7.7, 0.7 Hz, 1H$_{Ar}$). ^{13}C-NMR (100 MHz, DMSO-d_6): δ$_C$ 122.19 (CH), 123.21 (CH), 125.91 (CH), 128.29 (C), 129.16 (CH), 130.58 (C), 132.18 (CH), 134.45 (C), 134.71 (CH), 136.21 (CH), 137.26 (C), 143.39 (C, 4-C). LRMS: m/z (ESI) = 284.0 (M + Na)$^+$.

7-Bromo-2-nitrodibenzo[b,d]thiophene 5,5-dioxide (**6a**). Compound **5a** (1.75 g, 6.7 mmol) was dissolved in H_2SO_4 (96%, 30 mL). N-Bromosuccinimide (1.28 g, 7.2 mmol) was added to this solution in several portions and the mixture was stirred at room temperature for 24 h. The suspension was poured into ice-water (90 mL) and stirred for 10 min. The precipitate was filtered off, washed with H_2O until neutral, followed by MeOH, and dried at 60 °C to obtain a yellowish solid (2.29 g). The raw material was recrystallized twice from MeCN to afford the pure title compound **6a** (1.45 g, 4.27 mmol, 63.8% yield, R_f = 0.30, cyclohexane/CHCl$_3$, 1:4) as a pale yellow powder, m.p. 281.0–282.5 °C. ^{1}H-NMR (400 MHz, DMSO-d_6): δ_H 8.08 (dd, J = 8.4, 1.8 Hz, 1H$_{Ar}$, 8-H), 8.31 (d, J = 8.4 Hz, 1H$_{Ar}$, 4-H), 8.41–8.47 (m, 3H$_{Ar}$, 9-H, 3-H, 6-H), 9.07 (d, J = 1.8 Hz, 1H$_{Ar}$, 1-H). ^{13}C-NMR (100 MHz, DMSO-d_6): δ_C 118.52 (CH), 123.72 (CH), 125.11 (C), 125.36 (CH), 125.85 (CH), 126.35 (CH), 128.41 (C), 132.19 (C), 137.90 (CH), 138.66 (C), 141.11 (C), 151.78 (C). LRMS: m/z (ESI) = 364.0, 362.0 (M + Na)$^+$.

7-Bromo-2-fluorodibenzo[b,d]thiophene 5,5-dioxide (**8a**). TMAF tetrahydrate (0.22 g, 1.3 mmol, Acros) was dried under an atmosphere of argon by azeotropic distillation with a mixture of DMSO (6 mL) and cyclohexane (12 mL) using a water separator for 6 h (bath temperature: 115–120 °C). The bath temperature was allowed to cool to 80 °C and the nitro derivative **6a** (0.34 g, 1 mmol) was added under stirring to the suspension of the dried TMAF in one portion. The mixture immediately turned dark brown and was stirred for additional 5 h at 95 °C. Reaction progress was monitored by TLC (cyclohexane/CHCl$_3$, 1:4). The portion of cyclohexane was evaporated. The residual oil was poured into water (60 mL) and extracted with CH_2Cl_2 (4 × 15mL).

The extracts were combined and washed with water (10 mL), dried (MgSO$_4$) and evaporated. The residual yellow solid (0.283 g) was purified by column chromatography (silica gel 15–40 µm, 6 g) with CH_2Cl_2 to afford the pure title compound **8a**, (0.23 g, 0.74 mmol, 73.7% yield, R_f = 0.39, cyclohexane/CHCl$_3$, 1:4) as a colorless powder, m.p. 266.5–268.5 °C. ^{1}H-NMR (300 MHz, DMSO-d_6): δ_H 7.52 (dt-like, J = 8.8, 8.7, 2.4 Hz, 1H$_{Ar}$, 3-H), 8.05 (dd, J = 8.3, 1.8 Hz, 1H$_{Ar}$, 8-H), 8.11 (dd, J = 8.6, 4.9 Hz, 1H$_{Ar}$, 4-H), 8.16 (d, J = 8.3 Hz, 1H$_{Ar}$, 9-H), 8.18 (dd, J = 9.2, 2.3 Hz, 1H$_{Ar}$, 1-H), 8.35 (d, J = 1.6 Hz, 1H$_{Ar}$, 6-H); ^{13}C-NMR (75 MHz, DMSO-d_6): δ_C 110.74 (d, $^2J_{CF}$ = 25.6 Hz, CH), 118.22 (d, $^2J_{CF}$ = 24.1 Hz, CH), 124.53 (s, C), 124.89 (d, $^3J_{CF}$ = 10.2 Hz, CH), 125.09 (s, CH), 125.16 (s, CH), 128.98 (d, $^4J_{CF}$ = 2.4 Hz, C), 132.82 (d, $^4J_{CF}$ = 3.0 Hz, C), 133.45 (d, $^3J_{CF}$ = 10.7 Hz, C), 137.49 (s, CH), 139.31 (s, C), 165.85 (d, $^1J_{CF}$ = 252.2 Hz, C); ^{19}F-NMR (282.36 MHz, DMSO-d_6): δ_F −103.8 (m, F$_{Ar}$, 2-F). LRMS: m/z (ESI) = 334.9, 336.9 (M + Na)$^+$.

2-Fluorodibenzo[b,d]thiophene 5,5-dioxide (**7a**). Similar to the preparation of **8a**, TMAF tetrahydrate (1.35 g, 8.18 mmol) was azeotropically dried with cyclohexane/DMSO and subsequently reacted with **5a** (1.525 g, 5.84 mmol). After work-up, the yellow solid (1.25 g) was purified by column chromatography (silica gel 15–40 µm, 18 g) with cyclohexane/CH_2Cl_2 (1:1→1:3) as eluent to yield the pure title compound **7a** (1.03 g, 4.4 mmol, 75.2% yield, R_f = 0.4, cyclohexane/CHCl$_3$, 1:4) as a colorless powder, m.p. 227–228 °C. ^{1}H-NMR (400 MHz, CDCl$_3$): δ_H 7.21 (dt, J = 8.45, 8.45, 2.27 Hz, 1H$_{Ar}$, H-3), 7.45 (dd, J = 8.37, 2.26 Hz, 1H$_{Ar}$, H-1), 7.57 (dt, J = 7.56, 7.53, 1.09 Hz, 1H$_{Ar}$, H-7), 7.66 (dt, J = 7.65, 7.58, 1.17 Hz, 1H$_{Ar}$, H-8), 7.75 (ddd, J = 7.70, 1.01, 0.57 Hz, 1H$_{Ar}$, 9-H), 7.82 (dd, J = 8.42, 4.92 Hz, 1H$_{Ar}$, H-4), 7.83 (ddd, J = 7.60, 1.09, 0.69 Hz, 1H$_{Ar}$, 6-H); ^{13}C-NMR (100 MHz, CDCl$_3$): δ_C 109.34 (d, $^2J_{CF}$ = 24.6 Hz, CH), 117.61 (d, $^2J_{CF}$ = 23.9 Hz, CH), 121.98 (s, CH), 122.40 (s, CH), 124.58 (d, $^3J_{CF}$ = 9.9 Hz, CH), 130.47 (d, $^4J_{CF}$ = 2.5 Hz, C), 131.23 (s, CH), 133.72 (d, $^4J_{CF}$ = 3.2 Hz, C), 134.13 (s, CH), 134.92 (d, $^3J_{CF}$ = 9.7 Hz, C), 138.68 (s, C), 166.37 (d, $^1J_{CF}$ = 255.0 Hz, C); ^{19}F-NMR (282.36 MHz, DMSO-d_6): δ_F ppm −103.63 (m. 1F, 2-F); LRMS: m/z (ESI) = 257.0, 258.0 (M + Na)$^+$.

3-Bromo-2-fluorodibenzo[b,d]thiophene 5,5-dioxide (**8c**) *and 3,7-dibromo-2-fluoro-dibenzo[b,d]-thiophene 5,5-dioxide* (**8d**). Similar to the preparation of compound **6a**, compound **7a** (0.56 g, 2.4 mmol) was reacted with NBS (0.43 g, 2.4 mmol) in H_2SO_4. The solid residue obtained upon work-up (0.75 g) consisted of a mixture of starting material (R_f = 0.32), mono- (R_f = 0.43) and dibrominated derivative (R_f = 0.56) as shown by TLC analysis (cyclohexane/$CHCl_3$, 1:4). The mixture was subjected to chromatographic purification (silica gel 15–40 µm, 28 g) with cyclohexane/$CHCl_3$ (1:3) to yield the 2-fluoro-3,7-dibromo derivative **8d** (0.17 g, 0.43 mmol, 18%) from the first fraction and the 2-fluoro-3-bromo derivative **8c** (0.35 g, 1.12 mmol, 46.5%) from the second fraction. Compound **8c**: m.p. 269–274.5 °C. ^{1}H-NMR (400 MHz, DMSO-d_6): δ_H 7.72 (t, J = 7.6 Hz, 1H_{Ar}, 7-H), 7.84 (t, J = 7.6 Hz, 1H_{Ar}, 8-H), 8.01 (d, J = 7.7 Hz, 1H_{Ar}, 6-H), 8.22 (d, J = 7.7 Hz, 1H_{Ar}, 9-H), 8.32 (d, J = 8.9 Hz, 1H_{Ar}, 1-H), 8.55 (d, J = 6.2 Hz, 1H_{Ar}, 4-H). ^{13}C-NMR (100 MHz, DMSO-d_6): δ_C 110.78 (d, $^2J_{CF}$ = 23.4 Hz, C), 111.57 (d, $^2J_{CF}$ = 26.5 Hz, CH), 122.07 (s, CH), 123.33 (s, CH), 127.71 (d, $^3J_{CF}$ = 1.5 Hz, CH), 129.16 (d, $^4J_{CF}$ = 2.1 Hz, C), 131.80 (s, CH), 133.30 (d, $^3J_{CF}$ = 9.9 Hz, C), 134.06 (d, $^4J_{CF}$ = 3.5 Hz, C), 134.77 (s, CH), 137.28 (s, C), 162.00 (d, $^1J_{CF}$ = 251.8 Hz, C). ^{19}F-NMR (376.4 MHz, DMSO-d_6): δ_F −98.68 (m, 1 F_{Ar}, 2-F). LRMS: *m/z* (ESI) = 336.9, 335.0 (M + Na)$^+$. Compound **8d**: m.p. 282.5–285 °C. ^{1}H-NMR (300 MHz, $CDCl_3$): δ_H 7.49 (d, J = 7.8 Hz, 1H_{Ar}, 1-H), 7.60 (d, J = 8.2 Hz, 1H_{Ar}, 9-H), 7.79 (dd, J = 8.2, 1.8 Hz, 1H_{Ar}, 8-H), 7.94 (d, J = 1.7 Hz, 1H_{Ar}, 6-H), 8.01 (d, J = 6.0 Hz, 1H_{Ar}, 4-H). ^{13}C-NMR (75 MHz, $CDCl_3$): δ_C 110.07 (d, $^2J_{CF}$ = 25.8 Hz, CH), 111.88 (d, $^2J_{CF}$ = 12.2 Hz, C), 123.28 (s, CH), 125.53 (s, C), 125.84 (s, CH), 128.11 (d, $^3J_{CF}$ = 1.9 Hz, CH), 128.75 (s, C), 132.68 (d, $^3J_{CF}$ = 8.8 Hz, C), 134.32 (d, $^4J_{CF}$ = 3.8 Hz, C), 137.41 (s, CH), 139.67 (s, C), 162.85 (d, J = 256.4 Hz, C). ^{19}F-NMR (282.4 MHz, $CDCl_3$) δ_F −98.68 (m, 1 F_{Ar}, 2-F). LRMS: *m/z* (ESI) = 414.9, 416.9, 413.0 (M + Na)$^+$.

7-(1,4-Diazabicyclo[3.2.2]nonan-4-yl)-2-fluorodibenzo[b,d]thiophene 5,5-dioxide (**10a**). A mixture of $Pd_2(dba)_3$ (11 mg, 0.012 mmol) and BINAP (15 mg, 0.024 mmol) in toluene (1.5 mL) was stirred for 30 min at 90 °C. The red-orange colored solution of the catalyst was allowed to cool (22 °C) and added to a mixture of 1,4-diazabicyclo[3.2.2]nonane (**9a**, 53 mg, 0.42 mmol) and **8a** (0.126 g, 0.40 mmol) in toluene (2 mL). Cs_2CO_3 (Alfa Aesar, Karlsruhe, Germany; previously dried for 2 h at 120 °C, 4 mbar; 0.39 g, 1.2 mmol) was then added, and the reaction mixture was stirred under an atmosphere of argon for 24 h at 90° C. After cooling to room temperature, the solid was filtered off and washed with CH_2Cl_2 (2 × 4 mL). The filtrate was evaporated and chromatographically purified (silica gel 15–40 µm, 8 g) with a gradient from $CHCl_3$ (100%) to $CHCl_3$/MeOH/NH_3 (aq) (100:8:0.8). The fractions containing the product were combined, evaporated and the solid residue was recrystallized from EtOH to afford the title compound **10a** (0.075 g, 0.21 mmol, 52% yield, R_f = 0.18, $CHCl_3$/MeOH/NH_3 (aq.), 100:10:1) as yellow crystals, m.p. 317–319 °C (dec.). ^{1}H-NMR (400 MHz, $CDCl_3$): δ_H 1.78 (qd-like, J = 9.7, 4.6 Hz, 2H, 6'-H_a, 9'-H_a), 2.11 (m, 2H, 6'-H_b, 9'-H_b), 2.99 (m, 2H, 7'-H_a, 8'-H_a), 3.09 (A-part of AA'BB', 2H, 2'-H_2), 3.14 (m, 2H, 7'-H_b, 8'-H_{ba}), 3.61 (B-part of AA'BB', 2H, 3'-H_2), 4.08 (m, not resolved, 1H, 5'-H), 6.89 (dd, J = 8.8, 2.5 Hz, 1H_{Ar}, 8-H), 7.00 (td-like, J = 8.5, 8.5, 2.2 Hz, 1H_{Ar}, 3-H), 7.10 (d, J = 2.5 Hz, 1H_{Ar}, 6-H), 7.24 (dd, J = 8.8, 2.2 Hz, 1H_{Ar}, 1-H), 7.49 (d, J = 8.7 Hz, 1H_{Ar}, 9-H), 7.70 (dd, J = 8.4, 4.9 Hz, 1H_{Ar}, 4-H). ^{13}C-NMR (100 MHz, $CDCl_3$): δ_C 26.77 (s, 2 CH_2), 44.69 (s, CH_2), 46.51 (s, 2 CH_2), 51.74 (s, CH), 57.05 (s, CH_2), 105.19 (s, CH), 107.66 (d, $^2J_{CF}$ = 25.1 Hz, CH), 114.70 (d, $^2J_{CF}$ = 24.3 Hz, CH), 117.03 (s, CH), 117.28 (d, $^4J_{CF}$ = 2.2 Hz, C), 123.15 (s, CH), 124.27 (d, $^3J_{CF}$ = 10.3 Hz, CH), 132.88 (d, $^4J_{CF}$ = 3.0 Hz, C), 136.19 (d, $^3J_{CF}$ = 9.6 Hz, C), 140.61 (s, C), 151.15 (s, C), 166.58 (d,

$^1J_{CF}$ = 253.6 Hz, C). ^{19}F-NMR (282.4 MHz, CDCl$_3$): δ$_F$ −104.5 (m, 1F$_{Ar}$, 2-F). HRMS *m/z* (ESI): calcd for C$_{19}$H$_{20}$FN$_2$O$_2$S (M + H)$^+$ 359.12240, found 359.12223.

3-(1,4-Diazabicyclo[3.2.2]nonan-4-yl)-6-fluorodibenzo[b,d]thiophene 5,5-dioxide (**10b**). In a procedure similar to the preparation of **10a**, compound **8b** (0.14 g, 0.45 mmol) and amine **9a** (0.059 g, 0.47 mmol) were reacted in the presence of Cs$_2$CO$_3$ (0.44 g, 1.35 mmol) and a catalyst made from Pd$_2$(dba)$_3$ (10.3 mg, 0.011 mmol) and BINAP (14 mg, 0.022 mmol). For purification the same protocol as for **10a** was followed to afford the title compound **10b** (0.077 g, 0.21 mmol, 48% yield, R$_f$ = 0.19, CHCl$_3$/MeOH/NH$_3$ (aq), 100:10:1) as yellow crystals, m.p. 272.5–274 °C (dec.). ^{1}H-NMR (300 MHz, CDCl$_3$): δ$_H$ 1.77 (qd-like, *J* = 9.7, 4.6 Hz, 2H, 6′-H$_a$, 9′-H$_a$), 2.11 (ddtd, *J* = 12.4, 9.7, 5.0, 2.6 Hz, 2H, 6′-H$_b$, 9′-H$_b$), 3.05–2.92 (m, 2H, 7′-H$_a$, 8′-H$_a$), 3.20–3.06 (m, A-part of AA′BB′, 4H, 7′-H$_b$, 8′-H$_{ba}$, 3′-H$_2$), 3.61 (t-like, *J* = 5.7 Hz, B-part of AA′BB′, 2H, 3′-H$_2$), 4.08 (m, not resolved, 1H, 5′-H), 6.89 (dd, *J* = 8.8, 2.5 Hz, 1H$_{Ar}$, 2-H), 6.97 (t, *J* = 8.4 Hz, 1H$_{Ar}$, 7-H), 7.09 (d, *J* = 2.5 Hz, 1H$_{Ar}$, 4-H), 7.35 (d, *J* = 7.6 Hz, 1H$_{Ar}$, 9-H), 7.50 (dt, *J* = 8.0, 5.2 Hz, 1H$_{Ar}$, 8-H), 7.53 (d, *J* = 8.7 Hz, 1H$_{Ar}$, 1-H). ^{13}C-NMR (75 MHz, CDCl$_3$): δ$_C$ 26.78 (s, 2 CH$_2$), 44.70 (s, CH$_2$), 46.52 (s, 2 CH$_2$), 51.76 (s, CH), 57.06 (s, CH$_2$), 105.09 (s, C$_{ArH}$), 115.17 (d, $^2J_{CF}$ = 19.4 Hz, C$_{ArH}$), 115.85 (d, $^4J_{CF}$ = 3.4 Hz, C$_{ArH}$), 117.09 (s, C$_{ArH}$), 117.65 (d, $^4J_{CF}$ = 2.8 Hz, C$_{Ar}$), 123.25 (s, C$_{ArH}$), 123.63 (d, $^2J_{CF}$ = 17.7 Hz, C$_{Ar}$), 135.83 (d, $^3J_{CF}$ = 2.7 Hz, C$_{Ar}$), 136.17 (d, $^3J_{CF}$ = 7.9 Hz, C$_{ArH}$), 140.42 (s, C$_{Ar}$), 151.12 (s, C$_{Ar}$), 158.05 (d, $^1J_{CF}$ = 257.4 Hz, C$_{Ar}$). ^{19}F-NMR (282.4 MHz, CDCl$_3$): δ$_F$ −115.66 (m, F$_{Ar}$, 6-F). HRMS *m/z* (ESI): calcd for C$_{19}$H$_{20}$FN$_2$O$_2$S (M + H)$^+$ 359.12240, found 359.12235.

3-(1,4-Diazabicyclo[3.2.2]nonan-4-yl)-2-fluorodibenzo[b,d]thiophene 5,5-dioxide (**10c**). In a procedure similar to the preparation of **10a**, compound **8c** (0.126 g, 0.4 mmol) and **9a** (0.053 g, 0.42 mmol) were reacted in the presence of Cs$_2$CO$_3$ (0.39 g, 1.2 mmol) and a catalyst mixture made from Pd$_2$(dba)$_3$ (11 mg, 0.012 mmol) and BINAP (14.9 mg, 0.024 mmol). For purification the same protocol for **10a** was followed to afford the title compound **10c** (0.063 g, 0.18 mmol, 44% yield, R$_f$ = 0.20, CHCl$_3$/MeOH/NH$_3$ (aq), 100:10:1) as pale yellow crystals, m.p. 305–309 °C (dec.). ^{1}H-NMR (300 MHz, CDCl$_3$): δ$_H$ 1.75 (m, 2H, 6′-H$_a$, 9′-H$_a$), 2.11 (m, 2H, 6′-H$_b$, 9′-H$_b$), 3.04 (m, 4H, 7′-H$_2$, 8′-H$_2$), 3.16 (t, *J* = 5.6 Hz, A-part of AA′BB′, 2H, 2′-H$_2$), 3.44 (t, *J* = 5.6 Hz, B-part of AA′BB′, 2H, 3′-H$_2$), 3.80 (m, 1H, 5′-H), 7.32 (s, 1H$_{Ar}$, 4-H), 7.36 (d, *J* = 5.0 Hz, 1H$_{Ar}$, 1-H), 7.43 (ddd, *J* = 7.7, 5.1, 3.4 Hz, 1H$_{Ar}$, 7-H), 7.55–7.63 (m, 2H$_{Ar}$, 8-H, 9-H), 7.75 (d, *J* = 7.6 Hz, 1H$_{Ar}$, 6-H). ^{13}C-NMR (75 MHz, CDCl$_3$): δ$_C$ 27.64 (s, 2 CH$_2$), 46.70 (s, 2 CH$_2$), 47.50 (d, $^4J_{CF}$ = 1.2 Hz, CH$_2$), 55.83 (d, $^4J_{CF}$ = 6.0 Hz, CH), 56.30 (s, CH$_2$), 109.97 (d, $^2J_{CF}$ = 24.9 Hz, CH), 111.82 (d, $^3J_{CF}$ = 5.5 Hz, CH), 120.89 (s, CH), 122.19 (s, CH), 123.72 (d, $^3J_{CF}$ = 9.4 Hz, C), 129.25 (s, CH), 131.40 (d, $^4J_{CF}$ = 2.3 Hz, C), 133.98 (d, $^4J_{CF}$ = 2.9 Hz, C), 134.05 (s, C), 137.87 (s, C), 142.90 (d, $^2J_{CF}$ = 10.5 Hz, C), 157.62 (d, $^1J_{CF}$ = 252.1 Hz, C). ^{19}F-NMR (282.4 MHz, CDCl$_3$): δ$_F$ −112.95 (m, 1 F$_{Ar}$, 2-F). HRMS *m/z* (ESI): calcd for C$_{19}$H$_{20}$FN$_2$O$_2$S (M+H)$^+$ 359.12240, found 359.12235.

6-Fluoro-3-(9-methyl-3,9-diazabicyclo[3.3.1]nonan-3-yl)dibenzo[b,d]thiophene 5,5-dioxide (**12b**). In a procedure similar to the preparation of **10a**, compound **8b** (0.13 g, 0.42 mmol) and amine **9b** (0.065 g, 0.46 mmol) were reacted in the presence of Cs$_2$CO$_3$ (0.28 g, 0.84 mmol) and a catalyst made from Pd$_2$(dba)$_3$ (11.5 mg, 0.013 mmol) and BINAP (15.7 mg, 0.0252 mmol). For purification the same protocol for **10a** was followed to afford the title compound **12b** (0.065 g, 0.175 mmol, 42% yield, R_f = 0.31, CHCl$_3$/MeOH/NH$_3$ (aq), 100:10:1) as yellow crystals, m.p. 264–272 °C (dec.). ^{1}H-NMR (300 MHz, CDCl$_3$): δ_H 1.52–1.64 (m, 3H, 7′-H$_a$, 6′-H$_a$, 8′-H$_a$), 1.93–2.15 (m, 3H, 7′-H$_b$, 6′-H$_b$, 8′-H$_b$), 2.59 (s, 3H, 9′-NCH$_3$), 3.01 (s, 2H, 1′-H, 5′-H), 3.39–3.51 (m, 4H, 2′-H$_2$, 4′-H$_2$), 6.97–7.03 (m, 2H$_{Ar}$, 7-H, 2-H), 7.22 (d, J = 2.4 Hz, 1H$_{Ar}$, 4-H), 7.39 (dd, J = 7.7, 0.5 Hz, 1H$_{Ar}$, 9-H), 7.52 (ddd, J = 8.2, 7.8, 5.2 Hz, 1H$_{Ar}$, 8-H), 7.59 (d, J = 8.7 Hz, 1H$_{Ar}$, 1-H). ^{13}C-NMR (75 MHz, CDCl$_3$): δ_C 18.99 (s, 1 C$_{sec}$), 27.93 (s, 2C$_{sec}$), 40.62 (s, CH$_3$), 47.40 (s, 2CH$_2$), 53.03 (s, 2CH), 105.41 (s, C$_{Ar}$H), 115.46 (d, $^2J_{CF}$ = 19.4 Hz, C$_{Ar}$H), 116.05 (d, $^4J_{CF}$ = 3.4 Hz, C$_{Ar}$H), 117.22 (s, C$_{Ar}$H), 118.83 (d, $^4J_{CF}$ = 2.6 Hz, C$_{Ar}$), 122.99 (s, C$_{Ar}$H), 123.77 (d, $^2J_{CF}$ = 18.1 Hz, C$_{Ar}$), 135.76 (d, $^3J_{CF}$ = 2.5 Hz, C$_{Ar}$), 136.20 (d, $^3J_{CF}$ = 7.8 Hz, 1C$_{Ar}$H), 140.10 (s, C$_{Ar}$), 152.27 (s, C$_{Ar}$), 158.06 (d, $^1J_{CF}$ = 257.4 Hz, C$_{Ar}$). ^{19}F-NMR (282 MHz, CDCl$_3$): δ −115.56 (dd-like, J = 8.4, 5.1 Hz, 1F, 6-F). HRMS m/z (ESI): calcd for C$_{20}$H$_{21}$FN$_2$O$_2$S (M + H)$^+$ 373.13805, found 373.13773.

7-(1,4-Diazabicyclo[3.2.2]nonan-4-yl)-2-nitrodibenzo[b,d]thiophene 5,5-dioxide (**11a**). In a procedure similar to the preparation of **10a**, compound **6a** (0.17 g, 0.5 mmol) and amine **9a** (0.064 g, 0.51 mmol) were reacted in the presence of Cs$_2$CO$_3$ (0.49 g, 1.5 mmol) and a catalyst mixture made from Pd$_2$(dba)$_3$ (18.3 mg, 0.02 mmol) and BINAP (25 mg, 0.04 mmol). For purification the same protocol for **10a** was followed to afford the title compound **11a** (0.10 g, 0.18 mmol, 53% yield, R_f = 0.21, CHCl$_3$/MeOH/NH$_3$ (aq), 100:10:1) as dark red crystals, m.p. 291–294 °C (dec.). ^{1}H-NMR (300 MHz, CDCl$_3$): δ_H 1.79 (qd-like, J = 9.6, 4.6 Hz, 2H, 6′-H$_a$, 9′-H$_a$), 2.12 (m, 2H, 6′-H$_b$, 9′-H$_b$), 2.99 (m, 2H, 7′-H$_a$, 8′-H$_a$), 3.10 (A-part of AA′BB′, 2H, 2′-H$_2$), 3.13 (m, 2H, 7′-H$_b$, 8′-H$_{ba}$), 3.65 (B-part of AA′BB′, 2H, 3′-H$_2$), 4.11 (m, not resolved, 1H, 5′-H), 6.94 (dd, J = 8.8, 2.6 Hz, 1H$_{Ar}$, 8-H), 7.11 (d, J = 2.5 Hz, 1H$_{Ar}$, 6-H), 7.63 (d, J = 8.8 Hz, 1H$_{Ar}$, 9-H), 7.85 (d, J = 8.3 Hz, 1H$_{Ar}$, 4-H), 8.15 (dd, J = 8.3, 2.0 Hz, 1H$_{Ar}$, 3-H), 8.35 (d, J = 1.9 Hz, 1H$_{Ar}$, 1-H). ^{13}C-NMR (75 MHz, CDCl$_3$): δ_C 26.70 (2 CH$_2$), 44.71 (CH$_2$), 46.47 (2 CH$_2$), 51.72 (CH), 57.02 (CH$_2$), 105.18 (CH), 115.16 (CH), 116.04 (C), 117.33 (CH), 122.59 (CH), 123.18 (CH), 123.67 (CH), 135.15 (C), 140.17 (C), 141.86 (C), 151.46 (C), 151.82 (C). HRMS m/z (ESI): calcd for C$_{19}$H$_{20}$N$_3$O$_4$S (M + H)$^+$ 386.11690, found 386.11704.

3.3. Manual Radiosynthesis of [^{18}F]10a

No-carrier-added (n.c.a.) [^{18}F]fluoride was produced via the ^{18}O(p,n)^{18}F nuclear reaction by irradiation of a [^{18}O]H$_2$O target (Hyox 18 enriched water; Rotem Industries Ltd, Mishor Yamin, Israel) on a Cyclone®18/9 (IBA RadioPharma Solutions, Louvain-la-Neuve, Belgium) with 18 MeV proton beam using a Nirta® [^{18}F]fluoride XL target or [^{18}O]H$_2$O recycled by the established in-house method [63]. Starting with 1–2 GBq of n.c.a. ^{18}F-fluoride, [^{18}F]F$^-$-containing anion resin was eluted with a 20 mg mL^{-1} aqueous solution of K$_2$CO$_3$ (1.78 mg, 12.9 mmol) and added to a 5 mL V-vial in the Discover PET wave microwave CEM® (CEM Corporation, Matthews, NC, USA) cavity in the presence of Kryptofix®222 (11.2 mg, 29.7 mmol) in 1 mL MeCN. The aqueous [^{18}F]fluoride was dried under vacuum and argon flow in the microwave cavity (75 W, 20 cycles) at 50–60 °C for 10–12 min. Additional aliquots of MeCN (2 × 1.0 mL) were added for azeotropic drying. Thereafter, the precursor (1 mg) was added to the reactive anhydrous K[^{18}F]F-Kryptofix®222/K$_2$CO$_3$-complex. The reaction parameters were optimized varying the reaction time, temperature, solvent (MeCN, DMF, DMSO), and heating condition (thermal *vs.*

microwave-assisted). Thereafter, the crude reaction mixture was diluted with H_2O/MeCN (4:1, v/v) and directly applied onto an isocratic semi-preparative HPLC for isolation of [^{18}F]**10a**. Fractions were collected, diluted with 30 mL of H_2O and sodium hydroxide was added to neutralize the solution (~100–200 µL of 1 M NaOH). Final purification was performed using a Sep-Pak® C18 light cartridge (Waters, Milford, MA, USA) followed by elution with 1 mL of ethanol. To obtain an injectable solution the solvent was concentrated under a gentle argon stream and [^{18}F]**10a** was formulated in a sterile isotonic saline solution (5%–10% EtOH, v/v). The identity of [^{18}F]**10a** was verified by radio-HPLC analysis of a sample of the radiotracer solution spiked with the non-radioactive reference **10a**. Radiochemical and chemical purities were assessed by radio-TLC and analytical HPLC. The mass determination for specific activity was determined on the base of a calibration curve carried out under the same analytical HPLC conditions.

3.4. Automated Radiosynthesis of [^{18}F]**10a** *and* [^{18}F]**10b** *([^{18}F]ASEM)*

Remote controlled radiosynthesis was performed using a TRACERLAB™ FX$_{FN}$ synthesizer (GE Healthcare, Waukesha, WI, USA) equipped with a PU-980 pump (JASCO, Gross-Umstadt, Germany), WellChrom K-2001 UV detector (KNAUER GmbH, Berlin, Germany), NaI(Tl)-counter and automated data acquisition (NINA software version 4.8 rev. 4; Nuclear Interface GmbH, Dortmund, Germany). For transfer to the automated module we started with activities in the range of 5–12 GBq of n.c.a. ^{18}F-fluoride. Based upon the conditions optimized manually, the nitro-to-fluoro displacement was achieved by adding the respective nitro precursor (1 mg) dissolved in anhydrous DMF (1 mL) to the K[^{18}F]F-Kryptofix®222/K_2CO_3-complex. The reaction mixture was stirred at 120 °C for 10 min. After cooling, the reaction mixture was diluted with 3 mL of H_2O/MeCN (4:1) and directly applied onto the semi-preparative Reprosil-Pur 120 C18-AQ HPLC column (250 × 20 mm, 10 µm) using a solvent composition of 35% MeCN/H_2O/0.05%TFA as eluent and a flow rate of 10 mL·min^{-1}. The desired product was collected in 40 mL of H_2O, neutralized with 1 M NaOH and directly transferred to a pre-activated Sep-Pak® C18 light cartridge. The cartridge was washed with 2 mL of water and [^{18}F]**10a** or [^{18}F]**10b** was obtained after elution with 1 mL of EtOH. Injectable solutions were obtained by partial evaporation of the solvent under a gentle argon stream at 70 °C and dilution in isotonic saline (5%–10% of EtOH, v/v). Radiochemical purity and specific activity were assessed following the chromatographic methods described in quality control.

3.5. Quality Control

To control the quality of the K[^{18}F]F-Kryptofix®222/K_2CO_3-complex, reactions using ethylene glycol ditosylate were performed randomly. Radio-TLC of [^{18}F]**10a** and [^{18}F]**10b** was performed on Polygram® ALOX N/UV254 plates (Macherey-Nagel GmbH & Co. KG) with dichloromethane/methanol (9:1, v/v). The spots of the references were visualized using UV light at 254 nm. Radiochemical purity and specific activity were determined using analytical radio-HPLC with a Reprosil-Pur C18-AQ column (250 × 4.6 mm, 5 µm) and 26% MeCN/H_2O/0.05% TFA as eluent at a flow rate of 1 mL·min^{-1}. Analytical radio-HPLC profiles for the *in vivo* metabolism analysis in plasma was assessed by using a gradient mode (0–10 min: 10% MeCN/20 mM $NH_4OAc_{aq.}$, 10–40 min: 10%→90% MeCN/20 mM $NH_4OAc_{aq.}$, 40–50 min:

90%→10% MeCN/20 mM NH$_4$OAc$_{aq.}$, 50–60 min: 10% MeCN/20 mM NH$_4$OAc$_{aq.}$) on a Reprosil-Pur C18-AQ (250 × 4.6 mm, 5 μm) column at a flow rate of 1 mL·min^{-1}.

3.6. Determination of in Vitro Stability and Lipophilicity (Log D$_{7.2}$)

In vitro radiochemical stability of [^{18}F]**10a** was investigated in 0.9% NaCl solution, PBS (pH 7.2), and 0.01 M Tris-HCl (pH 7.4 at 21 °C) at 40 °C and in EtOH at room temperature for up to 90 min. Samples were taken at 15, 30, 60, and 90 min of incubation time and analyzed by radio-TLC and radio-HPLC. Log D$_{7.2}$ of [^{18}F]**10a** was experimentally determined in n-octanol/phosphate-buffered saline (PBS; 0.01 M, pH 7.2) at room temperature by the shake-flask method in multiple distribution. Measurement was performed twice in triplicate.

3.7. Biological Evaluation

3.7.1. Animals

Male Sprague-Dawley rats (275–300 g) were obtained from Charles River Laboratories, and female piglets (German Landrace pigs DL × Large White/Pietrain; mean weight 15.8 ± 0.8 kg, mean age 8 weeks) as well as female CD-1 mice (10–12 weeks, 22–26 g) were obtained from Medizinisch-Experimentelles Zentrum, Universität Leipzig, and all animals were provided with food and water ad libitum. All of the animal experiments were performed in accordance with the European Communities Council Directive of 24th November 1986 (86/609/EEC), and experiments using female piglets and female CD-1 mice were approved by the Animal Care and Use Committee of Saxony (TVV 08/13). Imaging experiments in rhesus monkeys were performed in accordance with a protocol approved by the Yale Institutional Animal Care and Use Committee.

3.7.2. Binding Assays

Affinity for Human α$_7$ nAChR

The affinity of the test compounds for human α$_7$ nAChR was determined by radioligand displacement experiments as previously described [12]. In brief, SH-SY5Y cells stably expressing human α$_7$ nAChR [64] were grown in DMEM/Ham'sF-12, supplemented with 10% FCS, stable glutamine, geneticin (100 μg·mL^{-1}), penicillin/streptomycin (100 U·mL^{-1}, 100 μg·mL^{-1}) at 37 °C in an atmosphere containing 5% CO$_2$. Cells were harvested by scraping, sedimented (800 rpm, 3 min), diluted with 50 mM TRIS–HCl, pH 7.4, and stored at −25°C until use. For determination of α$_7$ nAChR affinity of reference compounds, frozen cell suspensions were thawed, homogenized by a 27-gauge needle, and diluted with incubation buffer (50 mM TRIS–HCl, pH 7.4, 120 mM NaCl, 5 mM KCl). Membrane suspension was incubated with [^{3}H]methyllycaconitine ([^{3}H]MLA; ~0.3 nM final concentration; specific activity of 2.220 GBq·mmol^{-1}; NEN Life Sciences Products, Boston, MA, USA) and various concentrations of the test compound. Nonspecific binding was determined by co-incubation with 300 μM (−)-nicotine tartrate. The incubation was performed at room temperature for 120 min and terminated by rapid filtration using Whatman GF/B glass-fibre filters, presoaked in 0.3% polyethyleneimine, and a 48-channel harvester (Biomedical Research and Development Laboratories, Gaithersburg, MD, USA), followed by 4×

washing with ice-cold 50 mM TRIS-HCl, pH 7.4. Filter-bound radioactivity was quantified by liquid scintillation counting. The 50% inhibition concentrations (IC_{50}) were estimated from the competition curves by nonlinear regression using GraphPadPrism software and the K_i values calculated according to the Cheng-Prusoff equation [65].

Affinity for Human $\alpha_4\beta_2$ and $\alpha_3\beta_4$ nAChR

The affinity of the test compounds for human $\alpha_4\beta_2$ and $\alpha_3\beta_4$ nAChR was determined as described above. In brief, HEK293 cells stably expressing human $\alpha_4\beta_2$ or $\alpha_3\beta_4$ nAChR [66] were cultured. The cells were harvested and processed as described above, and membrane suspension was incubated with ($\pm$)-[^{3}H]epibatidine (0.4 to 0.6 nM final concentration; specific activity of 1250 GBq·mmol^{-1}; PerkinElmer Human Health, Rodgau, Germany) and various concentrations of the test compound. Nonspecific binding was determined in the presence of 300 µM ($-$)-nicotine tartrate. Incubation of the samples, separation of free from receptor-bound radioligand, and data evaluation were performed as described above.

Affinity for Subunit-Specific Antibodies Immobilized Native Rat $\alpha_6\beta_2$* nAChR

The specificity of the anti-α_6 antibody was tested as described [67]. The receptors were prepared and immobilized as previously described [68] by the subunit-specific antibody. The inhibition of [^{3}H]epibatidine binding to the immobilized subtypes by the test compounds was measured by pre-incubating increasing concentrations of the compounds for 1 h at room temperature, followed by incubation overnight at 4 °C with 0.1 nM ($\pm$)-[^{3}H]epibatidine (specific activity of 1250 GBq·mmol^{-1}; PerkinElmer Human Health, Rodgau, Germany). Incubations were performed in a buffer containing 50 mM Tris-HCl, pH 7, 150 mM NaCl, 5 mM KCl, 1 mM MgCl$_2$, 2.5 mM CaCl$_2$, 2 mg·mL^{-1} BSA, and 0.05% Tween 20. Specific ligand binding was defined as total binding minus the binding in the presence of 100 nM ($\pm$)-epibatidine dihydrochloride hydrate. After incubation, the wells were washed seven times with ice-cold PBS containing 0.05% Tween 20, the bound [^{3}H]epibatidine released by adding 200 µL of 2 M NaOH to each well and after incubation for 2 h determined by means of liquid scintillation in a beta counter.

Saturation binding analysis revealed an affinity of ($\pm$)-[^{3}H]epibatidine of 25 pM (CV 20%) for the nAChRs immobilized by anti-α_6 antibody from rat striatum, which contains two major $\alpha_6\beta_2$* subtypes ($\alpha_6\beta_2\beta_3$ and $\alpha_6\alpha_4\beta_2\beta_3$) in very similar proportions.

The experimental data obtained from the binding experiments were analyzed by means of a non-linear least square procedure using the LIGAND program [69]. The K_i values of the test compounds were also determined by means of the LIGAND program by simultaneously fitting the data obtained from three to five independent saturation and competition binding experiments.

Affinity for Human 5-HT$_3$ Receptor

The affinity of the test compounds for human 5-HT$_3$ receptor was evaluated by a contract research organization (Eurofins Panlabs, Inc., Taipei, Taiwan). All compounds were tested at 1 µM in radioligand binding assays using human recombinant HEK-293 cells for the percentage inhibition of the binding by the 5-HT$_3$ specific radioligand [^{3}H]GR-65630 (working concentration = 0.69 nM, K_D = 0.20 nM). In addition, for compound **10a** a K_i value was estimated based on the inhibition of the radioligand binding at seven concentrations of the test compound (20 µM–5 nM).

Affinity for Other Target Proteins

The activity of compound **10a** was evaluated by a contract research organization (Eurofins Panlabs, Inc.). The test concentration was 1 µM and radioligand binding assays for dopamine transporter (DAT), serotonin transporter (SERT), norepinephrine transporter (NET), monoamine transporter (VMAT), choline transporter, and α$_1$ nAChR performed according to the manufacturer's protocols. Responses were judged as significant at ≥50% inhibition or stimulation of the binding of the specific radioligand.

3.7.3. PET Studies in Piglets

Six female animals were used in this study with anesthesia and surgery performed as described previously [70]. In brief, the animals received initially a premedication with midazolam (1 mg·kg^{-1} i.m.) followed by induction of anesthesia with 3% isoflurane in 70% N$_2$O/30% O$_2$. All incision sites were infiltrated with 1% lidocaine, and the anesthesia was maintained with 1.5% isoflurane throughout the surgical procedure. A central venous catheter was introduced through the left external jugular vein and used for the administration of the radiotracer and drugs, and for volume substitution with heparinized (50 IE·mL^{-1}) lactated Ringer's solution (2 mL·kg^{-1}·h^{-1}). An endotracheal tube was inserted by tracheotomy for artificial ventilation (Servo Ventilator 900C, Siemens-Elema, Solna, Sweden) after immobilization with pancuronium bromide (0.2 mg·kg^{-1}·h^{-1}). The artificial ventilation was adjusted to obtain normoxia and normocapnia (Radiometer ABL 500, Copenhagen, Denmark). Polyurethane catheters (i.d., 0.5 mm) were advanced through the left and the right femoral arteries into the abdominal aorta to withdraw arterial blood samples for regular monitoring of blood gases and for radiotracer input function measurements and metabolite analyses. The body temperature was monitored by a rectal temperature probe and maintained at ~38 °C by a heating pad. After the surgical procedure has been completed, anesthesia was maintained with 0.5% isoflurane in 70% N$_2$O/30% O$_2$, and the animals were allowed to stabilize for 1 h before imaging procedure.

PET Scanning Protocol

PET imaging was performed according to the protocol described recently [21]. In brief, a clinical tomograph (ECAT Exact HR+; Siemens Healthcare GmbH, Erlangen, Germany) was used with animals lying prone with the head placed in a custom-made head holder. For attenuation and scatter correction, transmission scans were acquired using three rotating ^{68}Ge rod sources. The radiotracer was injected as an intravenous (i.v.) bolus (10 mL saline with 340–494 MBq [^{18}F]**10a**) by a syringe pump within 2 min followed by flushing with 10 mL heparinized saline (50 IE·mL^{-1}). PET scanning started at the time of injection (0 min) and dynamic emission data were acquired for a total of 240 min. Three animals were investigated under baseline conditions, and another three under blocking conditions with administration of 3 mg·kg^{-1} i.v. of the α$_7$ nAChR partial agonist NS6740 at 3 min before radiotracer injection followed by a continuous infusion throughout the scan (1 mg·kg^{-1}·h^{-1}).

Arterial blood was sampled using a peristaltic pump during the first 20 min of the scan followed by manual sampling at 25, 30, 40, 50, 60, 90, 120, 150, 180, 210, and 240 min after injection. Plasma was obtained by centrifugation (13,000 rpm, 2 min at 21 °C) and total radioactivity concentration was measured using a gamma-counter (1470 Wizard; PerkinElmer, Shelton, CT, USA) cross-calibrated to the PET scanner. In addition, arterial whole blood was sampled manually at 4, 16, 30, 60, and 120 min p.i. and plasma obtained for metabolite analyses to determine the fraction of non-metabolized [^{18}F]**10a**.

Arterial Input Function

The plasma input function was corrected for the presence of radioactive metabolites by determination of the parent fraction as described previously [12]. Plasma samples were de-proteinized by addition of ice-cold acetonitrile (500 µL MeCN/100 µL plasma) followed by centrifugation (13,000 rpm, 10 min). Aliquots of the original plasma sample and of the supernatant obtained after centrifugation were counted for radioactivity, and the supernatant was analyzed by radio-HPLC following the chromatographic methods described in quality control. The parent fraction as a function of time was fitted iteratively with the Levenberg-Marquardt algorithm, including decay correction, to calculate metabolite corrected input function [71].

Quantification of PET Data

After correction for attenuation, scatter, decay and scanner-specific dead time, images were reconstructed by filtered back-projection using a Hanning-filter of 4.9 mm FWHM into 40 frames of increasing length. A summed PET image of the 240-min scan was reconstructed for each animal and used for alignment with a T1-weighed MR image of a 6-week-old farm-bred pig as described previously [72]. The time-activity curves (TACs) were calculated for the following volumes of interest (VOIs): frontal cortex, temporal cortex, parietal cortex, occipital cortex, hippocampus, striatum (defined as mean radioactivity in caudate and putamen), cerebellum, thalamus, middle cortex, ventral cortex, midbrain, pons, and colliculi. Radioactivity in all VOIs was calculated as the average of radioactivity concentration (kBq·cm^{-3}) in the left and right sides. To generate standardized uptake values (SUVs) the TACs of the individual VOIs were normalized to the injected dose and corrected for animal weight (in kBq·g^{-1}). Kinetic analysis of the time-activity curves (TACs) was performed using an "in house" data analysis tool [73].

3.7.4. PET Study in Rhesus Monkey

One male rhesus monkey (*Macaca mulatta*, 6 years old, body weight = 10 kg) underwent two PET imaging sessions with [^{18}F]**10a** and [^{18}F]**10b**, respectively. The scans were conducted 19 days apart. The monkey was initially sedated by intramuscular ketamine at 10 mg·kg^{-1} at ~2 h before the PET scan and anesthetized by 1.5%–3% isoflurane throughout the imaging session. PET scan was performed using a FOCUS-220 PET scanner (Siemens Preclinical Solutions, Knoxville, TN, USA). After a transmission scan, 167 MBq of [^{18}F]**10a** (specific activity of 362 GBq·µmol^{-1}) or 185 MBq of [^{18}F]**10b** (specific activity of 159 GBq·µmol^{-1}) were administered intravenously. Emission data were acquired in list mode for 240 min. Raw list-mode PET data was histogrammed (frames of 6 × 30 s; 3 × 1 min; 2 × 2 min;

46 × 5 min to scan termination) and reconstructed with 2D filtered back projection, using a 0.15 cm^{-1} Shepp filter and including corrections for scanner normalization, detector dead time, randoms, and radiation scatter and attenuation. The PET images were then registered to MR images, acquired with a Siemens 3 T Trio scanner, with a 6-parameter rigid body registration. The MR native space was then normalized using nonlinear affine registration to a high-resolution rhesus monkey atlas using BioImage Suite 3.01 (http://www.bioimagesuite.org/index.html). Time-activity curves were extracted by mapping atlas-defined regions to PET native space using the optimal transformation matrices calculated in the registration and normalization steps. Regions extracted included amygdala, caudate, cerebellum, cingulate, cortical regions (frontal, occipital, and temporal cortex), hippocampus, putamen, thalamus, and pons.

3.7.5. Toxicity Studies in Rats

The acute toxicity of **10a** was assessed when administered by a single intravenous injection to rats followed by an observation period of 2 or 15 day (Hameln rds, Modra, Slovak Republic). The study was performed with 4 test groups, including 1 control and 3 dose groups (6.2, 62 and 620 µg·kg^{-1}), with 60 males and 60 females Wistar rats divided into two experiment cohorts: Day 2 (40 males and 40 females) and Day 15 (20 males and 20 females). The animals were weighed and allocated to the test groups based on their body weights. The animals were sacrificed in two time periods (day 2 and day 15 after test item administration) and then examined macroscopically and histopathologically. The following parameters were evaluated: mortality, clinical observation, body weights, food consumption, haematology (white blood cells, red blood cells, hemoglobin, hematocrit, mean corpuscular volume, mean corpuscular hemoglobin, platelets, lympohcytes, neutrophils, eosinophils, basophils, monocytes), clinical chemistry (alkaline phosphatase, aspartate aminotransferase, alanine aminotransferase, glucose, cholesterol, triacylglycerols, creatinine, urea, bilirubin, albumin, calcium, phosphorus, sodium, potassium chloride), pathology and histopathology.

4. Conclusions

We have designed and synthesized a number of compounds based on a new pharmacophore and tested their *in vitro* binding affinity and selectivity for α7 nAChR over other receptor subtypes. Among the new ligands, compound **10a** was found to have high α7 nAChR binding affinity and selectivity, and was selected for radiolabeling with ^{18}F and evaluation as PET imaging radioligand for α7 nAChR *in vivo*. [^{18}F]**10a** was produced in high radiochemical yield and purity both manually and in an automated synthesis. PET imaging experiments in piglets and monkey demonstrated that [^{18}F]**10a** possessed appropriate pharmacokinetic properties and displayed specific and displaceable binding to α7 nAChR in the brain. Taken together, the new radioligand [^{18}F]**10a** ([^{18}F]DBT-10) appears to be a promising agent for PET imaging and quantification of α7 nAChR availability in the primate brain.

Supplementary Materials

Supplementary materials including protocols for the preparation of intermediate compounds and NMR spectra of compounds **10, 10a, 10b, 10c, 11a, 12a, 12b, 13b** and **14b** can be accessed at: http://www.mdpi.com/1420-3049/20/10/18387/s1.

Acknowledgments

The Deutsche Forschungsgemeinschaft is acknowledged for financial support (Project DE 1165/2-3). We thank the staff of the Institute of Analytical Chemistry, Department of Chemistry and Mineralogy of the University of Leipzig, for the NMR spectra, Karsten Franke, Helmholtz-Zentrum Dresden-Rossendorf (HZDR), and the cyclotron staff of the Department of Nuclear Medicine of the University Hospital Leipzig for providing [^{18}F]fluoride, as well as Tina Spalholz, HZDR, and Juliane Schaller, HZDR, for technical assistance. Support to DanPET AB from CCJobs (Creating Competitive Jobs), FBÖ TransTechTrans and NRU, COGNITO, Copenhagen, is gratefully acknowledged.

Author Contributions

Rodrigo Teodoro, Barbara Wenzel, Jörg Steinbach, Marianne Patt, Ming-Qiang Zheng and Yiyun Huang designed and conducted the radiochemical experiments. Matthias Scheunemann and Dan Peters conceived and performed the chemical syntheses. Francesca Maria Fasoli, Cecilia Gotti and Winnie Deuther-Conrad conducted the *in vitro* binding studies. Winnie Deuther-Conrad, Cornelius K. Donat, Peter Brust, Osama Sabri, Ming-Qiang Zheng, Ansel Hillmer and Yiyun Huang designed and conducted the PET imaging experiments. Rodrigo Teodoro, Barbara Wenzel, Peter Brust, Mathias Kranz, Winnie Deuther-Conrad, Cecilia Gotti, Ansel Hillmer and Yiyun Huang analyzed the data. Rodrigo Teodoro, Matthias Scheunemann, Winnie Deuther-Conrad, Peter Brust, Osama Sabri, Ansel Hillmer, Ming-Qiang Zheng and Yiyun Huang wrote and revised the manuscript.

Conflicts of Interest

The authors declare no conflict of interest.

References

1. Dineley, K.T.; Pandya, A.A.; Yakel, J.L. Nicotinic ACh receptors as therapeutic targets in CNS disorders. *Trends Pharmacol. Sci.* **2015**, *36*, 96–108.
2. Weiland, S.; Bertrand, D.; Leonard, S. Neuronal nicotinic acetylcholine receptors: From the gene to the disease. *Behav. Brain Res.* **2000**, *113*, 43–56.
3. Briggs, C.A.; Gronlien, J.H.; Curzon, P.; Timmermann, D.B.; Ween, H.; Thorin-Hagene, K.; Kerr, P.; Anderson, D.J.; Malysz, J.; Dyhring, T.; *et al.* Role of channel activation in cognitive enhancement mediated by α7 nicotinic acetylcholine receptors. *Br. J. Pharmacol.* **2009**, *158*, 1486–1494.
4. Papke, R.L. Merging old and new perspectives on nicotinic acetylcholine receptors. *Biochem. Pharmacol.* **2014**, *89*, 1–11.
5. Palma, E.; Conti, L.; Roseti, C.; Limatola, C. Novel approaches to study the involvement of α7-nAChR in human diseases. *Curr. Drug Targets* **2012**, *13*, 579–586.
6. Horenstein, N.A.; Leonik, F.M.; Papke, R.L. Multiple pharmacophores for the selective activation of nicotinic α7-type acetylcholine receptors. *Mol. Pharmacol.* **2008**, *74*, 1496–1511.
7. Mazurov, A.; Hauser, T.; Miller, C.H. Selective α7 nicotinic acetylcholine receptor ligands. *Curr. Med. Chem.* **2006**, *13*, 1567–1584.

8. Kim, S.W.; Ding, Y.S.; Alexoff, D.; Patel, V.; Logan, J.; Lin, K.S.; Shea, C.; Muench, L.; Xu, Y.; Carter, P.; et al. Synthesis and positron emission tomography studies of C-11-labeled isotopomers and metabolites of GTS-21, a partial α7 nicotinic cholinergic agonist drug. *Nucl. Med. Biol.* **2007**, *34*, 541–551.
9. Toyohara, J.; Ishiwata, K.; Sakata, M.; Wu, J.; Nishiyama, S.; Tsukada, H.; Hashimoto, K. In vivo evaluation of α7 nicotinic acetylcholine receptor agonists [^{11}C]A-582941 and [^{11}C]A-844606 in mice and conscious monkeys. *PLoS ONE* **2010**, *5*, e8961.
10. Hashimoto, K.; Nishiyama, S.; Ohba, H.; Matsuo, M.; Kobashi, T.; Takahagi, M.; Iyo, M.; Kitashoji, T.; Tsukada, H. [^{11}C]CHIBA-1001 as a novel PET ligand for α7 nicotinic receptors in the brain: A PET study in conscious monkeys. *PLoS ONE* **2008**, *3*, e3231.
11. Ettrup, A.; Mikkelsen, J.D.; Lehel, S.; Madsen, J.; Nielsen, E.Ø.; Palner, M.; Timmermann, D.B.; Peters, D.; Knudsen, G.M. ^{11}C-NS14492 as a novel PET radioligand for imaging cerebral α7 nicotinic acetylcholine receptors: In vivo evaluation and drug occupancy measurements. *J. Nucl. Med.* **2011**, *52*, 1449–1456.
12. Deuther-Conrad, W.; Fischer, S.; Hiller, A.; Stergaard Nielsen, E.; Brunicardi Timmermann, D.; Steinbach, J.; Sabri, O.; Peters, D.; Brust, P. Molecular imaging of α7 nicotinic acetylcholine receptors: Design and evaluation of the potent radioligand [^{18}F]NS10743. *Eur. J. Nucl. Med. Mol. Imaging* **2009**, *36*, 791–800.
13. Rötering, S.; Scheunemann, M.; Fischer, S.; Hiller, A.; Peters, D.; Deuther-Conrad, W.; Brust, P. Radiosynthesis and first evaluation in mice of [^{18}F]NS14490 for molecular imaging of α7 nicotinic acetylcholine receptors. *Bioorg. Med. Chem.* **2013**, *21*, 2635–2642.
14. Briggs, C.A.; Schrimpf, M.R.; Anderson, D.J.; Gubbins, E.J.; Grønlien, J.H.; Håkerud, M.; Ween, H.; Thorin-Hagene, K.; Malysz, J.; Li, J. et al. α7 nicotinic acetylcholine receptor agonist properties of tilorone and related tricyclic analogues. *Br. J. Pharmacol.* **2008**, *153*, 1054–1061.
15. Schrimpf, M.R.; Sippy, K.B.; Briggs, C.A.; Anderson, D.J.; Li, T.; Ji, J.; Frost, J.M.; Surowy, C.S.; Bunnelle, W.H.; Gopalakrishnan, M. et al. SAR of α7 nicotinic receptor agonists derived from tilorone: Exploration of a novel nicotinic pharmacophore. *Bioorg. Med. Chem. Lett.* **2012**, *22*, 1633–1638.
16. Teodoro, R.; Deuther-Conrad, W.; Rötering, S.; Scheunemann, M.; Patt, M.; Donat, C.K.; Wenzel, B.; Peters, D.; Sabri, O.; Brust, P. Comparative evaluation of two novel fluorine-18 PET radiotracers for the α7 nicotinic acetylcholine receptor. *J. Nucl. Med.* **2014**, *55*, 1099.
17. Gao, Y.; Kellar, K.J.; Yasuda, R.P.; Tran, T.; Xiao, Y.; Dannals, R.F.; Horti, A.G. Derivatives of dibenzothiophene for positron emission tomography imaging of α7-nicotinic acetylcholine receptors. *J. Med. Chem.* **2013**, *56*, 7574–7589.
18. Horti, A.G.; Gao, Y.; Kuwabara, H.; Wang, Y.; Abazyan, S.; Yasuda, R.P.; Tran, T.; Xiao, Y.; Sahibzada, N.; Holt, D.P. et al. ^{18}F-ASEM, a radiolabeled antagonist for imaging the α7-nicotinic acetylcholine receptor with PET. *J. Nucl. Med.* **2014**, *55*, 672–677.
19. Horti, A.G. Development of [^{18}F]ASEM, a specific radiotracer for quantification of the α7-nAChR with positron-emission tomography. *Biochem. Pharmacol.* **2015**, doi:10.1016/j.bcp.2015.07.030.

20. Wong, D.F.; Kuwabara, H.; Pomper, M.; Holt, D.P.; Brasic, J.R.; George, N.; Frolov, B.; Willis, W.; Gao, Y.; Valentine, H.; et al. Human brain imaging of α7 nAChR with [^{18}F]ASEM: A new PET radiotracer for neuropsychiatry and determination of drug occupancy. *Mol. Imaging Biol.* **2014**, *16*, 730–738.
21. Deuther-Conrad, W.; Fischer, S.; Hiller, A.; Becker, G.; Cumming, P.; Xiong, G.; Funke, U.; Sabri, O.; Peters, D.; Brust, P. Assessment of α7 nicotinic acetylcholine receptor availability in juvenile pig brain with [^{18}F]NS10743. *European J. Nucl. Med. Mol. Imaging* **2011**, *38*, 1541–1549.
22. Toma, L.; Quadrelli, P.; Bunnelle, W.H.; Anderson, D.J.; Meyer, M.D.; Cignarella, G.; Gelain, A.; Barlocco, D. 6-Chloropyridazin-3-yl derivatives active as nicotinic agents: Synthesis, binding, and modeling studies. *J. Med. Chem.* **2002**, *45*, 4011–4017.
23. Eibl, C.; Tomassoli, I.; Munoz, L.; Stokes, C.; Papke, R.L.; Gündisch, D. The 3,7-diazabicyclo [3.3.1]nonane scaffold for subtype selective nicotinic acetylcholine receptor (nAChR) ligands. Part 1: The influence of different hydrogen bond acceptor systems on alkyl and (hetero)aryl substituents. *Bioorg. Med. Chem.* **2013**, *21*, 7283–7308.
24. Papke, R.L.; Schiff, H.C.; Jack, B.A.; Horenstein, N.A. Molecular dissection of tropisetron, an α7 nicotinic acetylcholine receptor-selective partial agonist. *Neurosci. Lett.* **2005**, *378*, 140–144.
25. Lehel, S.; Madsen, J.; Ettrup, A.; Mikkelsen, J.D.; Timmermann, D.B.; Peters, D.; Knudsen, G.M. [^{11}C]NS-12857: A novel PET ligand for α7-nicotinergic receptors. *J. Labelled Compd. Rad.* **2009**, *52*, S379.
26. O'Donnell, C.J.; Peng, L.; O'Neill, B.T.; Arnold, E.P.; Mather, R.J.; Sands, S.B.; Shrikhande, A.; Lebel, L.A.; Spracklin, D.K.; Nedza, F.M. Synthesis and SAR studies of 1,4-diazabicyclo [3.2.2]nonane phenyl carbamates—Subtype selective, high affinity α7 nicotinic acetylcholine receptor agonists. *Bioorg. Med. Chem. Lett.* **2009**, *19*, 4747–4751.
27. Slowinski, F.; Ben Ayad, O.; Vache, J.; Saady, M.; Leclerc, O.; Lochead, A. Synthesis of bridgehead-substituted azabicyclo[2.2.1]heptane and -[3.3.1]nonane derivatives for the elaboration of α7 nicotinic ligands. *J. Org. Chem.* **2011**, *76*, 8336–8346.
28. Wawzonek, S.; Thelen, P.J. Preparation of *N*-methylgranatanine. *J. Am. Chem. Soc.* **1950**, *72*, 2118–2120.
29. Lambert, F.L.; Ellis, W.D.; Parry, R.J. Halogenation of aromatic compounds by *N*-bromo- and *N*-chlorosuccinimide under ionic conditions. *J. Org. Chem.* **1965**, *30*, 304–306.
30. Butts, C.P.; Eberson, L.; Hartshorn, M.P.; Radner, F.; Robinson, W.T.; Wood, B.R. Regiochemistry of the reaction between dibenzothiophene radical cation and nucleophiles or nitrogen dioxide. *Acta Chem. Scand.* **1997**, *51*, 839–848.
31. Gilman, H.; Nobis, J.F. Some aminodibenzothiophenes. *J. Am. Chem. Soc.* **1949**, *71*, 274–276.
32. Cullinane, N.M.; Davies, C.G.; Davies, G.I. Substitution derivatives of diphenylene sulphide and diphenylenesulphone. *J. Chem. Soc.* **1936**, 1435–1437, doi:10.1039/JR9360001435.
33. Adams, D.J.; Clark, J.H. Nucleophilic routes to selectively fluorinated aromatics. *Chem. Soc. Rev.* **1999**, *28*, 225–231.
34. Boechat, N.; Clark, J.H. Fluorodenitrations using tetramethylammonium fluoride. *J. Chem. Soc. Chem. Commun.* **1993**, *11*, 921–922.
35. Manna, S.; Maity, S.; Rana, S.; Agasti, S.; Maiti, D. Ipso-nitration of arylboronic acids with bismuth nitrate and perdisulfate. *Org. Lett.* **2012**, *14*, 1736–1739.

36. Molander, G.A.; Cavalcanti, L.N. Nitrosation of aryl and heteroaryltrifluoroborates with nitrosonium tetrafluoroborate. *J. Org. Chem.* **2012**, *77*, 4402–4413.
37. Wu, X.F.; Schranck, J.; Neumann, H.; Beller, M. Convenient and mild synthesis of nitroarenes by metal-free nitration of arylboronic acids. *Chem. Commun.* **2011**, *47*, 12462–12463.
38. Xiao, Y.; Hammond, P.S.; Mazurov, A.A.; Yohannes, D. Multiple interaction regions in the *orthosteric* ligand binding domain of the α_7 neuronal nicotinic acetylcholine receptor. *J. Chem. Inf. Model.* **2012**, *52*, 3064–3073.
39. Brunzell, D.H.; McIntosh, J.M.; Papke, R.L. Diverse strategies targeting α_7 homomeric and $\alpha_6\beta_2$* heteromeric nicotinic acetylcholine receptors for smoking cessation. *Ann. N. Y. Acad. Sci.* **2006**, *1327*, 27–45.
40. Baddick, C.G.; Marks, M.J. An autoradiographic survey of mouse brain nicotinic acetylcholine receptors defined by null mutants. *Biochem. Pharmacol.* **2011**, *82*, 828–841.
41. Gotti, C.; Guiducci, S.; Tedesco, V.; Corbioli, S.; Zanetti, L.; Moretti, M.; Zanardi, A.; Rimondini, R.; Mugnaini, M.; Clementi, F.; *et al.* Nicotinic acetylcholine receptors in the mesolimbic pathway: Primary role of ventral tegmental area $\alpha_6\beta_2$* receptors in mediating systemic nicotine effects on dopamine release, locomotion, and reinforcement. *J. Neurosci.* **2010**, *30*, 5311–5325.
42. Quik, M.; Polonskaya, Y.; Gillespie, A.; Jakowec, M.; Lloyd, G.K.; Langston, J.W. Localization of nicotinic receptor subunit mRNAs in monkey brain by *in situ* hybridization. *J. Comp. Neurol.* **2000**, *425*, 58–69.
43. Han, Z.Y.; Le Novere, N.; Zoli, M.; Hill, J.A., Jr.; Champtiaux, N.; Changeux, J.P. Localization of nAChR subunit mRNAs in the brain of Macaca mulatta. *European J. Neurosci.* **2000**, *12*, 3664–3674.
44. Exley, R.; Maubourguet, N.; David, V.; Eddine, R.; Evrard, A.; Pons, S.; Marti, F.; Threlfell, S.; Cazala, P.; McIntosh, J.M.; *et al.* Distinct contributions of nicotinic acetylcholine receptor subunit α_4 and subunit α_6 to the reinforcing effects of nicotine. *Proc. Natl. Acad. Sci. USA* **2011**, *108*, 7577–7582.
45. Faure, P.; Tolu, S.; Valverde, S.; Naudé, J. Role of nicotinic acetylcholine receptors in regulating dopamine neuron activity. *Neuroscience* **2014**, *282*, 86–100.
46. Gotti, C.; Marks, M.J.; Millar, N.S.; Wonnacott, S. Nicotinic acetylcholine receptors: α_6. Available online: http://www.guidetopharmacology.org/GRAC/ObjectDisplayForward?objectId = 467&familyId=76&familyType=IC (accessed on 2 October 2015).
47. Le Novere, N.; Grutter, T.; Changeux, J.P. Models of the extracellular domain of the nicotinic receptors and of agonist- and Ca^{2+}-binding sites. *Proc. Natl. Acad. Sci. USA* **2002**, *99*, 3210–3215.
48. Ding, M.; Ghanekar, S.; Elmore, C.S.; Zysk, J.R.; Werkheiser, J.L.; Lee, C.M.; Liu, J.; Chhajlani, V.; Maier, D.L. [^{3}H]Chiba-1001(methyl-SSR180711) has low *in vitro* binding affinity and poor *in vivo* selectivity to nicotinic alpha-7 receptor in rodent brain. *Synapse* **2012**, *66*, 315–322.
49. Mazurov, A.A.; Kombo, D.C.; Akireddy, S.; Murthy, S.; Hauser, T.A.; Jordan, K.G.; Gatto, G.J.; Yohannes, D. Novel nicotinic acetylcholine receptor agonists containing carbonyl moiety as a hydrogen bond acceptor. *Bioorg. Med. Chem. Lett.* **2013**, *23*, 3927–3934.
50. Girard, B.M.; Merriam, L.A.; Tompkins, J.D.; Vizzard, M.A.; Parsons, R.L. Decrease in neuronal nicotinic acetylcholine receptor subunit and PSD-93 transcript levels in the male mouse MPG after cavernous nerve injury or explant culture. *Am. J. Physiol.* **2013**, *305*, F1504–F1512.

51. Glushakov, A.V.; Voytenko, L.P.; Skok, M.V.; Skok, V. Distribution of neuronal nicotinic acetylcholine receptors containing different alpha-subunits in the submucosal plexus of the guinea-pig. *Auton. Neurosci. Basic Clin.* **2004**, *110*, 19–26.
52. Teodoro, R.; Wenzel, B.; Oh-Nishi, A.; Fischer, S.; Peters, D.; Suhara, T.; Deuther-Conrad, W.; Brust, P. A high-yield automated radiosynthesis of the alpha-7 nicotinic receptor radioligand [^{18}F]NS10743. *Appl. Radiat. Isot.* **2015**, *95*, 76–84.
53. Jacobson, O.; Chen, X. PET designated flouride-18 production and chemistry. *Curr. Top. Med. Chem.* **2010**, *10*, 1048–1059.
54. Ravert, H.T.; Holt, D.P.; Gao, Y.; Horti, A.G.; Dannals, R.F. Microwave-assisted radiosynthesis of [^{18}F]ASEM, a radiolabeled α7-nicotinic acetylcholine receptor antagonist. *J. Labelled Compd. Radiopharm.* **2015**, *58*, 180–182.
55. Egleton, R.D.; Brown, K.C.; Dasgupta, P. Angiogenic activity of nicotinic acetylcholine receptors: Implications in tobacco-related vascular diseases. *Pharmacol. Ther.* **2009**, *121*, 205–223.
56. Cooke, J.P.; Bitterman, H. Nicotine and angiogenesis: A new paradigm for tobacco-related diseases. *Ann. Med.* **2004**, *36*, 33–40.
57. Pena, V.B.; Bonini, I.C.; Antollini, S.S.; Kobayashi, T.; Barrantes, F.J. α7-Type acetylcholine receptor localization and its modulation by nicotine and cholesterol in vascular endothelial cells. *J. Cell. Biochem.* **2011**, *112*, 3276–3288.
58. Innis, R.B.; Cunningham, V.J.; Delforge, J.; Fujita, M.; Gjedde, A.; Gunn, R.N.; Holden, J.; Houle, S.; Huang, S.C.; Ichise, M.; et al. Consensus nomenclature for *in vivo* imaging of reversibly binding radioligands. *J. Cereb. Blood Flow Metab.* **2007**, *27*, 1533–1539.
59. Cunningham, V.J.; Rabiner, E.A.; Slifstein, M.; Laruelle, M.; Gunn, R.N. Measuring drug occupancy in the absence of a reference region: The Lassen plot re-visited. *J. Cereb. Blood Flow Metab.* **2010**, *30*, 46–50.
60. Brust, P.; Peters, D.; Deuther-Conrad, W. Development of radioligands for the imaging of α7 nicotinic acetylcholine receptors with positron emission tomography. *Curr. Drug Targets* **2012**, *13*, 594–601.
61. Harwood, L.M. "Dry-column" flash chromatography. *Aldrichimica Acta* **1985**, *18*, 25.
62. Schrimpf, M.; Sippy, K.; Ji, J.; Li, T.; Frost, J.; Briggs, C.; Bunnelle, W. Amino-Substituted Tricyclic Derivatives and Methods of Use. U.S. Patent 20,050,234,031 A1, 20 October 2005.
63. Rötering, S.; Franke, K.; Zessin, J.; Brust, P.; Füchtner, F.; Fischer, S.; Steinbach, J. Convenient recycling and reuse of bombarded [^{18}O]H$_2$O for the production and the application of [^{18}F]F. *Appl. Radiat. Isot.* **2015**, *101*, 44–52.
64. Charpantier, E.; Wiesner, A.; Huh, K.H.; Ogier, R.; Hoda, J.C.; Allaman, G.; Raggenbass, M.; Feuerbach, D.; Bertrand, D.; Fuhrer, C. α7 neuronal nicotinic acetylcholine receptors are negatively regulated by tyrosine phosphorylation and Src-family kinases. *J. Neurosci.* **2005**, *25*, 9836–9849.
65. Cheng, Y.; Prusoff, W.H. Relationship between the inhibition constant (K_1) and the concentration of inhibitor which causes 50 per cent inhibition (IC$_{50}$) of an enzymatic reaction. *Biochem. Pharmacol.* **1973**, *22*, 3099–3108.
66. Michelmore, S.; Croskery, K.; Nozulak, J.; Hoyer, D.; Longato, R.; Weber, A.; Bouhelal, R.; Feuerbach, D. Study of the calcium dynamics of the human α$_4$β$_2$, α$_3$β$_4$ and α$_1$β$_1$γδ nicotinic acetylcholine receptors. *Naunyn-Schmiedeberg's Arch. Pharmacol.* **2002**, *366*, 235–245.

67. Grady, S.R.; Moretti, M.; Zoli, M.; Marks, M.J.; Zanardi, A.; Pucci, L.; Clementi, F.; Gotti, C. Rodent habenulo-interpeduncular pathway expresses a large variety of uncommon nAChR subtypes, but only the αβ_4* and α$\beta_3\beta_4$* subtypes mediate acetylcholine release. *J. Neurosci.* **2009**, *29*, 2272–2282.
68. Pucci, L.; Grazioso, G.; Dallanoce, C.; Rizzi, L.; De Micheli, C.; Clementi, F.; Bertrand, S.; Bertrand, D.; Longhi, R.; De Amici, M.; *et al.* Engineering of α-conotoxin MII-derived peptides with increased selectivity for native α$_6\beta_2$* nicotinic acetylcholine receptors. *FASEB J.* **2011**, *25*, 3775–3789.
69. Munson, P.J.; Rodbard, D. Ligand: A versatile computerized approach for characterization of ligand-binding systems. *Anal. Biochem.* **1980**, *107*, 220–239.
70. Brust, P.; Patt, J.T.; Deuther-Conrad, W.; Becker, G.; Patt, M.; Schildan, A.; Sorger, D.; Kendziorra, K.; Meyer, P.; Steinbach, J.; *et al. In vivo* measurement of nicotinic acetylcholine receptors with [^{18}F]norchloro-fluoro-homoepibatidine. *Synapse* **2008**, *62*, 205–218.
71. Gillings, N.M.; Bender, D.; Falborg, L.; Marthi, K.; Munk, O.L.; Cumming, P. Kinetics of the metabolism of four PET radioligands in living minipigs. *Nucl. Med. Biol.* **2001**, *28*, 97–104.
72. Brust, P.; Zessin, J.; Kuwabara, H.; Pawelke, B.; Kretzschmar, M.; Hinz, R.; Bergman, J.; Eskola, O.; Solin, O.; Steinbach, J.; *et al.* Positron emission tomography imaging of the serotonin transporter in the pig brain using [^{11}C](+)-McN5652 and S-[^{18}F]fluoromethyl-(+)-McN5652. *Synapse* **2003**, *47*, 143–151.
73. Brust, P.; Hinz, R.; Kuwabara, H.; Hesse, S.; Zessin, J.; Pawelke, B.; Stephan, H.; Bergmann, R.; Steinbach, J.; Sabri, O. *In vivo* measurement of the serotonin transporter with (*S*)-([^{18}F]fluoromethyl)-(+)-McN5652. *Neuropsychopharmacology* **2003**, *28*, 2010–2019.

Sample Availability: Samples of the compounds **10a,b** and **14b** are available from the authors.

© 2015 by the authors; licensee MDPI, Basel, Switzerland. This article is an open access article distributed under the terms and conditions of the Creative Commons Attribution license (http://creativecommons.org/licenses/by/4.0/).

Molecules **2015**, *20*, 14860-14878; doi:10.3390/molecules200814860

ISSN 1420-3049
www.mdpi.com/journal/molecules

Article

Development of a Single Vial Kit Solution for Radiolabeling of ^{68}Ga-DKFZ-PSMA-11 and Its Performance in Prostate Cancer Patients

Thomas Ebenhan [1,2], Mariza Vorster [1], Biljana Marjanovic-Painter [3], Judith Wagener [3], Janine Suthiram [1,3], Moshe Modiselle [1], Brenda Mokaleng [1], Jan Rijn Zeevaart [4] and Mike Sathekge [1,*]

1. University of Pretoria & Steve Biko Academic Hospital, Crn Malherbe and Steve Biko Rd, Pretoria 0001, South Africa; E-Mails: Thomas.ebenhan@gmail.com (T.E.); marizavorster@gmail.com (M.V.); janine.suthiram@necsa.co.za (J.S.); modisellemoshe@yahoo.co.uk (M.M.); tshelom@gmail.com (B.M.)
2. School of Health Sciences, Catalysis and Peptide Research Unit, E-Block 6th Floor, Westville Campus, University Road, Westville, Durban 3630, South Africa
3. The South African Nuclear Energy Corporation (Necsa), Building P1600, Radiochemistry, Pelindaba, Brits 0240, South Africa; E-Mails: biljana.marjanovic-painter@necsa.co.za (B.M.-P.); judith.wagener@necsa.co.za (J.W.)
4. Department of Science and Technology, Preclinical Drug Development Platform, North West University, 11 Hoffman St, Potchefstroom 2520, South Africa; E-Mail: janrijn.zeevaart@necsa.co.za

* Author to whom correspondence should be addressed; E-Mail: Mike.Sathekge@up.ac.za or sathekgemike@gmail.com; Tel.: +27-12-354-1794; Fax: +27-12-354-1219.

Academic Editor: Svend Borup Jensen

Received: 31 May 2015 / Accepted: 3 August 2015 / Published: 14 August 2015

Abstract: Prostate-specific membrane antigen (PSMA), a type II glycoprotein, is highly expressed in almost all prostate cancers. By playing such a universal role in the disease, PSMA provides a target for diagnostic imaging of prostate cancer using positron emission tomography/computed tomography (PET/CT). The PSMA-targeting ligand Glu-NH-CO-NH-Lys-(Ahx)-HBED-CC (DKFZ-PSMA-11) has superior imaging properties and allows for highly-specific complexation of the generator-based radioisotope Gallium-68 (^{68}Ga). However, only module-based radiolabeling procedures are currently available. This study intended to develop a single vial kit solution to radiolabel buffered DKFZ-PSMA-11 with

^{68}Ga. A ^{68}Ge/^{68}Ga-generator was utilized to yield ^{68}GaCl$_3$ and major aspects of the kit development were assessed, such as radiolabeling performance, quality assurance, and stability. The final product was injected into patients with prostate cancer for PET/CT imaging and the kit performance was evaluated on the basis of the expected biodistribution, lesion detection, and dose optimization. Kits containing 5 nmol DKFZ-PSMA-11 showed rapid, quantitative ^{68}Ga-complexation and all quality measurements met the release criteria for human application. The increased precursor content did not compromise the ability of ^{68}Ga-DKFZ-PSMA-11 PET/CT to detect primary prostate cancer and its advanced lymphatic- and metastatic lesions. The ^{68}Ga-DKFZ-PSMA-11 kit is a robust, ready-to-use diagnostic agent in prostate cancer with high diagnostic performance.

Keywords: PSMA; prostate cancer; PET/CT; ^{68}Ga-HBED-CC-(Ahx)Lys-NH-CO-NH-Glu; ^{68}Ga-DKFZ-PSMA-11; ^{68}Ga-PSMA^{HBED}

1. Introduction

Prostate cancer (PC) is the most commonly diagnosed cancer in men globally and is the second leading cause of death from malignancy among men in USA and other countries [1]. Currently, the available conventional diagnostic procedures, such as preventive blood diagnostics of tumour marker levels like prostate specific antigen (PSA), or inconvenient methods such as the digital rectal prostate examination or transrectal ultrasound (TRUS), are debatable. Non-invasive methods include sonographic Doppler techniques [2], computed tomography (CT), or magnetic resonance imaging (MRI) [3] of the abdomen and pelvis. These procedures frequently reveal PC but are lacking accuracy for detection of PC staging as well as in recurrent PC [4]. A hybrid imaging approach whereby CT is combined with positron emission tomography (PET) using Fluorine-18 fluorodeoxyglucose (FDG) has more significance, as the glucose metabolism can be instrumented to detect numerous malignancies. However, FDG-PET bears limitations towards the detection and localization of slow-growing primary prostate cancer and initial staging of disease with a tumour uptake level that can overlap with those in normal tissue and benign prostatic hyperplasia [5].

Published reports have highlighted the advantages of using ^{11}C- or ^{18}F-radiolabeled derivatives of acetate or choline, ^{18}F-labeled testosterone-derivatives as a ligand to androgen receptors, as well as PET radiotracers targeting prostate-specific membrane antigen (PSMA), prostate stem cell antigen- or gastrin-releasing peptide receptor [6–8]. ^{11}C-Choline-PET/CT and ^{18}F-Choline-PET/CT have been performed for years, however, in a considerable number of cases the metabolism of the tumor has not increased enough for the choline analogs to detect PC, especially in recurrent PC [9–11]—confirming the need for a more universal imaging agent. In the assessment of metastatic PC, it is becoming increasingly clear that ligands targeting PSMA may be a superior alternative to the aforementioned prostate cancer PET imaging agents. It is a type II membrane glycoprotein that is significantly overexpressed during all stages of the androgen-insensitive or the metastatic cancer of the prostate compared to other PSMA-expressing tissues, such as kidney, proximal small intestine, or salivary glands [12–15]. In 2008 M. Pomper and colleagues initiated the evaluation of radio-halogenated, technetium- and rhenium-labeled urea-based inhibitors of PSMA [16,17] to develop more potent imaging agents for prostate cancer. The PSMA-inhibiting peptide-based motif "glutamate-urea-lysin" (Glu-NH-CO-NH-Lys-(Ahx)) was discovered as a novel pharmacological entity. PSMA ligand derivatives were subsequently conjugated to 1,4,7,10-tetraazacyclododecane-1,4,7,10-tetraacetic acid (DOTA) to allow for ^{68}Ga-complexation, but notable concerns about the structural changes were addressed [18], as the DOTA macrocycle in the immediate vicinity decreased the binding affinity [19]. Alternatively, the novel and seldom used acyclic chelator agent N,N'-bis-[2-hydroxy-5-(carboxyethyl)benzyl]ethylenediamine-N,N'-diacetic acid (HBED-CC) was introduced to allow interaction with the hydrophobic part of the S1-subunit binding site of the PSMA protein and to facilitate rapid radiolabeling with metal-radioisotopes such as Gallium-68 at ambient temperatures. To date, a clinical pilot investigation with this ^{68}Ga-labeled PSMA derivative (Glu-NH-CO-NH-Lys-(Ahx)-(^{68}Ga)Ga(HBED-CC)) further denoted as ^{68}Ga-DKFZ-PSMA-11) suggested that it detects PC relapses and metastases with higher contrast as compared to ^{18}F-labeled choline [20]. The conjugation of HBED-CC enabled the research group to produce highly-specific activities of ^{68}Ga-DKFZ-PSMA-11 [21] and clinical experience show that the degree of accuracy of ^{68}Ga-DKFZ-PSMA-11 PET/CT will have a huge impact on the management of patients with PC and will address an important unmet need in this field [22].

From a medical point of view, compared to ^{18}F-FDG, ^{68}Ga allows shorter image acquisition, which has significant financial implications and demonstrates a more cost-effective approach with immediate benefit for patient care. It should also be noted that the ^{68}Ga half-life of 67.6 min often matches well with the pharmacokinetic of peptides and other structures, which makes it an attractive labelling option for novel diagnostic applications [23,24]. Particularly from an economic viewpoint, using generator-based PET-radiopharmaceuticals like ^{68}Ga has advantages over radioisotopes such as ^{11}C, which are produced in a cyclotron ^{68}Ga be conveniently yielded daily and GMP-compliant using mild acidic conditions (*i.e.*, 0.1 to 1 M HCl) and immediately used for radiolabeling.

Recently, examples for kit formulation strategies using single vial productions have arisen, such as for ^{68}Ga-radiolabeling of DOTA-peptides [25], NODAGA-conjugated compounds [26], and buffered citrate [27]. ^{68}Ga-DKFZ-PSMA-11 PET/CT will inevitably be a sensitive tool to support clinical trials and patient care using state-of-the-art diagnostic technologies and will significantly advance diagnosis and treatment management of patients with recurrent PC. To the best of our knowledge only elaborate cassette- and module-based radiolabeling techniques have been published [21]. In view of the suitability of the precursor we set out to develop a single vial kit radiolabeling procedure for ^{68}Ga-DKFZ-PSMA-11.

We report here the formulation of a freeze-dried kit containing DKFZ-PSMA-11 and sodium acetate trihydrate. We critically assessed the following aspects concerning kit development: radiolabeling with ^{68}Ga eluted from a certified generator, performance, safety, quality, sterility, and stability. Changes that were made to manufacture the DKFZ-PSMA-11 kit included an increase in the precursor content. The final product (having passed all necessary quality control requirements) was injected into patients with PC at various stages of the disease prior to PET/CT imaging. Kit performance was evaluated based on the expected biodistribution, lesion detection, and dose optimization.

2. Results and Discussion

2.1. ^{68}Ge/^{68}Ga Generator

1.85 GBq loaded ^{68}Ge/^{68}Ga-generators were utilized to yield 1.55 GBq ± 0.44 GBq of ^{68}Ga-activity, eluted as a 2–3 mL batch. Routinely, 1.0 mL generator eluate (457 MBq ± 210 MBq (n = 21)) was added to the DKFZ-PSMA-11 for labeling. Thus, two time-lagged radiosyntheses could be performed from one generator elution facilitating personalized administration (alternatively the full 2–3 mL batch could be used with an upscale protocol). Decayed total generator eluates and ^{68}Ga-DKFZ-PSMA-11 samples were routinely measured for residual ^{68}Ge. The maximum ^{68}Ge amount in the total eluate was 0.00074%. The average amount of ^{68}Ge found in the final product solution was 0.000057% ± 0.000015% (n = 8)). These levels were well below the limits outlined for ^{68}Ge-breakthrough (0.001% according to the Phar. Eur. 8.0). The contents of co-eluted ionic metal impurities as analyzed by ICP-OES was highest for the 246-day-old generator (13.7 ppm); this level was calculated below the limit of 20 ppm for the final product for all tested metals (Al^{3+}, Sn^{2+}, Fe^{3+}, Cu^{2+} and Zn^{2+}).

2.2. Preliminary Assessment of the Radiolabeling Parameters

Prior to the kit formulation, certain parameters were optimized to achieve quantitative complexation of ^{68}Ga to DKFZ-PSMA-11. The relevant results are summarized in Figure 1A, showing that under moderate acidic conditions a minimum of 5 nmol DKFZ-PSMA-11 was required in combination with 10 min incubation at RT to warrant a 95%–100% complexation of ^{68}Ga. Trace amounts of uncomplexed ^{68}Ga were determined with either HPLC or ITLC. Using 15%–25% ethanolic saline solution allowed recovering 84%–95% ^{68}Ga-DKFZ-PSMA-11 (Figure 1B) with C18 purification. Vortex stirring action caused no significant increase of the radiochemical yield (p = 0.338) or shortening of the incubation time, but is advised to be carried out as described (Material and Methods). An assessment of the radiolabeling depending on changes of pH, incubation temperature, and the choice of buffering solution was voided for this study. Ambient temperature labeling was a prerequisite for the kit formulation and former studies by Eder *et al.* suggested the optimal pH of 4.2 [19]; all crude radiolabeling solutions met the pH range of 4.0–4.5.

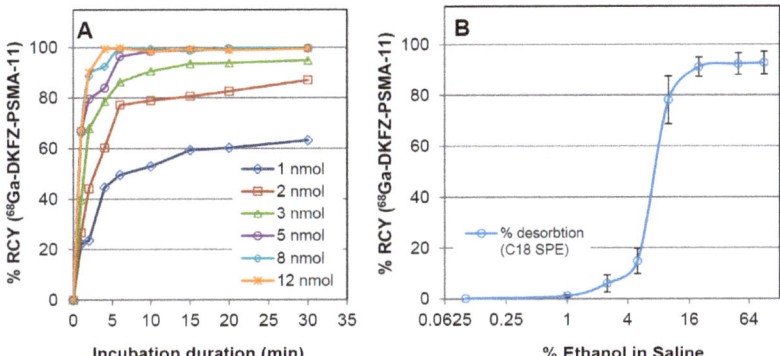

Figure 1. Results from the preliminary assessment of ^{68}Ga-DKFZ-PSMA-11 radiolabeling for studying: (**A**) precursor molarity as a function of incubation duration and (**B**) rising ethanol concentration required to desorb the purified product from a C18-SepPak light cartridge unit. The % RCY and % recovery of ^{68}Ga-DKFZ-PSMA-11 are displayed (determination of percentage activity of the tracer peak using ITLC). Samples were incubated at RT at pH 4.0–4.5. Mean values (±sem) of three independent experiments are displayed (error bars in A representing sem of 3.7%–12.4% are voided for more transparent presentation).

2.3. In-House Kit Vial Formulation of DKFZ-PSMA-11

The 5 nmol DKFZ-PSMA-11 were buffered with sodium acetate. The kit pellet was presented in a homogeneous-solid powder form, thus, no pellet bulking agents were added to the kit. After supplementing the 1.0 mL ^{68}Ga (0.6 N HCl) to the kit vial, the pellet dissolved rapidly to yield a clear particle-free solution. Table 1 summarizes the pre-release tests in comparison to the outcome of the ^{68}Ga-DKFZ-PSMA-11 kit-based radiosynthesis, indicating that all quality control tests met the prerequisite criteria.

Table 1. Overview of quality control tests compared to the release criteria for safe administration of ^{68}Ga-DKFZ-PSMA-11 to humans.

Quality Control Test	Specification	Test Results
Eluate fraction yield (MBq)	≥300 MBq/1 mL	332–1039
^{68}Ge breakthrough (total eluate batch 10 mL)	≤0.001% over 9 months	max: 0.00074% mean: 0.0003% ± 0.0001%
Cationic impurities (Zn, Fe, Cu, Sn, Al)	≤50 ppm/1 mL	pass
Product yield (MBq)	≥200 MBq/1 mL	310 ± 52
Visual inspection	Clear colourless, particle-free	pass
Radiochemical identity ITLC	R_f ^{68}Ga-DKFZ-PSMA-11 = 0.75 ± 0.2	0.73–0.77
Radiochemical identity HPLC	Retention time = 5.3 ± 0.5 min	5.18–5.58
Chemical identity HPLC$_{(UV214nm)}$	Retention time = 4.9 ± 0.5 min	4.83–5.08
Radiochemical purity	≥95%	99.6
pH for injection	physiological (6.0–7.6)	6.5–7.0
Sterile filter integrity	≥3.5 bar	5.9 ± 0.9 ($n = 7$)
Radionuclide identity	67.7 ± 5 min	65.1–69.8 ($n = 6$)
Residual ^{68}Germanium (2–5 mL)	≤0.001%	0.000057 ± 0.000015
Sterility	Sterile (fungal/anaerobe/aerobe)	pass
Total product endotoxins	max: 20 EU	pass

This study considered 28 kit based radiosyntheses. Eighteen kits were handled during the investigation of the purification step, here called Stage 1 (Stage 1, see Section 3.5.) and, building on those results, 10 additional kits were applied in Stage 2 where a true one-vial-one-step-radiolabeling approach was explored (Stage 2, see Section 3.5.). The average duration from generator elution to sterile dispensing of ^{68}Ga-DKFZ-PSMA was 41 ± 8 min ($n = 12$, phase 1) and 25 ± 4 min ($n = 5$, phase 2, $p = 0.068$) providing up to 800 MBq (calculated yield of two staggered synthesis) of product for injection and for quality-control purposes. The % recovery of ^{68}Ga-DKFZ-PSMA-11 using 20% ethanolic saline solution in Stage 1 was 78% ± 13% ($n = 14$). The final product was supplied in 8–10 mL with an average product yield range of 300–650 MBq for imaging purposes. In 16 out of 18 DKFZ-PSMA-11 kits (89%), the amount of uncomplexed ^{68}Ga was calculated ≤2.44%, whereas 2 kit radiosyntheses showed uncomplexed ^{68}Ga-levels of 6.2% and 6.9% (C18 SepPak purification was applied to all 18 radiosyntheses in phase 1). The % RCP after purification was calculated ≥98.9% ($n = 13$). The results achieved during the kit investigation (Stage 1) were a prerequisite to performing radiolabeling, voiding the purification step (Stage 2). At a pH value of 4.0 the kits containing 5 nmol DKFZ-PSMA-11 complexed the ^{68}Ga significantly better ($p < 0.01$) than the reference kit vials containing 2 nmol. The kit preparation of DKFZ-PSMA-11 did not compromise the radiolabeling parameters (Figure 2). HPLC-analysis showed free ^{68}Ga-release from the

C18-column at *ca.* 2–3 min and compound retention until 5.58 min. The UV-signal intensity for the compound identification was found at slightly earlier times (4.8–5.1 min) due to the consecutive alignment of the radio-HPLC detector (Figure 3 A,B). The DKFZ-PSMA-11 kits radiolabeled during phase 2 showed a % RCP of 96.8%–99.9% ($n = 7$) after 20 min incubation at RT ($p > 0.01$). Both radiolabeling procedures caused a 3%–16% loss of radioactivity to glass vial surfaces and disposal material such as the C18 cartridge. In light of the achieved results, calculations were made to achieve further optimization for a resourceful and economical protocol. The decay-corrected % RCY was very high for both methods applied, amounting to 87% ± 5% and 90% ± 3% ($p = 0.198$) for Stage 1 and Stage 2, respectively. On the basis of the presented results, 0.5 mL of the nine-months-old generator eluted ^{68}Ga (~240 MBq) would suffice as starting material to yield an appropriate single patient dose of 150–160 MBq after 25 min production. The ^{68}Ga eluted from a 30-days-old generator could be used in 0.25 mL aliquots/patient using kit adjustments, respectively. This might be the key aspect to an economic commercialization of the kit technique. Alternatively, if the PET imaging capacity will allow timely image acquisition of multiple patients, the kit vial can be adjusted to perform an upscaled radiolabeling procedure that serves for 4–5 injections. However, it should be noted that GMP complaint quality controls should be repeated accordingly.

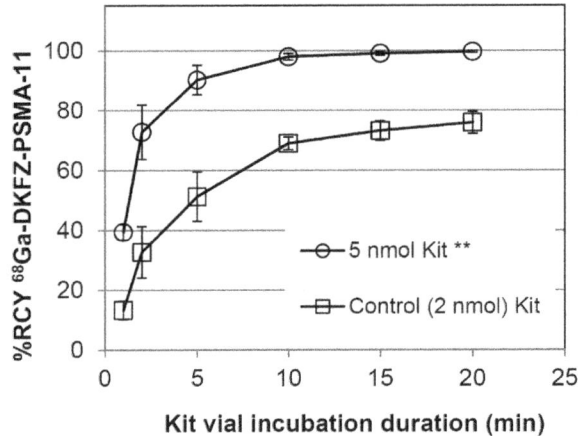

Figure 2. Kit radiolabeling of ^{68}Ga-DKFZ-PSMA-11 (Stage 2) as a function of incubation duration (ITLC-analysis mobile phase: Methanol/Saline 20:80 (v/v)). No purification was carried out post radiolabeling. Mean values (±sem) are displayed. ** Student *t*-tests returned a *p* value ≤ 0.01, for the % RCY of 5 nmol DKFZ-PSMA-11-containing kits ($n = 4$) *vs.* the control ($n = 3$).

2.4. Quality Assessment of the ^{68}Ga-DKFZ-PSMA-11 Kit

2.4.1. Appearance and Sterility

Optical inspection showed a clear, colourless product solution, which was free of particles or sediments. Decayed batches of ^{68}Ga-DKFZ-PSMA-11 were successfully tested free of anaerobe and aerobe bacteria, as well as fungal growth. Moreover, none of the sterile filters failed the integrity (bubble point burst) test ≥ 62 psi (*ca.* 4.1 bar).

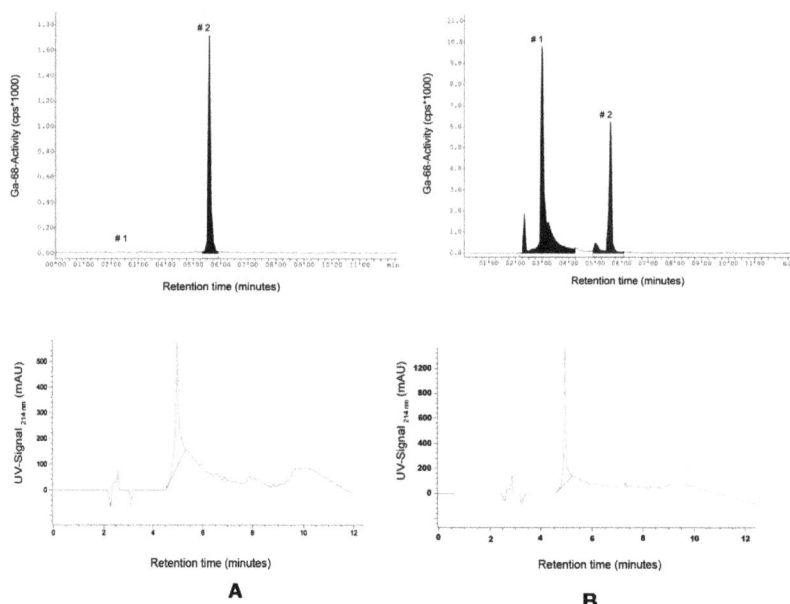

Figure 3. Exemplary HPLC chromatograms as recovered from the radioactivity channel (**top panel**) detecting (#1) free ^{68}Ga and (#2) ^{68}Ga-DKFZ-PSMA-11. Bottom panel showing DKFZ-PSMA-11 UV-signal detected simultaneously (λ: 214 nm). Samples were analyzed: (**A**) immediately after adding ^{68}Ga-activity; and (**B**) after 20 min incubation duration.

2.4.2. Radionuclidic Identity and pH Value

The ^{68}Ga-samples tested for half-life (radionuclidic) identification yielded results of 66.8 ± 2.9 min and could be differentiated from ^{68}Ge-samples (265.7 ± 4.9 min; $n = 4$). The pH value was *ca.* 4.0 for the reaction mixture and in physiological range for all product solutions for injection.

2.4.3. Radiochemical Stability

The radiochemical integrity of the DKFZ-PSMA-11-complexed ^{68}Ga was found to be >98% in 20% ethanolic saline solution at 37 °C over the 240-min duration observed ($n = 3$). Free ^{68}Ga-levels were found 0.3% ± 0.05%, 0.7% ± 0.18%, and 1.9% ± 0.31% for 60, 120, and 240 min, respectively. This high thermodynamic stability would allow for multi-patient production and resourceful optimization of the kit production as addressed earlier.

2.4.4. Long-Term Storage and Radiolabeling Reproducibility

There is limited information from former studies about the longitudinal performance of the iThemba LABS generator. As shown in Figure 4A, the reformulated kits give consistently high radiochemical yields when tested with 1.0 mL radiogallium eluted from a freshly manufactured generator (dark-grey column). No significant difference in the % RCY was found when the DKFZ-PSMA-11-kits were labeled with 1.0 mL eluate of the 1st batch, 2nd batch (light gray columns) of a nine-months-old generator or purified post labeling by SepPack C18 light (open-white column). The %RCY amounted to 90% ± 3.4%, 88% ± 5.1%, 86% ± 4.8%, and 94% ± 1.3%, for the aforementioned batches, respectively. Particular consideration was taken regarding the potential influence of the kit storage condition The DKFZ-PSMA-11 kit performance due to altered long-term storage (Figure 4B) showed % RCY of 97% ± 0.7%, 96% ± 0.9%, and 77% ± 7.2 % ($p = 0.008$) for vials that were frozen at −50 °C, cooled at 2 °C to 8 °C, and placed at RT, respectively. To date the frozen kits (4 months) perform stable radiosyntheses of DKFZ-PSMA-11 and the stability is monitored continuously. The single vial kit approach for the tracer preparation satisfies the necessity of a standardized pharmaceutical product with controlled quality and wide availability.

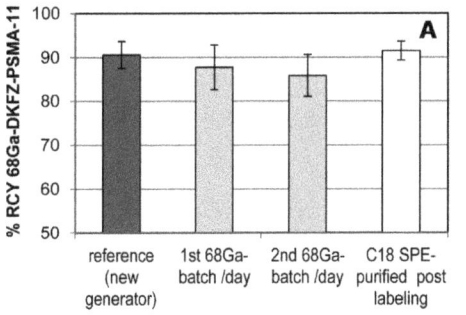
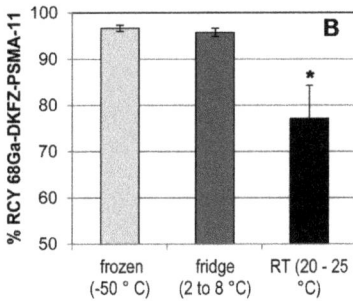

Figure 4. ^{68}Ga-DKFZ-PSMA-11 kit radiolabeling performance: (**A**) effect of purity of the eluted ^{68}Ga; and (**B**) long-term storage at different temperatures. Kit contents were labeled at RT for 15 min at pH 4.0–4.5 and analysed using radio-ITLC. Mean values (± sem) of two to nine independent experiments are displayed. * Statistical significance tests returned *p* values ≤0.05 for the % RCY of DKFZ-PSMA-11 kits incubated at RT *vs.* % RCY of frozen and/or cooled kit vials.

2.5. Clinical PET/CT—^{68}Ga-DKFZ-PSMA-11 Kit Performance in Prostate Cancer Patients

Fifteen male patients (up to 92 yr-old, history of rising PSA-levels, with and without ^{18}F-FDG-PET or ^{99m}Tc-MDP-SPECT (technetium-99m methylene diphosphonate/single photon emission computed tomography) scan prior to this imaging study) were considered for studying the kit performance in various patients to address localization of primary tumours, lymph node involvement and metastasis. Figure 5 demonstrates the image findings in a patient with limited disease at presentation (age: 63-years-old; weight: 125 kg; PSA = 146 µg/L; ^{99m}Tc-MDP-SPECT negative for lesions). The organ-distribution of ^{68}Ga-DKFZ-PSMA-11 consists of uptake in expected organs such as in the lacrimal glands, salivary glands, liver and spleen, minimal bowel excretion, and main excretion via the kidneys into the urinary bladder. The image quality and the ability to delineate the tumour mass can be considered comparable when using a low dose (Figure 5A) SUV$_{max}$ = 27.3)) instead of a high dose (Figure 5B) SUV$_{max}$ = 23.7)) in the same patient. This is an example how a dose of 97 MBq has sufficiently detected the primary lesion without significantly compromising the image quality. Figure 6 is a typical example of a patient with a limited history of the disease at presentation (age: 92-yr-old; weight = 65 kg; history of rising PSA levels post-surgery (orchidectomy)). No ^{18}F-FDG-PET or ^{99m}Tc-MDP-SPECT was carried out prior to ^{68}Ga-DKFZ-PSMA-11-PET, which showed an advanced disease state with an involvement of lymph nodes in the pelvis area (SUV$_{max}$ = 18.3); the low dose of 44.4 MBq injected localized the pathologic tissues adequately. In Figure 7, a patient had a surgical history of bilateral orchidectomy and left nephrectomy. He represents a typical example of advanced prostate cancer (age: 63-yr-old; weight: 77 kg; PSA = 291 μg/L) with a positive ^{99m}Tc-MDP-SPECT scan for bone metastases (Figure 7A). ^{68}Ga-DKFZ-PSMA-11 PET/CT confirmed prostate cancer recurrence (SUV$_{max}$ = 21.6) including multiple soft tissue and bone metastatic lesions with SUV$_{max}$ values ranging from 8.6–22.6 (Figure 7B).

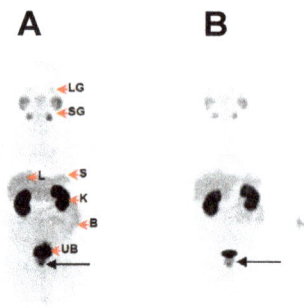

Figure 5. ^{68}Ga-DKFZ-PSMA-11-PET images showing pathology of primary prostate cancer (**black arrow**) in a 63-yr-old patient (weight = 125 kg; PSA = 146 μg/L; ^{99m}Tc-MDP-SPECT was negative for lesions) at 60 min following administration of (**A**) a low dose of ^{68}Ga-DKFZ-PSMA-11 (97 MBq) and (**B**) a high dose of ^{68}Ga-DKFZ-PSMA-11 (325 MBq), respectively. Images were obtained on a Siemens Biograph 40 PET/CT scanner and displayed in anterior projection. The normal bio-distribution (**red arrow heads**) consists of uptake in the lacrimal glands (LG), salivary glands (SG), liver (L) and spleen (S), minimal bowel excretion (B), and main excretion via the kidneys (K) into the urinary bladder (UB).

Figure 6. PET image displayed in anterior view of ^{68}Ga-DKFZ-PSMA-11 in a 92-yr-old male with limited disease at presentation (weight = 65 kg; history of rising PSA levels post orchidectomy for prostate; no prior ^{18}F-FDG-PET or ^{99m}Tc-MDP-SPECT carried out). Prostate cancer is demonstrated (**black arrow**) including an intense accumulation of ^{68}Ga-DKFZ-PSMA-11 in the left internal iliac node (**red arrow head**) at 60 min post-injection.

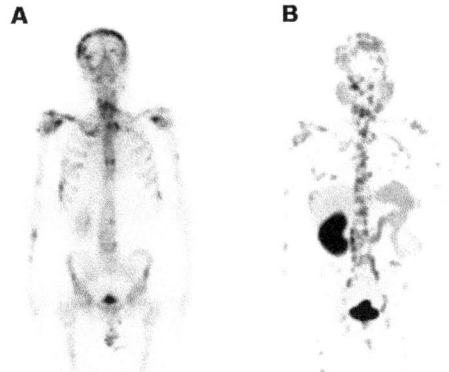

Figure 7. Whole body image projections of a 63-yr-old patient presenting with advanced disease (weight = 77 kg; PSA = 291 µg/L) (**A**) ^{99m}Tc-MDP-SPECT was positive for bone metastases; (**B**) PET/CT imaging detected widely distributed skeletal and visceral metastases at 60 min after injection of a low dose of ^{68}Ga-DKFZ-PSMA-11 (70 MBq). Images were obtained on a Siemens Biograph 40 PET/CT scanner and are displayed in anterior projection, showing multiple skeletal and soft tissue lesions.

2.6. Discussion

Nuclear medical PET/CT-diagnostics using ^{18}F-FDG, or ^{11}C or ^{18}F labeled derivatives of choline and acetate is based on cellular glucose accumulation or tumour proliferation rate, which is found to be low in most forms of PC [28]. However, alternative PET imaging with ligands targeting receptors such as PSMA that are overexpressed independently from tumour proliferation seem superior also because PSMA ligands exhibit the ability to be internalized by receptor-mediated endocytosis [29]. This binding, in combination with a suitable radioisotope, allows for highly selective PET imaging with intense tumour-to-background ratios as a result of enhanced tumour cell retention of the tracer molecule. In 2009 scientists described the pseudo-peptide structure Glutamine-Urea-Lysin to be a PSMA-targeting inhibitor that clears rapidly from the circulation, leading to images with clear contrast [30,31]. As ^{68}Ga was envisaged as the preferred radioisotope, the first challenge occurring was to successfully complex it with DOTA without hampering the targeting abilities. The limited binding of DOTA-PSMA was overcome by the conjugation of Glutamine-Urea-Lysin to the hetero-bifunctional chelator HBED-CC [21,32] to form DKFZ-PSMA-11, which warranted efficient radiolabeling with ^{68}Ga at ambient temperature. The high specificity of DKFZ-PSMA-11 to complex ^{68}Ga led us to set up a convenient kit labeling procedure. Additionally, PC imaging using the kit manufacturing technique is tested for its performance in PC patients.

To our knowledge, this is the first report on this novel single vial kit assessment involving a SnO$_2$-based ^{68}Ge/^{68}Ga generator presenting ^{68}Ga-DKFZ-PSMA-11-PET images with high lesion-to-background ratios. The major challenge was to address an appropriate precursor concentration and radiolabel it with ^{68}Ga while meeting all safety and purity requisites, without compromising the straightforward radiolabeling. In the past five years, ^{68}Ge/^{68}Ga-generators have emerged as a reliable, source for daily ^{68}Ga, possibly detaching PET-radiopharmaceutical development from a cyclotron environment. We managed to utilise a fractionated batch of ^{68}Ga that was eluted from a commercially available SnO$_2$-based generator (IDB

Holland, Netherlands). The amount of elutable ^{68}Ga-activity over the generator's nine-month-life span was excellent and the breakthrough of ^{68}Ge and co-eluted metal impurities were of no concern for these radiopharmaceutical productions (Table 1). If the DKFZ-PSMA-11-peptide-labeling formulation can be supplied in a GMP-compliant kit form in analogy to all the conventional ^{99m}Tc-radiotracers, it can be made available to all the PET/CT facilities in South Africa and beyond.

The concept of utilizing PET-kits was first addressed by E. Deutsch 1993 in the *Journal of Nuclear Medicine*, where he suggests that in order to satisfy the governmental regulatory and manufacturing requirements one significant approach would entail the development of a family of ^{68}Ga-radiopharmaceuticals that can be prepared from cold kits and a ^{68}Ge/^{68}Ga-generator [33]. Nowadays, we have singular kit vial solutions becoming an asset to PET radiopharmaceutical research, as exemplified by a universal technique to radiolabel proteins [26] or macro-aggregated albumin [34] with ^{68}Ga. In 2013 we have successfully reported a procedure of a one-step-aseptic technique to radiolabel ACD-A kits with ^{68}Ga for imaging of tuberculosis [27,35]. Many problems concerning the regulation and manufacture of PET radiopharmaceuticals can be alleviated in their future development. However, careful, critical attention to detail and rigorous quality-assurance protocols are essential if complex radiolabeling procedures are to be successfully embedded at a hospital radiopharmacy [36].

The kit-derived ^{68}Ga-DKFZ-PSMA-11 was injected into patients diagnosed with PC. PET/CT images demonstrated the expected bio-distribution with the most intense tracer accumulation noted in the kidneys and salivary glands. The lacrimal glands, liver, spleen, and the small and large intestines showed moderate-to-low uptake (Figure 5), which confirmed what other groups have demonstrated [14,37]. Moreover, our amended approach to work with an increased mass of DKFZ-PSMA-11 (to achieve quantitative ^{68}Ga-complexation) did not hamper the pharmacological ability of the molecule; a clear delineation of the tumour mass through the PET scan was determined, even with less than 100 MBq administered, as a result of the high specific activity. The low injected doses in this study ($n = 6$) were lower than doses administered to study dosimetric aspects of ^{68}Ga-DKFZ-PSMA-11 [38]. These doses exhibit less radiation burden (effective dose of 3 mSv), delivering organ doses that are comparable to (kidneys), or lower than, those delivered by ^{18}F-FDG. Following low-dose tracer administration, we were able to detect primary tumours (Figure 5) and early metastatic disease with pelvic lymph node involvement (Figure 6), as well as widespread metastatic disease (Figure 7). It should be noted that this clinical investigation is merely a performance test to address the radiopharmaceutical amendments in the production according to GMP and GCP. This study cannot motivate the application of low doses routinely for this tracer, but it might be interesting to implement this objective in a larger scaled investigation. Low dose tracer administrations might bear the risk of overlooking critical lesions or leaving malignant loci undetected post therapy. If accurately investigated, the use of lower injected doses offers significant advantages; it helps to reduce patient radiation exposure, while allowing for the optimization of costs and patient throughput. Cyclotron-independent PET tracers also allow for the possibility of PET/CT centres in rural areas.

Based on these results, the performance of the ^{68}Ga-DKFZ-PSMA-11 kit appears similar to published findings with ^{68}Ga-DKFZ-PSMA-11, which were produced in a module- or cassette-like procedure [21,22]. We have also reported a case study where ^{68}Ga-DKFZ-PSMA-11 PET/CT was successfully imaging metastatic breast cancer [39] and the tracer was also found to accumulate in metastatic renal cell carcinoma [40].

The ^{68}Ga-DKFZ-PSMA-11 kit is a safe and useful, ready-to-use diagnostic agent in PC with high diagnostic performance. A multi-patient dose can be produced and dispensed in less than 30 min at RT, featuring high thermodynamic stability, and a high degree of reproducibility and robustness towards the storage environment and the quality of the eluted radiogallium. The simplicity of the method provides a highly convenient and easy-to-integrate ^{68}Ga-tool to tracer production in the hospital radiopharmacy. Providing an immaculate generator performance, this simplified technique makes post-purification obsolete and usable by radiography personnel. The latter approach may be suitable for implementation of other ^{68}Ga-radiopharmaceuticals for a more elegant way of translational research.

3. Material and Methods

3.1. Chemicals and Materials

If not stated otherwise, only pharmacological-grade solvents were used in the procedures. DKFZ-PSMA-11 was purchased from ABX advanced biochemical compounds (Biomedizinische Forschungsreagenzien GmbH, Radeberg, Germany) in GMP-compliant grade, supplied as trifluoroacetate salt. A 30% solution of ultrapure grade hydrochloric acid (HCl), trifluoroacetic acid and methanol were purchased from Fluka Analytical (Steinheim, Germany). High-performance liquid chromatography (HPLC) grade water (resistivity = 18.2 MΩ cm) was produced in-house by a Simplicity 185 Millipore system (Cambridge, MA, USA). All other solvents were purchased in at least HPLC grade from Sigma Aldrich (Steinheim, Germany). Certified sterile pyrogen-free sealed borosilicate glass vials (5–30 mL) were provided by NTP Radioisotopes (Pty) Ltd. (Pelindaba, South Africa) and were utilized for kit production, generator elution, and sterile saline dispension. Silica gel ITLC paper was purchased from Agilent Technologies (Forest Lakes, CA, USA). Sterile filters were obtained from Millipore (Millipore, New York, NY, USA).

3.2. ^{68}Ge/^{68}Ga Generator

^{68}Ga (89%; EC β$^+$ max. 1.9 MeV) was yielded from two consecutive tin-dioxide-based ^{68}Ge/^{68}Ga generators (iThemba LABS, Somerset West, South Africa). The eluate fractionation and purification was carried out as previously reported [27]. A Jelco 22G × 1 polymer catheter (Smiths Medical, Croydon, South Africa) was utilized to warrant metal-free radiogallium transfer. Routinely, inductively coupled plasma optical emission spectroscopy (ICP-OES) was carried out to detect trace metal impurities. The levels of co-eluted ^{68}Ge were routinely surveyed in the total generator eluate and in retained ^{68}Ga-DKFZ-PSMA-11 solutions. After at least 48 h, ^{68}Ge was measured indirectly in a CRC 25 ionization chamber (Capintec Inc., Ramsey, NJ, USA), by detecting radiogallium that was generated *in situ* by ^{68}Ge impurities. ^{68}Ge samples of known activity were used as references. Low ^{68}Ge-levels were detected using a single probe well counter (Biodex Inc., Shirley, New York, NY, USA) or a calibrated gamma spectrometer as previously described [27]. The percentage ^{68}Ge was calculated by as follows:

$$68\text{Ge} - \text{impurity (\%)} = \frac{68\text{Ge} - \text{activity (Bq)}}{68\text{Ga} - \text{activity (Bq)}} \times 100\% \tag{1}$$

3.3. Preliminary Assessment of Radiolabeling Parameters

In order to achieve the highest labeling efficiency in optimal time, the following labeling parameters were optimized: DKZF-PSMA-11 molarity given a constant ^{68}Ga pH value (buffered with sodium acetate trihydrate to yield pH 4.0–4.5) and incubation duration (with or without applying vortex stirring of the sample). Therefore, 1 mL of the ^{68}Ga eluate was pre-mixed into vials containing 1–12 nmol DKFZ-PSMA-11 and 98 mg sodium acetate trihydrate salt and incubated at room temperature (RT) up to 30 min; one set of samples underwent vortex stirring action every 5 min for 15 s. At 2, 4, 6, 10, 15, 20 and 30 min, 4 µL per sample was extracted for analysis. The labeling efficiency for all crude samples was determined by ITLC as described. Potential impurities and un-chelated ^{68}Ga [41] were purified using Sep-Pak light C18 solid-phase-extraction (Waters, Eschborn, Germany).

3.4. In-House Kit Vial Formulation of DKFZ-PSMA-11

Sterile kit vials were produced at Radiochemistry; The South African Nuclear Energy Corporation, using 10 mL vials provided by NTP Radioisotopes. A stock solution of 50 nmol/mL DKFZ-PSMA-11 (resultant mass ^{68}Ga-DKFZ-PSMA-11: 5.07 µg (MW$_{DKFZ-PSMA-11}$ = 947.0 g/mol + MW$_{68Ga}$ = 69.7 g/mol) was prepared by dissolving the peptide in Millipore water. Aliquots of 100 µL of the peptide stock solution were mixed with 98 mg sodium acetate trihydrate and vortexed vigorously to yield a homogeneous, gel-like consistency. Alternatively, 5 nmol DKFZ-PSMA-11 were supplemented with 250 µL of 392 mg/mL sodium acetate trihydrate solution and vortexed. The kit vials were immediately placed in an ultra-low freezer (Bio-Freezer, Thermo Fisher Scientific, Waltham, MA, USA) for a minimum of four h (−50 °C) and subsequently transferred to the Alpha 1–5 laboratory freeze dryer (Christ, Osterode am Harz, Germany) where lyophilisation was carried out overnight under Argon atmosphere at 0.05 mbar. The vials were sealed and routinely stored at 2–8 °C.

3.5. ^{68}Ga Radiolabeling of DKFZ-PSMA-11 Kits

Eighteen kit vials containing 5 nmol buffered DKFZ-PSMA-11 were mixed with 1 mL ^{68}GaCl$_3$ solution and allowed to incubate at RT for 15 min with gentle vortexing for at least 15 s in five-min intervals. Thereafter the radiolabeling was carried as following: Stage 1 involves the kit investigation phase where the need for a purification step was studied by passing the reaction mixture through a SepPack C18 light as published [42], recovering the product with ethanolic saline solution. Stage 2 involves a true one-vial-one-step-radiolabeling approach followed by supplementing the kit vial with 1.5 mL of 2.5 M sodium acetate trihydrate and 3 mL of saline, to yield a physiological pH. Before dispensing of ^{68}Ga-DKFZ-PSMA-11, the solution was run through a 0.22 µm membrane using a low protein-binding filter. An aliquot of >2 mL was retained for further quality assessment of the kits. Radiochemical purity (% RCP) was determined by gradient radio-high-pressure-liquid-chromatography (HPLC) and the decay-corrected radiochemically recovered yield (% RCY) was determined by radio-thin layer chromatography (ITLC) as optimized from previously described procedures [21,43]. A reverse-phase HPLC column (Zorbax SB C18, 4.6 mm × 250 mm × 5 µm; Agilent Technologies, CA, USA) with a linear A–B gradient (0% B to 100% B in 6 min) at a flow rate of 1 mL/min was utilized for quantification analysis. Solvent A consisted of 0.1% aqueous trifluoroacetic acid (TFA) and solvent B was 0.1% TFA in acetonitrile. The HPLC system

(Agilent 1200 series HPLC instrument coupled to 6100 Quadruple MS detector, Agilent Technologies Inc., Wilmington, DE, USA), equipped with a diode array detector (DAD) and radioactive detector (Gina Star, Raytest, Straubenhardt, Germany) measured UV absorbance at 214 nm. For ITLC quantification silica gel impregnated chromatographic paper (1 × 10 cm, Agilent Technologies, Forrest Lake, CA, USA) was employed as the stationary phase. The paper was spotted with reaction mixture, dried and exposed for 5–8 min to the mobile phase (Saline/MeOH 80:20 v/v). The chromatograms were visualized by ITLC radio-chromatography imaging (VSC-201, Veenstra Ind., Oldenzaal, The Netherlands) using a gamma radiation detector (Scionix 25B25/1.5-E2, Bunnik, The Netherlands). The obtained chromatograms allowed for peak identification and performing "area under the curve" analysis for percentage quantification (Genie2000 software, Veenstra Ind., Oldenzaal, The Netherlands). Free radiogallium remained close to the baseline (R_f: 0.05) whereas the ^{68}Ga-DKFZ-PSMA-11 peak was detected at R_f of 0.8–0.95.

3.6. Quality Assessment of the ^{68}Ga-DKFZ-PSMA-11 Kit

Assessment of the quality of the cold or radiolabeled kits included sterility, pH value of the ^{68}Ga reaction mixture and product solutions, percentage radiolabeling efficiency (% LE), radiochemical stability, radionuclidic identity and labeling reproducibility with different generator eluate purities, and storage stability.

3.6.1. Appearance and Sterility

After optical and light microscopic inspection of particles and change of colour, aliquots of the sterile filtered sample were analysed by the NHLS microbiological laboratory at Steve Biko Academic Hospital for bacterial growth of aerobe, anaerobe, and fungal species as carried out previously [27]. Sterile filters were tested for pressure integrity (bubble point test).

3.6.2. Radionuclidic Identity and pH Value

Radionuclidic identity was measured by 120 min decay analysis for determination of ^{68}Ga half-life using a CRC 25 ionization chamber (Capintec Inc., Ramsey, NJ, USA). The pH value was assessed using a narrow range pH paper (pH Fix 0-7, Macherey-Nagel, Düren, Germany) technique.

3.6.3. Radiochemical Stability

As multi-patient doses can potentially be achieved from an upscale radiopharmaceutical production the stability of ^{68}Ga-DKFZ-PSMA-11 was evaluated over three h of incubation at 37 °C and subsequent analysis for potential free ^{68}Ga recurrence using radio-ITLC.

3.6.4. Long-Term Storage and Radiolabeling Reproducibility

Verification of the kit shelf-life was carried out considering two aspects A) storage stability (aliquots for the kit formulation were kept at either −50 °C, 2–8 °C, or at room temperature for 30–60 days prior to radiolabeling and B) robustness towards purity changes of the generator-eluted ^{68}GaCl$_3$: kits contents were mixed with equal ^{68}Ga-batches yielded from a freshly-manufactured generator and compared with those of an outdated generator: (1) a routinely eluted generator batch, (2) the consecutive batch of the same day yielded >4 h later, and (3) a C18-purified ^{68}Ga-DKFZ-PSMA-11 batch, respectively. Radiolabeling was carried out according to the above mentioned protocol. The % RCY and % RCP was determined by HPLC and ITLC as described earlier.

3.7. Clinical PET/CT—^{68}Ga-DKFZ-PSMA-11 Kit Performance in Prostate Cancer Patients

The study was conducted in accordance with the Declaration of Helsinki. The University of Pretoria's Research Ethics Committee granted approval for this study and written or verbal consent was obtained from each participant prior to tracer injection. Patients with histologically confirmed PC were included with referral indications for initial staging, restaging or suspected recurrence.

Image Acquisition, Reconstruction and Analysis

Image acquisition and reconstruction was carried out as previously reported [27]. Briefly, no special patient preparation was required and all patients were imaged on a Siemens Biograph 40-slice PET/CT scanner according to standard protocol. Intravenous contrast was injected unless a contra-indication existed and all patients were imaged from vortex to mid-thigh at 60 min post-injection of low doses of ^{68}Ga-DKFZ-PSMA-11 (44-126 MBq). Images were independently analysed by two trained physicians determining the maximum standard uptake values (SUV_{max}) in lesions or targeted organs.

3.8. Statistical Analysis

If necessary, data was normalized by a log10 transformation before statistical analysis. If not stated otherwise, analytical datasets were expressed as mean and standard error of mean (sem). Dependency between two parameters was analysed by the Spearman correlation to provide the correlation coefficient (r). Significance of two mean values was calculated by *Student-t-test* (paired and unpaired comparison). For all statistical tests, the level of significance (p) was set at <0.05 (two-tailed) where * $p < 0.05$, ** $p < 0.01$, *** $p < 0.001$ *vs.* references or controls.

4. Conclusions

We managed to produce ^{68}Ga-DKFZ-PSMA-11 from a freeze-dried one-vial kit that meets all QC criteria for PET imaging. An efficient technique for the routine preparation of ^{68}Ga-DKFZ-PSMA-11 was presented, showing highly-specific, reproducible radiolabeling that accommodated the acidic conditions needed to elute ^{68}Ga from a SnO_2-based ^{68}Ge/^{68}Ga generator. Up to 800 MBq highly pure (>98%) ^{68}Ga-DKFZ-PSMA-11 was provided for patient administration within 20–30 min, and the localization of primary PC and lymph node involvement, as well as advanced metastatic PC scenarios, seemed not to be compromised by the kit-manufacturing protocol. Moreover, ^{68}Ga-DKFZ-PSMA-11 PET/CT is desirable for an effective stratification of patients undergoing theranostic radioligand therapy with ^{177}Lu-labeled PSMA-ligands.

Acknowledgments

The work related to this study was funded by the Department of Science and Technology, The South African Nuclear Energy Corporation and the Nuclear Technologies in Medicine and the Biosciences

Initiative. Meltem Ocak from Istanbul University is thanked for assistance in the initiation of this study. The personnel of the Nuclear Medicine Department are thanked for the kind patient preparation and handling prior and after the PET/CT imaging. Mrs. Barbara English is thanked for editing of English language and style.

Author Contributions

T.E. was the principal investigator on this project, planning all studies and tests carried out preliminary radiolabeling prior to kit formulation and preparing of the manuscript drafts. B.M., B.M.-P., J.W. and J.S.: assistance on freeze-dried-kit procedures, radiolabeling, HPLC/ITLC analyses and all QC method validations. J.R.Z.: supervision and assistance on the release criteria and critical review of the results. M.M.: assistance on the patient population, patient data acquisition and image analysis and reporting of patients' disease histories. M.V. and M.S.: supervision and provision of clinical competence, revision and enhancement of the intellectual contents of the manuscript. All co-authors contributed in critical revision of the manuscript and in approving the final content of the manuscript.

Conflicts of Interest

The authors declare no conflict of interest.

References

1. Center, M.M.; Jemal, A.; Lortet-Tieulent, J.; Ward, E.; Ferlay, J.; Brawley, O.; Bray, F. International variation in prostate cancer incidence and mortality rates. *Eur. Urol.* **2012**, *61*, 1079–1092.
2. Kravchick, S.; Cytron, S.; Peled, R.; Altshuler, A.; Ben-Dor, D. Using gray-scale and two different techniques of color Doppler sonography to detect prostate cancer. *Urology* **2003**, *61*, 977–981.
3. Hricak, H.; Dooms, G.C.; Jeffrey, R.B.; Avallone, A.; Jacobs, D.; Benton, W.K.; Narayan, P.; Tanagho, E.A. Prostatic carcinoma: Staging by clinical assessment, CT, and MR imaging. *Radiology* **1987**, *162*, 331–336.
4. Gronberg, H. Prostate cancer epidemiology. *Lancet* **2003**, *361*, 859–864.
5. Jadvar, H. Imaging evaluation of prostate cancer with ^{18}F-fluorodeoxyglucose PET/CT: Utility and limitations. *Eur. J. Nucl. Med. Mol. Imaging* **2013**, *40 (Suppl. 1)*, 5–10.
6. Apolo, A.B.; Pandit-Taskar, N.; Morris, M.J. Novel tracers and their development for the imaging of metastatic prostate cancer. *J. Nucl. Med.* **2008**, *49*, 2031–2041.
7. Jadvar, H. Molecular imaging of prostate cancer: PET radiotracers. *Am. J. Roentgenol.* **2012**, *199*, 278–291.
8. Castellucci, P.; Jadvar, H. PET/CT in prostate cancer: Non-choline radiopharmaceuticals. *Q. J. Nucl. Med. Mol. Imaging* **2012**, *56*, 367–374.
9. Jadvar, H. Can choline PET tackle the challenge of imaging prostate cancer? *Theranostics* **2012**, *2*, 331–332.

10. Afshar-Oromieh, A.; Zechmann, C.M.; Malcher, A.; Eder, M.; Eisenhut, M.; Linhart, H.G.; Holland-Letz, T.; Hadaschik, B.A.; Giesel, F.L.; Debus, J.; et al. Comparison of PET imaging with a ^{68}Ga-labeled PSMA ligand and ^{18}F-choline-based PET/CT for the diagnosis of recurrent prostate cancer. *Eur. J. Nucl. Med. Mol. Imaging* **2014**, *41*, 11–20.
11. Vorster, M.; Modiselle, M.; Ebenhan, T.; Wagener, C.; Sello, T.; Zeevaart, J.R.; Moshokwa, E.; Sathekge, M.M. Fluorine-18-fluoroethylcholine PET/CT in the detection of prostate cancer: A South African experience. *Hell. J. Nucl. Med.* **2015**, *18*, 53–59.
12. Demirkol, M.O.; Acar, O.; Ucar, B.; Ramazanoglu, S.R.; Saglican, Y.; Esen, T. Prostate-specific membrane antigen-based imaging in prostate cancer: Impact on clinical decision making process. *Prostate* **2015**, *75*, 748–757.
13. Osborne, J.R.; Akhtar, N.H.; Vallabhajosula, S.; Anand, A.; Deh, K.; Tagawa, S.T. Prostate-specific membrane antigen-based imaging. *Urol. Oncol.* **2013**, *31*, 144–154.
14. Afshar-Oromieh, A.; Malcher, A.; Eder, M.; Eisenhut, M.; Linhart, H.G.; Hadaschik, B.A.; Holland-Letz, T.; Giesel, F.L.; Kratochwil, C.; Haufe, S.; et al. PET imaging with a [^{68}Ga]gallium-labeled PSMA ligand for the diagnosis of prostate cancer: Biodistribution in humans and first evaluation of tumour lesions. *Eur. J. Nucl. Med. Mol. Imaging* **2013**, *40*, 486–495.
15. Buhler, P.; Wolf, P.; Elsasser-Beile, U. Targeting the prostate-specific membrane antigen for prostate cancer therapy. *Immunotherapy* **2009**, *1*, 471–481.
16. Chen, Y.; Foss, C.A.; Byun, Y.; Nimmagadda, S.; Pullambhatla, M.; Fox, J.J.; Castanares, M.; Lupold, S.E.; Babich, J.W.; Mease, R.C.; et al. Radiohalogenated prostate-specific membrane antigen (PSMA)-based ureas as imaging agents for prostate cancer. *J. Med. Chem.* **2008**, *51*, 7933–7943.
17. Banerjee, S.R.; Foss, C.A.; Castanares, M.; Mease, R.C.; Byun, Y.; Fox, J.J.; Hilton, J.; Lupold, S.E.; Kozikowski, A.P.; Pomper, M.G. Synthesis and evaluation of technetium-99m- and rhenium-labeled inhibitors of the prostate-specific membrane antigen (PSMA). *J. Med. Chem.* **2008**, *51*, 4504–4517.
18. Banerjee, S.R.; Pullambhatla, M.; Byun, Y.; Nimmagadda, S.; Green, G.; Fox, J.J.; Horti, A.; Mease, R.C.; Pomper, M.G. ^{68}Ga-labeled inhibitors of prostate-specific membrane antigen (PSMA) for imaging prostate cancer. *J. Med. Chem.* **2010**, *53*, 5333–5341.
19. Eder, M.; Schafer, M.; Bauder-Wust, U.; Hull, W.E.; Wangler, C.; Mier, W.; Haberkorn, U.; Eisenhut, M. ^{68}Ga-complex lipophilicity and the targeting property of a urea-based PSMA inhibitor for PET imaging. *Bioconjugate Chem.* **2012**, *23*, 688–697.
20. Afshar-Oromieh, A.; Haberkorn, U.; Eder, M.; Eisenhut, M.; Zechmann, C.M. [^{68}Ga]Gallium-labeled PSMA ligand as superior PET tracer for the diagnosis of prostate cancer: Comparison with ^{18}F-FECH. *Eur. J. Nucl. Med. Mol. Imaging* **2012**, *39*, 1085–1086.
21. Eder, M.; Neels, O.; Muller, M.; Bauder-Wust, U.; Remde, Y.; Schafer, M.; Hennrich, U.; Eisenhut, M.; Afshar-Oromieh, A.; Haberkorn, U.; et al. Novel preclinical and radiopharmaceutical aspects of [^{68}Ga]Ga-PSMA-HBED-CC: A new PET tracer for imaging of prostate cancer. *Pharmaceuticals (Basel)* **2014**, *7*, 779–796.
22. Afshar-Oromieh, A.; Avtzi, E.; Giesel, F.L.; Holland-Letz, T.; Linhart, H.G.; Eder, M.; Eisenhut, M.; Boxler, S.; Hadaschik, B.A.; Kratochwil, C.; et al. The diagnostic value of PET/CT imaging with the ^{68}Ga-labeled PSMA ligand HBED-CC in the diagnosis of recurrent prostate cancer. *Eur. J. Nucl. Med. Mol. Imaging* **2015**, *42*, 197–209.

23. Al-Nahhas, A.; Win, Z.; Szyszko, T.; Singh, A.; Khan, S.; Rubello, D. What can gallium-68 PET add to receptor and molecular imaging? *Eur. J. Nucl. Med. Mol. Imaging* **2007**, *34*, 1897–1901.
24. Velikyan, I. Prospective of ^{68}Ga-radiopharmaceutical development. *Theranostics* **2013**, *4*, 47–80.
25. Mukherjee, A.; Pandey, U.; Chakravarty, R.; Sarma, H.D.; Dash, A. Development of single vial kits for preparation of ^{68}Ga-labeled peptides for PET imaging of neuroendocrine tumours. *Mol. Imaging Biol.* **2014**, *16*, 550–557.
26. Wangler, C.; Wangler, B.; Lehner, S.; Elsner, A.; Todica, A.; Bartenstein, P.; Hacker, M.; Schirrmacher, R. A universally applicable ^{68}Ga-labeling technique for proteins. *J. Nucl. Med.* **2011**, *52*, 586–591.
27. Vorster, M.; Mokaleng, B.; Sathekge, M.M.; Ebenhan, T. A modified technique for efficient radiolabeling of ^{68}Ga-citrate from a SnO$_2$-based ^{68}Ge/^{68}Ga generator for better infection imaging. *Hell. J. Nucl. Med.* **2013**, *16*, 193–198.
28. Hain, S.F.; Maisey, M.N. Positron emission tomography for urological tumours. *BJU Int.* **2003**, *92*, 159–164.
29. Goodman, O.B., Jr.; Barwe, S.P.; Ritter, B.; McPherson, P.S.; Vasko, A.J.; Keen, J.H.; Nanus, D.M.; Bander, N.H.; Rajasekaran, A.K. Interaction of prostate specific membrane antigen with clathrin and the adaptor protein complex-2. *Int. J. Oncol.* **2007**, *31*, 1199–1203.
30. Hillier, S.M.; Maresca, K.P.; Femia, F.J.; Marquis, J.C.; Foss, C.A.; Nguyen, N.; Zimmerman, C.N.; Barrett, J.A.; Eckelman, W.C.; Pomper, M.G.; *et al.* Preclinical evaluation of novel glutamate-urea-lysine analogues that target prostate-specific membrane antigen as molecular imaging pharmaceuticals for prostate cancer. *Cancer Res.* **2009**, *69*, 6932–6940.
31. Maresca, K.P.; Hillier, S.M.; Femia, F.J.; Keith, D.; Barone, C.; Joyal, J.L.; Zimmerman, C.N.; Kozikowski, A.P.; Barrett, J.A.; Eckelman, W.C.; *et al.* A series of halogenated heterodimeric inhibitors of prostate specific membrane antigen (PSMA) as radiolabeled probes for targeting prostate cancer. *J. Med. Chem.* **2009**, *52*, 347–357.
32. Schafer, M.; Bauder-Wust, U.; Leotta, K.; Zoller, F.; Mier, W.; Haberkorn, U.; Eisenhut, M.; Eder, M. A dimerized urea-based inhibitor of the prostate-specific membrane antigen for ^{68}Ga-PET imaging of prostate cancer. *EJNMMI Res.* **2012**, *2*, doi:10.1186/2191-219X-2-23.
33. Deutsch, E. Clinical PET: Its time has come? *J. Nucl. Med.* **1993**, *34*, 1132–1133.
34. Amor-Coarasa, A.; Milera, A.; Carvajal, D.; Gulec, S.; McGoron, A.J. Lyophilized kit for the preparation of the PET perfusion agent [^{68}Ga]-MAA. *Int. J. Mol. Imaging* **2014**, *2014*, doi:10.1155/2014/269365.
35. Vorster, M.; Maes, A.; van de Wiele, C.; Sathekge, M.M. ^{68}Ga-citrate PET/CT in Tuberculosis: A pilot study. *Q. J. Nucl. Med. Mol. Imaging* **2014**.
36. Breeman, W.A.; de Blois, E.; Sze Chan, H.; Konijnenberg, M.; Kwekkeboom, D.J.; Krenning, E.P. ^{68}Ga-labeled DOTA-peptides and ^{68}Ga-labeled radiopharmaceuticals for positron emission tomography: Current status of research, clinical applications, and future perspectives. *Semin. Nucl. Med.* **2011**, *41*, 314–321.
37. Kabasakal, L.; Demirci, E.; Ocak, M.; Akyel, R.; Nematyazar, J.; Aygun, A.; Halac, M.; Talat, Z.; Araman, A. Evaluation of PSMA PET/CT imaging using a ^{68}Ga-HBED-CC ligand in patients with prostate cancer and the value of early pelvic imaging. *Nucl. Med. Commun.* **2015**, *36*, 582–587.

38. Herrmann, K.; Bluemel, C.; Weineisen, M.; Schottelius, M.; Wester, H.J.; Czernin, J.; Eberlein, U.; Beykan, S.; Lapa, C.; Riedmiller, H.; et al. Biodistribution and radiation dosimetry for a novel probe targeting prostate specific membrane antigen for imaging and therapy (^{68}Ga-PSMA I & T). *J. Nucl. Med.* **2015**, *56*, 855–861.
39. Sathekge, M.; Modiselle, M.; Vorster, M.; Mokgoro, N.; Nyakale, N.; Mokaleng, B.; Ebenhan, T. ^{68}Ga-PSMA imaging of metastatic breast cancer. *Eur. J. Nucl. Med. Mol. Imaging* **2015**, *42*, 1482–1483.
40. Demirci, E.; Ocak, M.; Kabasakal, L.; Decristoforo, C.; Talat, Z.; Halac, M.; Kanmaz, B. ^{68}Ga-PSMA PET/CT imaging of metastatic clear cell renal cell carcinoma. *Eur. J. Nucl. Med. Mol. Imaging* **2014**, *41*, 1461–1462.
41. Ebenhan, T.; Chadwick, N.; Sathekge, M.M.; Govender, P.; Govender, T.; Kruger, H.G.; Marjanovic-Painter, B.; Zeevaart, J.R. Peptide synthesis, characterization and ^{68}Ga-radiolabeling of NOTA-conjugated ubiquicidin fragments for prospective infection imaging with PET/CT. *Nucl. Med. Biol.* **2014**, *41*, 390–400.
42. Rossouw, D.D.; Breeman, W.A. Scaled-up radiolabeling of DOTATATE with ^{68}Ga eluted from a SnO$_2$-based ^{68}Ge/^{68}Ga generator. *Appl. Radiat. Isot.* **2012**, *70*, 171–175.
43. Baur, B.; Solbach, C.; Andreolli, E.; Winter, G.; Machulla, H.J.; Reske, S.N. Synthesis, radiolabeling and *in vitro* characterisation of the Gallium-68-, Yttrium-90- and Lutetium-177-labeled PSMA ligand, CHX-A''-DTPA-DUPA-Pep. *Pharmaceuticals (Basel)* **2014**, *7*, 517–529.

Sample Availability: DKFZ-PSMA-11 kit vials are available from the authors.

© 2015 by the authors; licensee MDPI, Basel, Switzerland. This article is an open access article distributed under the terms and conditions of the Creative Commons Attribution license (http://creativecommons.org/licenses/by/4.0/).

Molecules **2015**, *20*, 13112-13126; doi:10.3390/molecules200713112

ISSN 1420-3049
www.mdpi.com/journal/molecules

Article

The Influence of the Combination of Carboxylate and Phosphinate Pendant Arms in 1,4,7-Triazacyclononane-Based Chelators on Their ^{68}Ga Labelling Properties

Gábor Máté [1], Jakub Šimeček [2,†], Miroslav Pniok [3], István Kertész [1], Johannes Notni [2], Hans-Jürgen Wester [2], László Galuska [1] and Petr Hermann [3,*]

[1] Department of Nuclear Medicine, Faculty of Medicine, University of Debrecen, Nagyerdei krt 98, H-4032 Debrecen, Hungary; E-Mails: mate.gabor@med.unideb.hu (G.M.); kertesz.istvan@med.unideb.hu (I.K.); galuska.laszlo@med.unideb.hu (L.G.)

[2] Lehrstuhl für Pharmazeutische Radiochemie, Technische Universität München, Walther-Meissner-Strasse 3, D-85748 Garching, Germany; E-Mails: jakubsimecek@seznam.cz (J.Š.); johannes.notni@tum.de (J.N.); h.j.wester@tum.de (H.-J.W.)

[3] Department of Inorganic Chemistry, Charles University in Prague, Hlavova 2030, 12840 Prague 2, Czech Republic; E-Mail: mirekpniok@gmail.com

† Current address: Scintomics GmbH, Lindach 4, Fürstenfeldbruck 82256, Germany.

* Author to whom correspondence should be addressed; E-Mail: petrh@natur.cuni.cz; Tel.: +420-221-951-263; Fax: +420-221-951-253.

Academic Editor: Svend Borup Jensen

Received: 10 May 2015 / Accepted: 13 July 2015 / Published: 21 July 2015

Abstract: In order to compare the coordination properties of 1,4,7-triazacyclononane (tacn) derivatives bearing varying numbers of phosphinic/carboxylic acid pendant groups towards ^{68}Ga, 1,4,7-triazacyclononane-7-acetic-1,4-bis(methylenephosphinic) acid (NOPA) and 1,4,7-triazacyclononane-4,7-diacetic-1-[methylene(2-carboxyethyl)phosphinic] acid (NO2AP) were synthesized using Mannich reactions with trivalent or pentavalent forms of *H*-phosphinic acids as phosphorus components. Stepwise protonation constants $\log K_{1-3}$ 12.06, 3.90 and 1.95, and stability constants with Ga^{III} and Cu^{II}, $\log K_{\text{GaL}}$ 24.01 and $\log K_{\text{CuL}}$ 16.66, were potentiometrically determined for NOPA. Both ligands were labelled with ^{68}Ga and compared with NOTA (tacn-*N*,*N'*,*N''*-triacetic acid) and NOPO, a TRAP-type [tacn-*N*,*N'*,*N''*-tris(methylenephosphinic acid)] chelator. At pH 3, NOPO and NOPA showed higher labelling efficiency (binding with lower ligand excess) at both room temperature and 95 °C, compared to NO2AP and NOTA. Labelling efficiency at pH = 0–3 correlated with a number of phosphinic acid pendants: NOPO >> NOPA > NO2AP >> NOTA; however, it was more apparent at 95 °C than at room temperature. By contrast, NOTA was found to be labelled more efficiently at pH > 4 compared to the ligands with phosphinic acids. Overall, replacement of a single phosphinate donor with a carboxylate does not challenge ^{68}Ga labelling of TRAP-type chelators. However, the presence of carboxylates facilitates labelling at neutral or weakly acidic pH.

Keywords: positron emission tomography; metal complexes; macrocyclic ligands; radiopharmaceuticals; tacn derivative; phosphinate complexes; gallium complexes; radiolabelling; PET tracer development; molecular imaging

1. Introduction

In analogy to ^{99m}Tc, the most commonly used radionuclide for single-photon emission tomography (SPECT) [1], the generator-produced radiometal ^{68}Ga with its favourable physical properties (89% β^+-emission; $t_{1/2}$ = 67.7 min; $E_{av}(\beta^+)$ = 740 keV) is a valuable resource for decentralised manufacturing of positron emission tomography (PET) radiopharmaceuticals [2–4]. For application in nuclear medicine, ^{68}Ga is attached to a biological vector as a complex with a suitable chelator that is conjugated to the targeting group, frequently through an additional linker.

Current ^{68}Ga-based PET is dominated by peptide conjugates of DOTA and NOTA (Figure 1), mainly due to the success of the corresponding radiolabelled octreotide analogues, such as ^{68}Ga-DOTATOC, ^{68}Ga-DOTATATE, or ^{68}Ga-DOTANOC for imaging of neuroendocrine tumours [5,6]. However, although ^{68}Ga^{3+} labelling of DOTA is feasible, this chelator has been mainly employed for ^{90}Y, ^{111}In, ^{152}Tb, ^{177}Lu, ^{212}Pb or ^{213}Bi radioisotopes, whose coordination requires higher coordination numbers [7]. Since the coordination chemistry of the radiometal and the chelator determines the labelling conditions [8], an extensive effort has recently been dedicated to the development of improved bifunctional chelators tailored for gallium(III) [9–18]. For the development of ^{68}Ga-based imaging agents, 1,4,7-triazacyclononane-based (tacn-based) NOTA-like bifunctional derivatives (3 [11], 2 [12], 4 or 5 [13], 1 [14,15]; Figure 1) have been shown as promising chelators for ^{68}Ga^{3+} ion. Compared to DOTA, the NOTA-like derivatives can also be labelled efficiently at lower ligand concentrations/excess and lower temperatures [19]. However, ^{68}Ga labelling of NOTA proved to be influenced to a considerable extent by metal contaminants present in the ^{68}Ge/^{68}Ga generator eluates, most notably by Zn^{2+}, the inevitable decay product of ^{68}Ga [20]. Among the open-chain chelators, despite the lower kinetic inertness of their metal complexes compared to those of macrocyclic ligands, several conjugates of ligands derived from **6** and **7** showed promising results in preclinical and clinical studies [21–23].

Previously, we have evaluated a number of 1,4,7-triazacyclononane-1,4,7-tris(methylenephosphinic acids) (TRAP ligands) for gallium(III) complexation/labelling [9,10,24–26]. The phosphinate ligands, **8** [27] and **9** [28], reported earlier, were compared to NOTA, DOTA and phosphinate chelators, **10** and **11** [25]. The TRAP-type chelators showed significantly improved labelling properties when compared with their acetic acid analogues. Apart from the feasibility of labelling at room temperature (RT) and at low chelator concentrations, the higher acidity of phosphinic acids allowed for labelling at acidic conditions (pH < 2), where formation of insoluble $^{68}Ga^{3+}$ hydroxide species is avoided [29]. Among the TRAP chelators, no statistically significant difference in labelling properties has been found; only labelling of the more lipophilic **9** resulted in slightly worse ^{68}Ga incorporation efficiency. The TRAP motif was also employed for a straightforward preparation of a PET/MRI bimodal contrast agent, combining TRAP and DOTA structures for Ga^{3+} and Gd^{3+} chelation, respectively [30]. More recently, excellent labelling properties have also been reported for the monoconjugable TRAP-type chelator NOPO [10,31,32] (Figure 2) which combines the pendant arm moieties of **10** and **11**. Interestingly, bringing the asymmetric element to the N-substitution pattern did not entail any loss of ^{68}Ga-labelling performance. Moreover, NOPO and **10** were found to be highly chemoselective for Ga^{3+}, even in the presence of high concentrations of contaminating metallic cations [20].

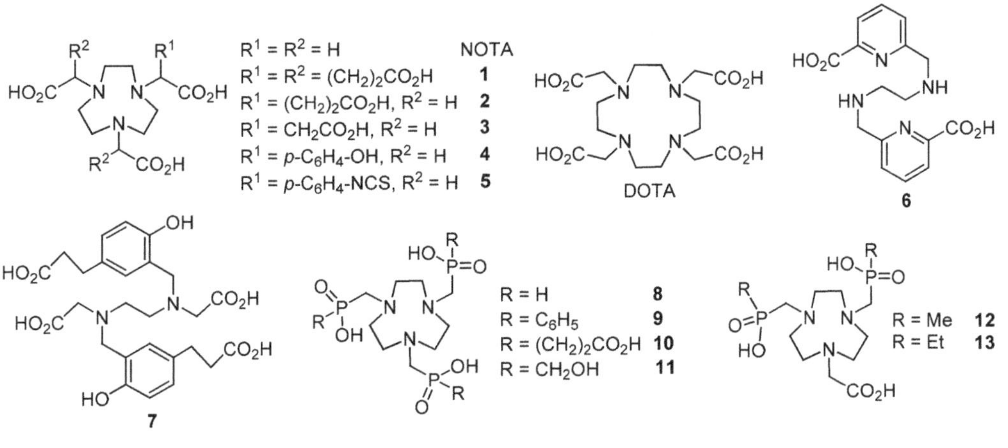

Figure 1. Macrocyclic and open-chain chelators for trivalent gallium.

In order to gain a better understanding of the factors responsible for the ^{68}Ga-labelling efficiency of TRAP chelators, we have now investigated two tacn-based bifunctional chelators with asymmetrical N-substitution patterns, involving both phosphinate and carboxylate coordination sites (NO2AP and NOPA, Figure 2).

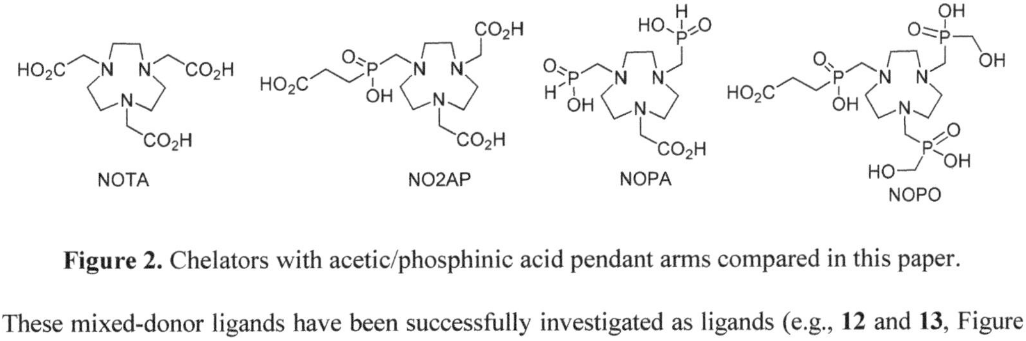

Figure 2. Chelators with acetic/phosphinic acid pendant arms compared in this paper.

These mixed-donor ligands have been successfully investigated as ligands (e.g., **12** and **13**, Figure 1) selective for Mg^{2+} over Ca^{2+} [33,34]. Their ^{68}Ga labelling performance was compared to that of NOTA and NOPO as representatives of symmetrically substituted carboxylate-type and phosphinate-type chelators.

2. Results and Discussion

2.1. Ligand Synthesis

Synthesis of NOPA was carried out according to the reaction sequence shown in Scheme 1. 1,4,7-Triazacyclononane was reacted with *N,N*-dimethylformamide dimethyl acetal to give aminal **14** [35] which was monoalkylated *in situ* [36,37], affording the ammonium salt **15** that crystallized from the reaction mixture. This one-pot alkylation followed by hydrolysis is—despite requiring several steps—simple and easy to carry out on a large scale. Compound **16** [33,38–40] was then obtained by alkaline hydrolysis of **15**. Moedritzer-Irani (phospho-Mannich) [41] reaction of **15** with phosphinic acid and paraformaldehyde readily afforded NOPA; similarly to the analogous reaction on *N*-monobenzylated tacn [32], the typical formation (according to NMR and MS spectra of the reaction mixture) of *N*-methylated by-products [42] in the last reaction step was suppressed by low reaction temperature. Pure NOPA was obtained in a zwitterionic form after simple purification on a strong cationic exchanger; surprisingly, separation of NOPA from the *N*-methylated by-product on cationic exchange resin was more efficient than that in previously published synthesis of the tris(phosphinic acid) ligand **8** [25].

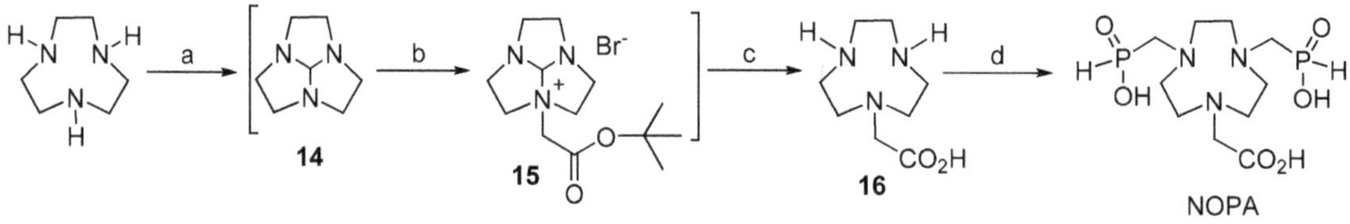

Scheme 1. NOPA synthesis. Reagents and conditions: (**a**) (MeO)$_2$CHNMe$_2$, dioxane, 105 °C, 4 h; (**b**) *t*BuO$_2$CCH$_2$Br, dioxane, room temp., 1 h; (**c**) NaOH, water/EtOH, reflux, 72 h, 89% based on tacn; (**d**) paraformaldehyde, H$_3$PO$_2$, water, room temp., 12 h, 63%.

Two synthetic pathways were evaluated for the preparation of NO2AP. In the first approach, reaction of the phosphinic acid **17** with tacn-1,7-diacetic acid (NO2A) and formaldehyde in conc. aq. HCl at elevated temperatures (50–70 °C) resulted in the formation of complex mixtures, difficult to separate mainly due to the formation of the *N*-methylated side products. Furthermore, the presence of the free acetic acid pendant arms discourages utilisation of the chelator for selective coupling to a primary amine group in e.g., peptides. Therefore, another route employing a precursor with ester protected *N*-acetates was investigated, in which the phosphite intermediate **18** was generated *in-situ* by reaction of acid **17** with hexamethyldisilazane (HMDSA). The latter intermediate was reacted with tacn-1,7-bis(*t*-butyl acetate) **19** under anhydrous conditions according to our previously reported synthetic procedure [32] to give ester **20** (Scheme 2) [34]. Comparing to the published synthesis (the esterified mixed acetate-phosphinate tacn derivatives have been prepared from the *t*-butyl ester of **16** or from **19** by reaction with paraformaldehyde and MeP(OEt)$_2$ or EtP(OEt)$_2$, respectively, in anhydrous solvents but the product was isolated in very low overall yields and after difficult purification procedures [34]), the latter procedure is characterized by simple purification and higher overall yield despite the seemingly more demanding synthetic protocols. The silyl groups were removed by treatment with methanol and the free chelator NO2AP was obtained by deprotection with trifluoroacetic acid. The reaction sequence confirmed that silylated phosphites derived from *H*-phosphinic acids are valuable, readily available reagents for the anhydrous variant of Mannich reaction. Utilization of the silylated phosphinic acids for the formation of the >N–CH2–P pendant arm might represent a feasible general approach for the synthesis of mixed and/or selectively protected phosphorylated polyazamacrocycles.

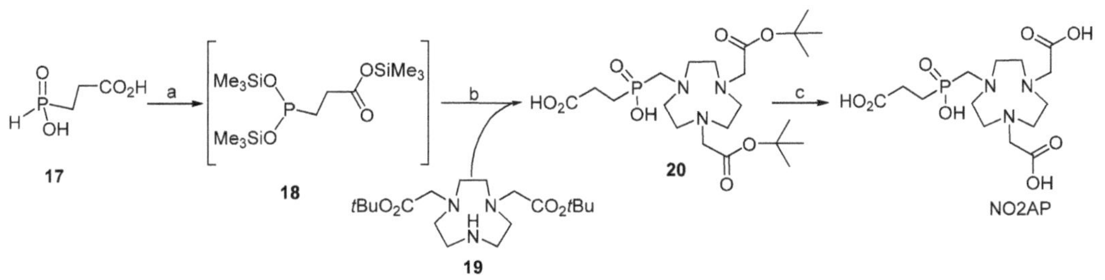

Scheme 2. Synthesis of NO2AP through ester **20**. Reagents and Conditions: (**a**) HMDSA, 130 °C, 24 h, quantitative; (**b**) (*i*) paraformaldehyde, HMDSA, 130 °C, 24 h; (*ii*) MeOH, HPLC purification; 46% (based on **19**); (**c**) CF3CO2H:CH2Cl2 1:1, room temp..

2.2. Equilibrium Studies

Protonation constants and gallium(III) complex stability constants of NOPA were determined by potentiometry (Table 1); for the species distribution diagram, see Figure 3. As expected, values of the protonation constants of NOPA were found to be between those of the mother ligands, NOTA and **8**, and, taking into account different experimental conditions, are in a good agreement with the data reported for its methyl- (**12**) and ethyl phosphinate (**13**) analogues (Figure 1) [33,34]. The first protonation constant is relatively high as it should correspond to protonation of the ring amine with the attached acetate moiety, whereas the second protonation constant should be connected with an amine substituted with methyl phosphinate group [34]. Gallium(III) complexation in acidic solution was very fast and complete complex formation was observed at the beginning of titrations at pH 1.5. In this region, formation of a protonated complex was observed (β_{HLGa} = 25.14(8), logK_a = 1.10). The Ga^{3+} complex stability constant was thus determined through competition with hydroxide anions in alkaline solution. Similarly to other tacn-based ligands [9,25,32], equilibration above pH ~ 6 was slow (more than two weeks) and "out-of-cell" titration method had to be used. Mixed hydroxido species were also found (β_{H-1LGa} = 16.04(5), logK_a = 8.00). As NOTA derivatives are now commonly used as ligands of choice for complexation of ^{64}Cu, stability constants for Cu^{2+}-NOPA system were determined as well. The respective complex (β_{LCu} = 16.66(2)) is formed even in very acidic solutions, which nevertheless contained 25% free Cu^{2+} at pH 1.7, enabling the stability constant determination; the chemical model also required a hydroxido species (β_{H-1LCu} = 5.36(2), logK_a = 11.30). Thermodynamic stabilities of the [Ga(NOPA)] and [Cu(NOPA)]− complexes correlate with the overall ligand basicity [43] (defined as basicity of the ring nitrogen atoms, logK_1 + logK_2) of NOPA and, thus, are between those for the NOTA and **8** complexes.

The protonated [Ga(HNOPA)]+ species should be the "*in-cage*" complex as the proton is probably attached to the phosphoryl oxygen atom of the coordinated phosphinate pendant arm [9,25]. Abundance of the [Ga(OH)(NOPA)]− species (Figure 3) is relatively high, and its possible formation during radiolabelling might explain lower radiolabelling yields at higher pH (see below).

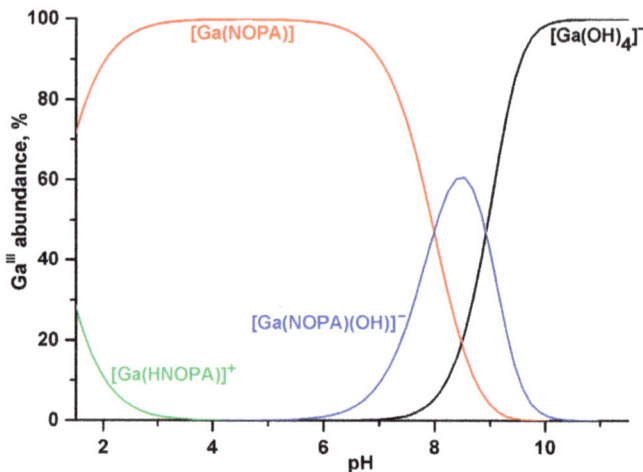

Figure 3. Species distribution diagram of the Ga^{3+}-NOPA system.

Table 1. Stepwise protonation (logK_n) and thermodynamic stability (logK_{GaL}) constants of free ligands and their gallium(III) complexes, respectively (25 °C, I = 0.1 M (Me$_4$N)Cl). Literature data are given for comparison.

Constant	Ligand					
	NOPA [a]	12 [b] [34]	8 [25]	10 [15]	NOPO [32]	NOTA [44]
logK_1	12.06 12.058(4)	11.7	10.48	11.48	11.96	13.17
logK_2	3.90 15.958(6)	4.24	3.28	5.44	5.22	5.74
logK_3	1.95 17.910(6)	2.10		4.84	3.77	3.22
logK_4				4.23	1.54	1.96
logK_5				3.45		
logK_6				1.66		
logK_{GaL} [c]	24.04 24.04(6)		21.91	26.24	25.0	29.60 [25]

[a] This work; experimentally determined overall protonation/stability constants (log$β_{hlm}$) are in italics; [b] 25 °C, I = 0.1 M KCl; [c] Equilibrium constant for reaction Ga^{3+} + L^{n-} ↔ [Ga(L)]$^{(n-3)-}$ where L^{n-} is the fully deprotonated ligand.

2.3. ^{68}Ga Radiolabelling

Radiolabelling of the chelators at pH 3 exhibited similar shapes and relations of the curves for 95 °C and 25 °C (Figure 4) while, as expected, increased chelator concentrations were required for labelling at ambient temperature. In all cases, the tris(phosphinate) ligand NOPO showed superior labelling compared to the mixed-pendant arm ligands and NOTA. Interestingly, the presence of a single carboxylate donor in NOPA did not significantly affect the labelling performance at pH 3 in comparison to NOPO. Likewise, the behaviour of the monophosphinate ligand NO2AP closely resembled that of NOTA at 95 °C. However at 25 °C, NO2AP showed slightly improved labelling efficiency compared to that of NOTA, although more than 90% radiolabelling yield was not reached, even at fairly high concentrations. Hence, in terms of chelator concentration required for ^{68}Ga labelling, the largest difference is observed between the chelators possessing one and two carboxylates or phosphinates. At both temperatures investigated, NOPA could be labelled with three-times better efficiency than NO2AP (comparing at 50% activity incorporation), while NOPO and NOTA are separated by a factor of ten. In addition, the data for NOPO showed a much better reproducibility than those for the other ligands. All this indicates that no less than three phosphinate donors are required to observe high indifference of the TRAP ligand to non-Ga^{3+} ions in the labelling solution, rooted in the exceptional gallium(III) selectivity.

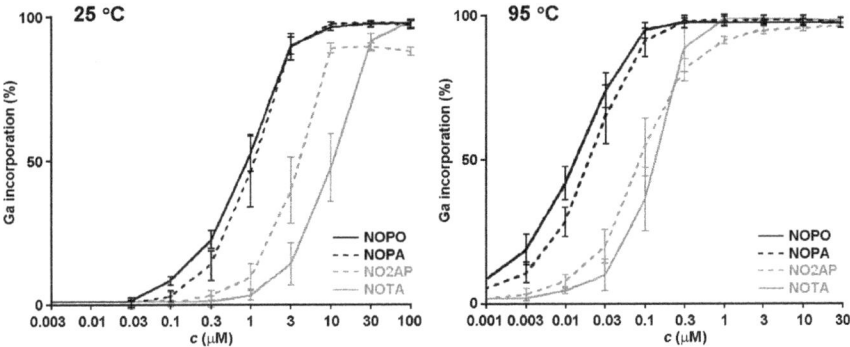

Figure 4. Labelling efficiency of the discussed chelators at 25 and 95 °C at different chelator concentrations (pH = 3, n = 3).

Since all the investigated compounds showed almost quantitative radiolabelling at 3 μM (95 °C) and 30 μM (25 °C), those concentrations were selected for further investigation of labelling efficiency at various pH (Figure 5). At 95 °C, an increasing number of phosphinate side arms mainly resulted in higher labelling yields at lower pH due to the high acidity of phosphinic acids. In accordance with previous results [19], NOPO could be labelled quantitatively already at pH 0.5 and even to a small extent at pH 0. In turn, NOTA showed better performance in the neutral and mildly acidic region. Above pH 8, none of the compounds was labelled anymore.

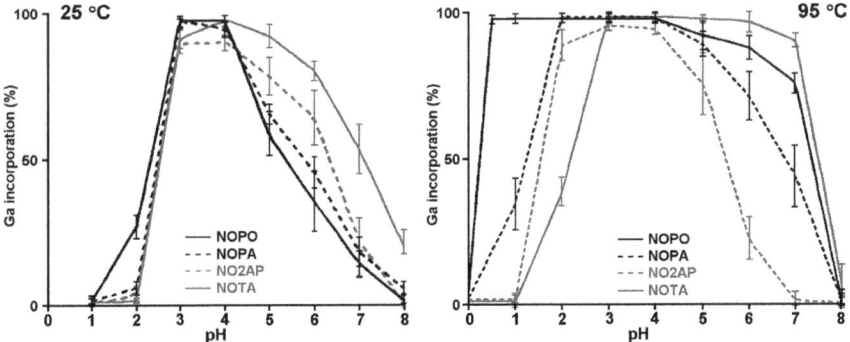

Figure 5. The ^{68}Ga activity incorporation into the discussed chelators at 25 °C and 95 °C at different pH values at constant ligand concentrations of 30 and 3 μM, respectively (n = 3).

At ambient temperature, labelling of all chelators was restricted to a much narrower pH region. While NOPO still performed slightly better at lower pH, NOPO, NOPA and NO2AP reached their optimum between pH 3 and 4. However, ^{68}Ga incorporation by the latter ligand again did not exceed 90%, while the first two ligands were labelled quantitatively. Above pH 4, labelling efficiency of NOPO was decreasing to a larger extent than that observed for the other chelators. By contrast, and similarly to the situation observed at 95 °C, NOTA performed better than the other ligands between pH 4 and 7, with an optimum at pH 4. Notably, some radioactivity can be clearly incorporated by NOTA even at pH 8.

Overall, radiolabelling results are in line with the previously obtained data on TRAP ligands. Due to the selectivity of phosphinate-containing tacn derivatives for gallium(III) [10,19,20], a lower ligand excess is required for efficient radiolabelling with an increasing number of phosphinate pendant arms. A similar decrease in ^{68}Ga incorporation due to presence of the acetate pendant arms has been very recently observed for a diacetate-phosphinate tacn derivative with the *P*-bound –CH$_2$CH(PO$_3$H$_2$)$_2$ group [45]. More phosphinate pendant arms also means a better incorporation of ^{68}Ga in more acidic solutions due to the higher acidity of phosphinic acids. On the other hand, ligands with more acetate pendant arms are more suitable for ^{68}Ga labelling at pH > 4–5. This might be caused by competition with the hydroxide anion, which is more pronounced for complexes exhibiting lower overall thermodynamic stability [25], *i.e.*, for the phosphinate-containing tacn derivatives (see e.g., Figure 3) than for all-carboxylate NOTA.

3. Experimental Section

3.1. General Information

NOPO [10] and NOTA [46] were synthesized by a published procedure. Ester **19** and 1,4,7-triazacyclononane (tacn) were purchased from CheMatech (Dijon, France). Characterization NMR spectra were recorded using Bruker (600 MHz), Varian UNITY Inova (400 MHz) or VNMRS (300 MHz) spectrometers. ^{1}H- and ^{13}C-NMR chemical shifts were referenced to *t*-BuOH as internal standard, and ^{31}P-NMR chemical shifts were referenced relative to 85% aq. H$_3$PO$_4$ as external standard. Electrospray mass spectra (ES-MS) spectra were measured with Varian Ion-trap 500 spectrometer in negative or positive modes. High-resolution mass spectra (HR-MS) were measured on UPLC/MS system consisting of Accela 1250 quaternary gradient pump coupled to LTQ Velos Pro/Orbitrap ELITE mass spectrometer (both Thermo, Waltham, MA, USA); samples were dissolved 50% aq. MeOH. Analytical experiments were performed on a HPLC system composed of a Beta 10 gradient pump (ECOM, Prague, Czech Republic) equipped with an active mixer Knauer A0285 and a Topaz dual-UV detector (ECOM), and on Luna RP8, 5 µm, 150 × 4.6-mm column (Phenomenex, Torrance, CA, USA) equipped with a Security Guard system (Phenomenex) holding a C8-cartridge. The mobile phase was continuously vacuum-degassed in a DG 3014 degasser (ECOM, Czech Republic). Semi-preparative HPLC was run with LCD 50K gradient pump (ECOM) and UV-Vis detector LCD2083 (ECOM) on a Luna RP8, 10 µm, 250 × 21.2-mm column (Phenomenex). For the radiolabelling studies, Ultrapur® water, HCl and NaOH were obtained from Merck KGaA (Darmstadt, Germany); all other materials used were commercially available and of analytical grade. At all cases, incorporation of ^{68}Ga was determined by radio-TLC on silica-impregnated chromatography paper (Agilent, Santa Clara, CA, USA) with 1 M aq. NH$_4$OAc:MeOH

1:1 as mobile phase; scanning and evaluation were performed with a MiniGITA Star TLC-scanner (Raytest, Straubenhardt, Germany).

3.2. Syntheses

3.2.1. Synthesis of (1,4,7-Triazacyclononan-1-yl)acetic Acid (**16**)

Tacn (4.00 g, 31 mmol) was dissolved in dioxane (30 mL) and *N,N*-dimethylformamide dimethyl acetal (4.40 g, 36.9 mmol) was added. The mixture was heated at 105 °C (in bath) for 4 h, then cooled to room temperature, and *t*-butyl bromoacetate (7.24 g, 37.1 mmol) was added dropwise. Immediately formed suspension was diluted by addition of dioxane (10 mL) and stirred at room temperature for 1 h. Diethyl ether (20 mL) was added and yellow microcrystalline solid was filtered off, washed with Et$_2$O and dissolved in solution of NaOH (5.00 g, 125 mmol) in 50% aq. EtOH (40 mL). The solution was refluxed for 72 h, then evaporated to dryness in vacuum and the residue was purified on Dowex 50 in H$^+$-form (column size ~3 × 20 cm). The column was washed with water and the product was eluted by 5% aq. NH$_3$. The fraction containing pure product was evaporated. The residue was dissolved in water (50 mL) and evaporated in vacuum to dryness; the procedure was repeated twice. The product was isolated as yellow oil (5.20 g, 89%) which solidified upon standing at 4 °C. ^{1}H-NMR (300 MHz, D$_2$O): δ (ppm) 2.82–2.95 (m, HO$_2$CCH$_2$NCH$_2$CH$_2$NH, 8H), 3.11 (s, HNCH$_2$CH$_2$NH, 4H), 3.30 (s, CH$_2$CO$_2$H, 2H). ^{13}C{^{1}H} NMR (75.4 MHZ, D$_2$O): δ (ppm) 43.88, 43.98, 50.09 (s 3×, ring CH$_2$), 58.26 (s, NCH$_2$CO$_2$H), 180.44 (s, CO$_2$H). MS (ESI, positive mode, *m/z*): 188.3 [M + H]$^+$. calc. for M (C$_8$H$_{17}$N$_3$O$_2$) 187.2.

3.2.2. Synthesis of 1,4,7-Triazacyclononane-7-(carboxymethyl)-1,4-bis(methylenephosphinic acid) (NOPA)

Compound **16** (6.20 g, 33.2 mmol) was dissolved in 50% aq. H$_3$PO$_2$ (36.3 mL, 33.2 mmol) and paraformaldehyde (1.96 g, 65.3 mmol) was added. The mixture in a closed flask was stirred at room temperature for 12 h and paraformaldehyde slowly dissolved. The mixture was evaporated in vacuum to dryness, dissolved in small amount of water and the solution was soaked on Dowex 50 in H$^+$-form (column size ~3 × 20 cm). The column was eluted by water and the first acidic fraction, containing phosphinic acid, was discarded, and the product was eluted in further neutral fractions. The fractions containing pure product were collected, evaporated in vacuum and finally freeze-dried to give transparent solid of NOPA (7.20 g, 63%). ^{1}H-NMR (300 MHz, D$_2$O): δ 3.34 (d, $^2J_{PH}$ = 9.9 Hz, NCH$_2$P, 4H), 3.39–3.56 (m, ring CH$_2$, 8H), 3.62 (s, ring CH$_2$, 4H), 3.91 (s, NCH$_2$CO$_2$H, 2H), 7.23 (d, $^2J_{PH}$ = 546 Hz, PH, 2H). ^{13}C{^{1}H} NMR (75 MHz, D$_2$O): δ 49.97 (s, ring CH$_2$), 51.83 (d, $^3J_{PC}$ = 5.0 Hz, ring CH$_2$), 52.07 (d, $^3J_{PC}$ = 3.8 Hz, ring CH$_2$), 56.35 (s, NCH$_2$CO$_2$H), 56.21 (d, $^2J_{PC}$ = 88.0 Hz, NCH$_2$P), 172.33 (s, CO$_2$H). ^{31}P-NMR (121 MHz, D$_2$O): δ 16.76 (d, $^1J_{PH}$ = 542 Hz). MS (ESI, positive, *m/z*): 366.6 [M + Na]$^+$, 344.0 [M + H]$^+$; calc. for M (C$_{10}$H$_{23}$N$_3$O$_6$P$_2$) 342.8. HR-MS (positive mode, *m/z*): 344.1143 [M + H]$^+$, calc. for C$_{10}$H$_{23}$N$_3$O$_6$P$_2$: 343.1062.

3.2.3. Synthesis of 1,4,7-Triazacyclononane-4,7-bis(*t*-butyloxycarbonylmethyl)-1-[methylene(2-carboxyethyl)phosphinic acid] (**20**)

(2-Carboxyethyl)phosphinic acid **17** (0.260 g, 1.9 mmol) [9,47] was dissolved in hexamethyl-disilazane (HMDS, 5 mL) in dry glassware under argon and the solution was heated at 140 °C (in oil bath) for 24 h to give intermediate **18**. Ester **19** (0.200 g, 0.56 mmol) was separately dissolved in HMDS (7 mL) and added into the cooled solution of **18**. Dried paraformaldehyde (0.050 g, 1.6 mmol) was added in one portion, flask was tightly closed and the reaction mixture was heated at 130 °C (in oil bath) for 24 h and then cooled to 25 °C. MeOH (5 mL) was slowly added to remove the trimethylsilyl groups. The reaction mixture was evaporated in vacuum to yield a yellow oil. It was divided into 200 mg portions and each portion was dissolved in water (1 mL), solution was filtered through a 0.5-μm syringe filter and purified using semi-preparative HPLC in gradient mode using solution A (20% MeCN, 20% 0.1 M aq. NH_4OAc and 60% H_2O) and B (33% MeCN, 20% 0.1 M aq. NH_4OAc and 47% H_2O); flow rate 20 mL/min, gradient: 100% of A to 100% of B in 19 min. The fraction containing pure product (r_t = 5.7 min) was collected, evaporated in vacuum and finally freeze-dried. Yield 0.130 g (46%, based on *t*Bu$_2$NO2A). ^{1}H-NMR (600 MHz, D_2O): δ (ppm) 1.49 (s, C$\underline{H}_3$, 18H), 1.87 (m, PC$\underline{H}_2$CH$_2$, 2H), 2.41 (m, PCH$_2$C$\underline{H}_2$, 2H), 2.89 (bs, ring C$\underline{H}_2$, 4H), 3.30 (d, $^2J_{PH}$=7.5 Hz, NC$\underline{H}_2$P, 2H), 3.12 (bs, ring C$\underline{H}_2$, 4H), 3.35 (bs, ring C$\underline{H}_2$, 4H), 3.63 (s, NC$\underline{H}_2$CO, 4H). ^{13}C{^{1}H} NMR (150 MHz, D_2O): δ (ppm) 27.63 (d, $^1J_{PC}$ = 72.0 Hz, P$\underline{C}$H$_2$CH$_2$), 28.03 (s, CH$_3$), 30.0 (d, $^2J_{PC}$ = 3.0 Hz, PCH$_2$$\underline{C}H_2$), 47.56 (s, ring $\underline{C}$H$_2$), 49.66 (s, ring $\underline{C}$H$_2$), 53.30 (s, ring $\underline{C}$H$_2$), 53.77 (d, $^1J_{PC}$ = 88.0 Hz, N$\underline{C}$H$_2$P), 56.62 (s, N$\underline{C}$H$_2$CO), 84.27 (s, $\underline{C}_q$), 172.98 (s, NCH$_2$$\underline{C}$O), 181.29 (d, $^3J_{PC}$ = 16.7 Hz, PCH$_2$CH$_2$$\underline{C}O_2$H). ^{31}P{^{1}H} NMR (121 MHz, D_2O): δ (ppm) 32.42 (s). MS (ESI, positive, *m/z*): 508.3 [M + H]$^+$, calc. for M ($C_{22}H_{42}N_3O_8P$) 507.6. HR-MS (positive mode, *m/z*): 508.2797 [M + H]$^+$, calc. for $C_{22}H_{42}N_3O_8P$ 507.2710.

3.2.4. Synthesis of 1,4,7-Triazacyclononane-4,7-bis(carboxymethyl)-1-[methylene(2-carboxyethyl)phosphinic acid] (NO2AP)

Ester **20** (48.2 mg, 0.095 mmol) was dissolved in dry CH_2Cl_2:TFA 1:1 (10 mL) and the solution was stirred in dark at room temperature for 12 h. Solvents were evaporated in vacuum and the crude product was dissolved in water and evaporated, and the procedure was repeated twice. The residue was dissolved in water and the solution was freeze-dried. Product yield 37.1 mg as the trifluoroacetate salt. ^{1}H-NMR (600 MHz, D_2O): δ (ppm) 2.13 (m, PC$\underline{H}_2$CH$_2$, 2H), 2.67 (m, PCH$_2$C$\underline{H}_2$, 2H), 3.45 (d, $^2J_{PH}$ = 5.7 Hz, NC$\underline{H}_2$P, 2H), 3.50–3.56 (m, ring C$\underline{H}_2$, 8H), 3.66 (s, ring C$\underline{H}_2$, 4H), 4.14 (s, NC$\underline{H}_2$CO, 4H). ^{13}C{^{1}H} NMR (150 MHz, D_2O): δ (ppm) 24.72 (d, $^1J_{PC}$ = 92.3 Hz, P$\underline{C}$H$_2$CH$_2$), 27.01 (s, PCH$_2$$\underline{C}H_2$), 51.44 (s, ring $\underline{C}$H$_2$), 52.19 (s, 2× ring $\underline{C}$H$_2$), 55.01 (d, $^1J_{PC}$ = 96.4 Hz, N$\underline{C}$H$_2$P), 57.39 (s, N$\underline{C}$H$_2$CO), 116.7 (q, $^1J_{CF}$ = 290.4 Hz), 163.1 (q, $^2J_{CF}$ = 36.5 Hz), 170.92 (s, NCH$_2$$\underline{C}$O), 177.14 (d, $^3J_{PC}$ = 13.5 Hz, PCH$_2$CH$_2$$\underline{C}O_2$H). ^{31}P{^{1}H} NMR (121 MHz, D_2O): δ (ppm) 43.77 (s). MS (ESI, positive, *m/z*): 396.1 [M + H]$^+$, calc. for M ($C_{14}H_{26}N_3O_8P$) 395.3. HR-MS (positive mode, *m/z*): 396.1534 [M + H]$^+$, calc. for $C_{14}H_{26}N_3O_8P$: 395.1457.

3.3. Potentiometry

Potentiometry was carried out (preparation of stock solutions and chemicals, electrode system calibration, titration procedures, equipment and data treatment) according to the previously published procedures [48]. The Ga(NO$_3$)$_3$ stock solution contained known amount of HNO$_3$ to protect it against hydrolysis. Protonation and stability constants were determined in 0.1 M (NMe$_4$)Cl at 25.0 °C and they are concentration constants. Protonation constants of NOPA (c_L = 0.004 M) and Cu-NOPA stability constants (c_L = c_{Cu} = 0.004 M) were determined by normal ("in-cell") titrations in pH range 1.6–12 with ≈40 points per titration and four parallel titrations. The stability constants in the Ga^{3+}–NOPA system were obtained by "out-of-cell" method as described previously (c_L = c_{Ga} = 0.004 M, pH range 1.5–11.5, 25 points per titration, two parallel titrations, equilibration time three weeks) [10,48]. The titration data were treated with OPIUM [49] program. Stability constants of gallium(III) hydroxide species and pK_w = 13.81 were taken from literature [50,51]. Throughout the text, the pH means −log[H$^+$].

3.4. ^{68}Ga Labelling

The labelling was done manually according to the procedure described in ref. [19]. Briefly, ^{68}Ga was eluted from a SnO$_2$-based ^{68}Ge/^{68}Ga-generator (iTHEMBA Labs, Cape Town, South Africa) with 1 M aq. HCl. A 1250-μL fraction containing the highest activity (≈70 MBq) was collected and buffered with 2-[4-(2-hydroxyethyl)-piperazin-1-yl]ethanesulfonic acid (HEPES; 800 μL, 2.7 M aq.). Aliquots of that solution (90 μL) were added to ligand stock solutions of appropriate concentration (10 μL, pH ≈3.0) and left to incubate at 95 °C or 25 °C for 5 min. For pH dependence experiments, pH was adjusted with aq. HCl and/or aq. NaOH.

4. Conclusions

A detailed comparison of a series of four tacn-based chelators with various phosphinic/carboxylic acid substitution patterns provided a better understanding of the structural factors governing metal ion complexation properties of this class of ligands. The presence of at least two phosphinic acid pendant arms is a key to the unique ^{68}Ga-labelling properties of TRAP-like chelators. Apparently, one phosphinate coordination site of the TRAP motif can be exchanged with a different donor, e.g. carboxylate, without compromising its affinity to gallium(III). On the other hand, the presence of carboxylate groups facilitates the complex formation at neutral or weakly acidic pH. Overall, our findings help with the fine-tuning of metal-binding properties of the pendant-armed 1,4,7-triazacyclononanes and, thus, provide a strong basis for future rational design of these ligands for medical applications.

Acknowledgments

Financial support by the Deutsche Forschungsgemeinschaft (grant NO822/4-1 and SFB 824, Projects B5 and Z1) and the Grant Agency of the Czech Republic (13-08336S) is gratefully acknowledged. G.M. acknowledges support from the European Social Fund in the framework of TÁMOP-4.2.4.A/ 2-11/1-2012-0001 National Excellence Program. The authors would like to thank Jana Havlíčková (Charles University, Prague) for potentiometry service. The work was done in the frame of TD1004 and TD1007 COST Actions.

Author Contributions

Ligand design (J.Š., P.H.), ligand syntheses (J.Š., M.P.), characterizations of ligands (J.Š., M.P.), radiochemical study design (G.M., J.Š., H.-J.W.), radiolabelling experiments (G.M., I.K.) and the data evaluation, both carried out at two organizations (G.M., J.Š., I.K., L.G., J.N.), manuscript preparation (G.M., P.H.).

Conflicts of Interest

The authors declare no conflict of interest.

References and Notes

1. Charkraborty, S.; Liu, S. ^{99m}Tc and ^{111}In-labeling of small biomolecules: Bifunctional chelators and related coordination chemistry. *Curr. Top. Med. Chem.* **2010**, *10*, 1113–1134.
2. Rösch, F. Past, present and future of ^{68}Ge/^{68}Ga generators. *Appl. Radiat. Isot.* **2013**, *76*, 24–30.
3. Banerjee, S.R.; Pomper, M.G. Clinical applications of gallium-68. *Appl. Radiat. Isot.* **2013**, *76*, 2–13.
4. Velikyan, I. Continued rapid growth in ^{68}Ga applications: Update 2013 to June 2014. *J. Label. Compd. Radiopharm.* **2015**, *58*, 99–121.
5. Geijer, H.; Breimer, L.H. Somatostatin receptor PET/CT in neuroendocrine tumours: Update on systematic review and meta-analysis. *Eur. J. Nucl. Med. Mol. Imaging* **2013**, *40*, 1770–1780.
6. Van Essen, M.; Sundin, A.; Krenning, E.P.; Kwekkeboom, D.J. Neuroendocrine tumours: The role of imaging for diagnosis and therapy. *Nat. Rev. Endocrinol.* **2014**, *10*, 102–114.
7. Ramogida, C.F; Orvig C. Tumour targeting with radiometals for diagnosis and therapy. *Chem. Commun.* **2013**, *49*, 4720–4739.
8. Price, E.W.; Orvig, C. Matching chelators to radiometals for radiopharmaceuticals. *Chem. Soc. Rev.* **2014**, *43*, 260–290.
9. Notni, J.; Hermann, P.; Havlíčková, J.; Kotek, J.; Kubíček, V.; Plutnar, J.; Loktionova, N.; Riss, P.J.; Rösch, F.; Lukeš, I. A triazacyclononane based bifunctional phosphinate ligand for preparation of multimeric ^{68}Ga PET tracers. *Chem. Eur. J.* **2010**, *16*, 7174–7185.
10. Šimeček, J.; Zemek, O.; Hermann, P.; Wester, H.J.; Notni, J. A monoreactive bifunctional triazacyclononane-phosphinate chelator with high selectivity for gallium-68. *ChemMedChem* **2012**, *7*, 1375–1378.
11. Andre, J.P.; Mäcke, H.R.; Zehnder, M.; Macko, L.; Akyel, K.G. 1,4,7-triazacyclononane-1-succinic acid-4,7-diacetic acid (NODASA): A new bifunctional chelator for radio gallium-labelling of biomolecules. *Chem. Commun.* **1998**, 1301–1302.
12. Eisenwiener, K.P.; Prata, M.I.M.; Buschmann, I.; Zhang, H.W.; Santos, A.C.; Wenger, S.; Reubi, J.C.; Mäcke, H.R. NODAGATOC, a new chelator-coupled somatostatin analogue labeled with [Ga-67/68] and [In-111] for SPECT, PET, and targeted therapeutic applications of somatostatin receptor (hsst2) expressing tumors. *Bioconjugate Chem.* **2002**, *13*, 530–541.

13. Riss, P.J.; Kroll, C.; Nagel, V.; Rösch, F. NODAPA-OH and NODAPA-(NCS)n: Synthesis, ^{68}Ga-radiolabelling and *in vitro* characterisation of novel versatile bifunctional chelators for molecular imaging. *Bioorg. Med. Chem. Lett.* **2008**, *18*, 5364–5367.
14. Singh, A.N.; Liu, W.; Hao, G.; Kumar, A.; Gupta, A.; Oz, O.K.; Hsieh, J.T.; Sun, X. Multivalent bifunctional chelator scaffolds for gallium-68 based positron emission tomography imaging probe design: Signal amplification via multivalency. *Bioconjugate Chem.* **2011**, *22*, 1650–1662.
15. Guerra Gomez, F.L.; Uehara, T.; Rokugawa, T.; Higaki, Y.; Suzuki, H.; Hanaoka, H.; Akizawa, H.; Arano, Y. Synthesis and evaluation of diastereoisomers of 1,4,7-triazacyclononane-1,4,7-tris-(glutaric acid) (NOTGA) for multimeric radiopharmaceuticals of galium. *Bioconjugate Chem.* **2012**, *23*, 2229–2238.
16. Waldron, B.P.; Parker, D.; Burchardt, C.; Yufit, D.S.; Zimny, M.; Rösch, F. Structure and stability of hexadentate complexes of ligands based on AAZTA for efficient PET labelling with gallium-68. *Chem. Commun.* **2013**, *49*, 579–581.
17. Boros, E.; Ferreira, C.L.; Cawthray, J.F.; Price, E.W.; Patrick, B.O.; Wester, D.W.; Adam, M.J.; Orvig, C. Acyclic chelate with ideal properties for Ga-68 PET imaging agent elaboration. *J. Am. Chem. Soc.* **2010**, *132*, 15726–15733.
18. Berry, D.J; Ma, Y.; Ballinger, J.R.; Tavaré, R.; Koers, A.; Sunassee, K.; Zhou, T.; Nawaz, S.; Mullen, G.E.D.; Hider, R.C.; *et al.* Efficient bifunctional gallium-68 chelators for positron emission tomography: Tris(hydroxypyridinone) ligands. *Chem. Commun.* **2011**, *47*, 7068–7070.
19. Notni, J.; Šimeček, J.; Hermann, P.; Wester, H.J. TRAP, a powerful and versatile framework for gallium-68 radiopharmaceuticals. *Chem. Eur. J.* **2011**, *17*, 14718–14722.
20. Šimeček, J.; Hermann, P.; Wester, H.J.; Notni, J. How is ^{68}Ga labeling of macrocyclic chelators influenced by metal ion contaminants in ^{68}Ge/^{68}Ga generator eluates? *ChemMedChem* **2013**, *8*, 95–103.
21. Eder, M.; Schäfer, M.; Bauder-Wüst, U.; Hull, W.E; Wängler, C.; Mier, W.; Haberkorn, U.; Eisenhut, M. ^{68}Ga-Complex lipophilicity and the targeting property of a urea-based PSMA inhibitor for PET imaging. *Bioconjugate Chem.* **2012**, *23*, 688–697.
22. Afshar-Oromieh, A.; Malcher, A.; Eder, M.; Eisenhut, M.; Linhart, H.G.; Hadaschik, B.A.; Holland-Letz, T.; Giesel, F.L.; Kratochwil, C.; Haufe, S.; *et al.* PET imaging with a [^{68}Ga] gallium-labelled PSMA ligand for the diagnosis of prostate cancer: Biodistribution in humans and first evaluation of tumour lesions. *Eur. J. Nucl. Med. Mol. Imaging* **2013**, *40*, 486–495.
23. Boros, E.; Ferreira, C.L.; Yapp, D.T.T.; Gill, R.K.; Price, E.W.; Adam, M.J.; Orvig, C. RGD conjugates of the H$_2$dedpa scaffold: Synthesis, labeling and imaging with ^{68}Ga. *Nucl. Med. Biol.* **2012**, *29*, 785–794.
24. Notni, J.; Pohle, K.; Wester, H.J. Be spoilt for choice with radiolabelled RGD peptides: Preclinical evaluation of ^{68}Ga-TRAP(RGD)$_3$. *Nucl. Med. Biol.* **2013**, *40*, 33–41.
25. Šimeček, J.; Schulz, M.; Notni, J.; Plutnar, J.; Kubíček, V.; Havlíčková, J.; Hermann, P. Complexation of metal ions with TRAP (1,4,7-triazacyclononane phosphinic acid) ligands and NOTA: Phosphinate-containing ligands as unique chelators for trivalent galium. *Inorg. Chem.* **2012**, *51*, 577–590.

26. Notni, J.; Šimeček, J.; Wester, H.J. Phosphinic acid functionalized polyazacycloalkane chelators for radiodiagnostics and radiotherapeutics: Unique characteristics and applications. *ChemMedChem* **2014**, *9*, 1107–1115 (corrigendum: *ibid.* **2014**, *9*, 2614).
27. Bazakas, K.; Lukeš, I. Synthesis and complexing properties of polyazamacrocycles with pendant *N*-methylenephosphinic acid. *J. Chem. Soc. Dalton Trans.* **1995**, doi:10.1039/DT9950001133.
28. Cole, E.; Parker, D.; Ferguson, G.; Gallagher, J.F.; Kaitner, B. Synthesis and structure of chiral metal complexes of polyazacycloalkane ligands incorporating phosphinic acid donors. *J. Chem. Soc. Chem. Commun.* **1991**, 1473–1475, doi:10.1039/C39910001473.
29. Hacht, B. Gallium(III) ion hydrolysis under physiological conditions. *Bull. Korean Chem. Soc.* **2008**, *29*, 372–376.
30. Notni, J.; Hermann, P.; Dregely, I.; Wester, H.J. Convenient synthesis of ^{68}Ga-labeled gadolinium(III) complexes: Towards bimodal responsive probes for functional imaging with PET/MRI. *Chem. Eur. J.* **2013**, *19*, 12602–12606.
31. Šimeček, J.; Notni, J.; Kapp, T.G.; Kessler, H.; Wester, H.J. Benefits of NOPO as chelator in gallium-68 peptides, exemplified by preclinical characterization of ^{68}Ga-NOPO–c(RGDfK). *Mol. Pharm.* **2014**, *11*, 1687–1695.
32. Šimeček, J.; Zemek, O.; Hermann, P.; Notni, J.; Wester, H.J. Tailored gallium(III) chelator NOPO: Synthesis, characterization, bioconjugation, and application in preclinical Ga-68-PET imaging. *Mol. Pharm.* **2014**, *11*, 3893–3903.
33. Van Haveren, J.; DeLeon, L.; Ramasamy, R.; van Westrenen, J.; Sherry, A.D. The design of macrocyclic ligands for monitoring magnesium in tissue by ^{31}P-NMR. *NMR Biomed.* **1995**, *8*, 197–205.
34. Huskens, J.; Sherry, A.D. Synthesis and characterization of 1,4,7-triazacyclononane derivatives with methylphosphinate and acetate side chains for monitoring free Mg^{II} by ^{31}P- and ^{1}H-NMR spectroscopy. *J. Am. Chem. Soc.* **1996**, *118*, 4396–4404.
35. Atkins, T.J. Tricyclic trisaminomethanes. *J. Am. Chem. Soc.* **1980**, *102*, 6364–6365.
36. Schulz, D.; Weyhermüller, T.; Wieghard, K.; Nuber, B. The monofunctionalized 1,4,7-triazacyclononane derivatives 1,4,7-triazacyclononane-*N*-acetate (L^1) and *N*-(2-hydroxybenzyl-1,4,7-triazacyclononane (HL^2) and their complexes with vanadium(IV)/(V). Localized and delocalized electronic structures in compounds containing the mixed valent [OV^{IV}-O-V^{V}O]$^{3+}$ core. *Inorg. Chim. Acta* **1995**, *240*, 217–229.
37. Warden, A.C.; Spiccia, L.; Hearn, M.T.W.; Boas, J.F.; Pilbrow, J.R. The synthesis, structure and properties of copper(II) complexes of asymmetrically functionalized derivatives of 1,4,7-triazacyclononane. *Dalton Trans.* **2005**, 1804–1813.
38. Studer, M.; Kaden, T.A. Metal complexes with macrocyclic ligands. Part XXV. One-step synthesis of mono-*N*-substituted azamacrocycles with a carboxylic group in the side-chain and their complexes with Cu^{2+} and Ni^{2+}. *Helv. Chim. Acta* **1986**, *69*, 2081–2086.
39. Warden, A.; Graham, B.; Hearn, M.T.W.; Spiccia, L. Synthesis of novel derivatives of 1,4,7-triazacyclononane. *Org. Lett.* **2001**, *3*, 2855–2858.
40. Kovács, Z.; Sherry, A.D. A general synthesis of mono- and disubstituted 1,4,7-triazacyclononanes. *Tetrahedron Lett.* **1995**, *36*, 9269–9272.

41. Moedritzer, K.; Irani, R.R. The direct synthesis of α-aminomethylphosphonic acids. Mannich-type reactions with orthophosphorous acid. *J. Org. Chem.* **1966**, *31*, 1603–1607.
42. Remore, D. Chemistry of phosphorous acid: new routes to phosphonic acids and phosphate esters. *J. Org. Chem.* **1978**, *43*, 992–996.
43. Lukeš, I.; Kotek, J.; Vojtíšek, J.; Hermann, P. Complexes of tetraazacycles bearing methylphosphinic/phosphonic acid pendant arms with copper(II), zinc(II) and lanthanides(III). A comparison with their acetic acid analogues. *Coord. Chem. Rev.* **2001**, *216–217*, 287–312.
44. Drahoš, B.; Kubíček, V.; Bonnet, C.S.; Hermann, P.; Lukeš, I.; Tóth, É. Dissociation kinetics of Mn^{2+} complexes of NOTA and DOTA. *Dalton Trans.* **2011**, *40*, 1945–1951.
45. Holub, J.; Meckel, M.; Kubíček, V.; Rösch, F.; Hermann, P. Gallium(III) complexes of NOTA-bis(phosphonate) conjugates as PET radiotracers for bone imaging. *Contrast Media Mol. Imaging* **2015**, *10*, 122–134.
46. Größ, S.; Elias, H. Kinetics and mechanism of complex formation: The reaction of nickel(II) with 1,4,7-triazacyclononane-*N*,*N'*,*N''*-triacetic acid. *Inorg. Chim. Acta* **1996**, *251*, 347–354.
47. Řezanka, P.; Kubíček, V.; Hermann, P.; Lukeš, I. Synthesis of a bifunctional monophosphinate DOTA derivative having free carboxylate group in the phosphorus side chain. *Synthesis* **2008**, 1431–1435.
48. Kubíček, V.; Havlíčková, J.; Kotek, J.; Tircsó, G.; Hermann, P.; Tóth, É.; Lukeš, I. Gallium(III) complexes of DOTA and DOTA-monoamide: Kinetic and thermodynamic studies. *Inorg. Chem.* **2010**, *49*, 10960–10969.
49. Kývala, M.; Lubal, P.; Lukeš, I. Determination of equilibrium constants with the OPIUM computer program. Proceedings of the IX. Spanish-Italian and Mediterranean Congress on Thermodynamics of Metal Complexes (SIMEC 98), Girona, Spain, 2–5 June 1998; p. 94. The Full Version of the OPIUM Program is Available (Free of Charge) on http://web.natur.cuni.cz/~kyvala/opium.html (accessed on 23 September 2014).
50. *NIST Standard Reference Database 46 (Critically Selected Stability Constants of Metal Complexes)*, Version 7.0; National Institute of Standards and Technology: Gaithersburg, MD, USA, 2003.
51. Baes, C.F., Jr.; Mesmer, R.E. *The Hydrolysis of Cations*; Wiley: New York, NY, USA, 1976.

Sample Availability: Samples are not available.

© 2015 by the authors; licensee MDPI, Basel, Switzerland. This article is an open access article distributed under the terms and conditions of the Creative Commons Attribution license (http://creativecommons.org/licenses/by/4.0/).

Molecules 2015, 20, 12863-12879; doi:10.3390/molecules200712863

OPEN ACCESS

molecules
ISSN 1420-3049
www.mdpi.com/journal/molecules

Article

Dual Radiolabeling as a Technique to Track Nanocarriers: The Case of Gold Nanoparticles

Clinton Rambanapasi [1,*], Nicola Barnard [1], Anne Grobler [1], Hylton Buntting [1], Molahlehi Sonopo [2], David Jansen [2], Anine Jordaan [3], Hendrik Steyn [4] and Jan Rijn Zeevaart [1]

[1] DST/NWU Preclinical Drug Development Platform, Faculty of Health Sciences, Potchefstroom Campus, North-West University, Potchefstroom 2531, South Africa; E-Mails: Nicola.Barnard@nwu.ac.za (N.B.); Anne.Grobler@nwu.ac.za (A.G.); 24861820@nwu.ac.za (H.B.); janrijn.zeevaart@necsa.co.za (J.R.Z.)

[2] Radiochemistry Department, South African Nuclear Energy Corporation (SOC) Ltd., P. O. Box 482, Pretoria 0001, South Africa; E-Mails: Molahlehi.Sonopo@necsa.co.za (M.S.); david.r.jansen@gmail.com (D.J.)

[3] Laboratory for Electron Microscopy, Chemical Resources Beneficiation Group, Potchefstroom Campus, North-West University, Potchefstroom 2531, South Africa; E-Mail: Anine.Jordaan@nwu.ac.za

[4] Statistical Consultation Services, Potchefstroom Campus, North-West University, Potchefstroom 2531, South Africa; E-Mail: Faans.Steyn@nwu.ac.za

* Author to whom correspondence should be addressed; E-Mail: 24089117@nwu.ac.za or crambanapasi@gmail.com; Tel.: +27-18-299-2281; Fax: +27-18-285-2233.

Academic Editor: Svend Borup Jensen

Received: 23 May 2015 / Accepted: 10 July 2015 / Published: 16 July 2015

Abstract: Gold nanoparticles (AuNPs) have shown great potential for use in nanomedicine and nanotechnologies due to their ease of synthesis and functionalization. However, their apparent biocompatibility and biodistribution is still a matter of intense debate due to the lack of clear safety data. To investigate the biodistribution of AuNPs, monodisperse 14-nm dual-radiolabeled [^{14}C]citrate-coated [^{198}Au]AuNPs were synthesized and their physico-chemical characteristics compared to those of non-radiolabeled AuNPs synthesized by the same method. The dual-radiolabeled AuNPs were administered to rats by oral or intravenous routes. After 24 h, the amounts of Au core and citrate surface coating were quantified using gamma spectroscopy for ^{198}Au and liquid scintillation for the ^{14}C. The Au core and citrate surface coating had different biodistribution profiles in the organs/tissues analyzed, and no oral absorption was observed. We conclude that the different components of the AuNPs system, in this case the Au core and citrate surface coating, did not remain intact, resulting in the different distribution profiles observed. A better understanding of the biodistribution profiles of other surface attachments or cargo of AuNPs in relation to the Au core is required to successfully use AuNPs as drug delivery vehicles.

Keywords: gold nanoparticles; dual radiolabeling; biodistribution profiles; Sprague Dawley rats

1. Introduction

The use of engineered nanomaterials, such as gold (Au) nanoparticles (AuNPs), promises to have a great impact on the field of nanomedicine and nanotechnologies. As a result, AuNPs have become an on-going area of research for a wide range of biomedical applications, such as plasmon-based labeling and imaging, diagnostics and therapeutics [1–3]. AuNPs' unique surface, electronic and optical properties, as well as their apparent biocompatibility [4] make them ideal drug delivery vectors [5–8]. However, their biocompatibility and toxicity have recently been questioned [9,10], and currently, there is no consensus on their biodistribution [4,11–13]. This can be attributed to the use of different methodologies with a diversity of objectives that do not collate easily into a single general conclusion. The lack of correlation between *in vitro* and *in vivo* toxicity results further complicates matters.

In biodistribution and toxicity studies, it is necessary to accurately determine the amount of Au in various tissues/organs. The quantification of the other components of an AuNP drug delivery vesicle, the surface coating and surface attachments or the cargo, can assist in the elucidation of potential toxicity mechanisms. Several techniques have been used to measure the content of Au in rodents, for example; inductively-coupled plasma mass spectroscopy (ICP-MS) [10,14–18], atomic absorption spectroscopy (AAS) [19], radioactive analysis (RA) using gamma spectroscopy [20–23] and instrumental neutron activation analysis (INAA) [24,25]. Gamma spectroscopy and INAA are preferred analytical techniques in biodistribution studies due to the lower limits of detection compared to AAS and ICP-MS. Gamma spectroscopy offers the added advantage of a quick and relatively simple sample preparation. However, all of these quantification methods mentioned lack the ability to track and quantify the other components/ surface attachments of AuNPs simultaneously *in vivo*.

Whilst the use of a single radiolabel is common [13,20,23], to the best of our knowledge, there are no published studies using dual radiolabeling to determine the biodistribution profiles of the different components in a multi-component systems for AuNPs. However, dual radiolabeling has been reported before to study the biodistribution of the components of a vaccine system (both adjuvant and antigen) [26] and for AuNPs with two radiolabels for use in single-photon emission computed tomography (SPECT) for bioimaging applications in diagnostics [27]. An approach similar to the one we are taking in this study was done for superparamagnetic iron using ^{59}Fe for the nanoparticle core and labeled surface attachments [28,29].

An understanding of the biodistribution profile of each component (Au core and any surface attachments) would be ideal, as this will enable any observed end organ toxicity to be attributed to the whole system or a part thereof. This can be achieved by radiolabeling each of the desired components;

in this case, the Au core and the citrate surface coating. The methods used by Hirn *et al.* of radiolabeling AuNPs by irradiation of a pellet of AuNPs (^{197}Au (n,γ) ^{198}Au) [20,23] cannot be used for dual radiolabeling of the Au core and the surface coating, as neutron activation only produces ^{198}Au. In this study, the Au was radiolabeled using ^{198}Au, while [^{14}C]citrate was used for the citrate surface coating.

The aim of the present study was to synthesize dual-radiolabeled AuNPs and to determine the biodistribution profiles of the Au core and citrate surface coating, while investigating the influence of the route of administration and the dose level. Well-characterized 14-nm AuNPs that were dual-radiolabeled were administered to healthy male Sprague Dawley rats intravenously and orally. The study served as a proof of principle that dual radiolabeling can be used to determine the biodistribution profiles of the different components of a multi-component system. Therefore, in this study, the acute biodistribution profiles of the Au core and citrate surface coating after oral and intravenous (*i.v.*) administrations are presented. Future studies must investigate the biodistribution of Au after multiple doses and assess its biopersistence, while focusing more on toxicity endpoints.

2. Results

2.1. Synthesis and Characterization of AuNPs

AuNPs were synthesized from both radioactive and non-radioactive precursors using the citrate reduction method. The UV spectra peaks in Figure 1 were similar for both radioactive and non-radioactive AuNPs.

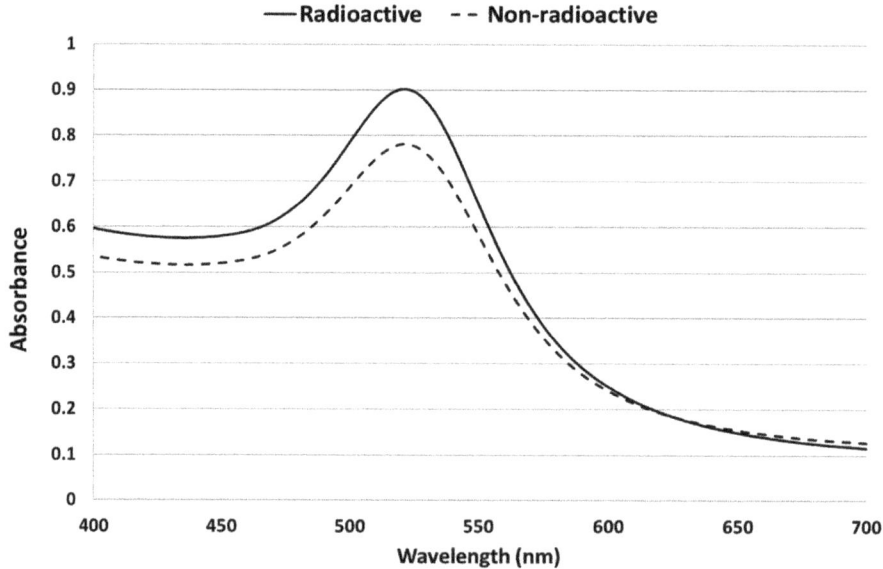

Figure 1. UV-Vis spectra of radioactive (**continuous line**) and non-radioactive (**dashed line**) AuNPs. The measurements were done after the synthesis of both samples.

The UV peak was around 520 nm as expected for this particle size range, whilst the dispersion quality was confirmed by the absence of absorbance at wavelengths greater than 600 nm [30].

From the TEM images and the particle size distribution plots (see Figure 2), it can be seen that the morphology and primary particles size distribution of 14 ± 1.2 nm and 14 ± 1.5 nm for radioactive and nonradioactive AuNPs, respectively, are similar/comparable. Hydrodynamic sizes of 25 nm for the non-radioactive preparation and 23 nm for the radioactive samples with zeta potentials of −50.9 mV and −48.9 mV were found, respectively.

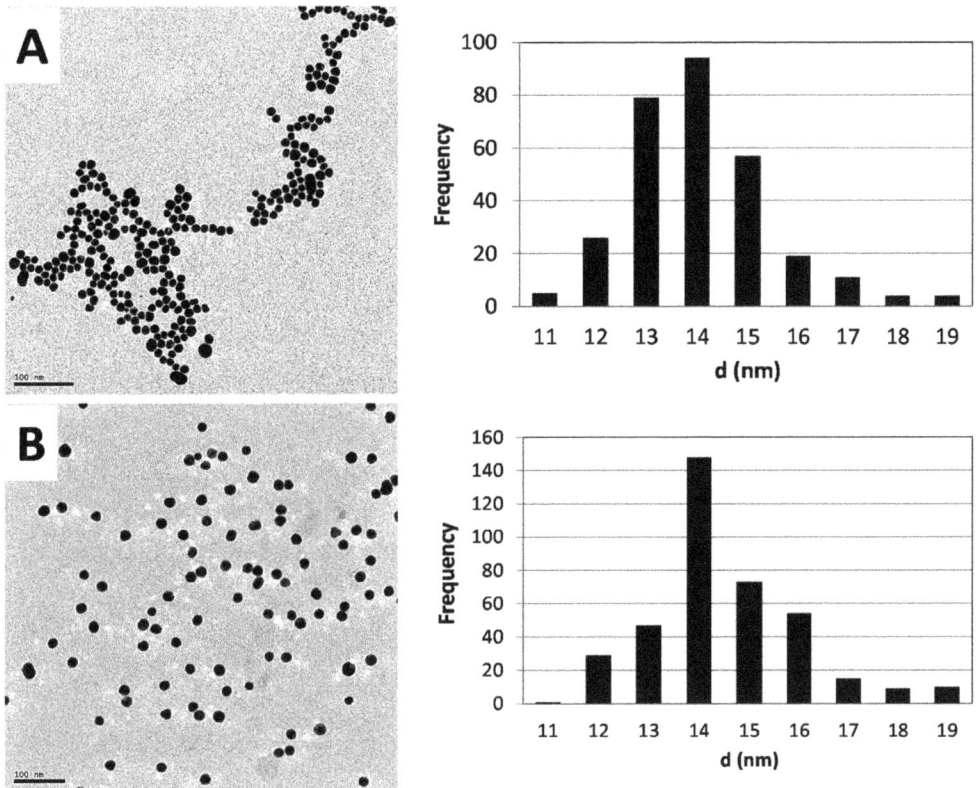

Figure 2. Morphology and size distribution profiles of the synthesized AuNPs fabricated from natural gold (**A**) and radioactive gold (**B**). On the left are TEM images, with the particle size distribution plots on the right.

The specific activity of the ^{198}Au was 108 GBq/g, with an activity concentration of 25.9 MBq/mL and an isotope ratio (^{198}Au:^{197}Au) of 1.45×10^{-5}. The specific activity of the ^{14}C was 52.43 MBq/g, with an activity concentration of 0.054 MBq/mL and an isotope ratio (^{14}C:^{12}C) of 0.8.

2.2. Biodistribution of Gold vs. Citrate in the Rat

2.2.1. Dosimetry

Table 1 summarizes the main dosimetric features of the [^{14}C]citrate-[^{198}Au]AuNPs used. The surface area and number of nanoparticles were calculated using the initial mass of Au used in the synthesis.

Table 1. Characteristics of the dual-radiolabeled AuNPs used in the study at the 2 dose levels used (high and low). The surface area of the AuNPs was calculated using the primary size determined using TEM.

		Dose	
		High	Low
Administered radioactivity per rat (MBq)	^{198}Au	12.95	1.22
	^{14}C	0.027	0.0027
Administered mass per rat (µg)	Au	90	9
	Citrate	520	52
Administered number of AuNPs per rat		3.27×10^{12}	3.327×10^{11}
Administered surface area (cm^2) of AuNPs per rat		20.16	2.02

2.2.2. Biodistribution Profiles

In order to ensure that the signal measured was only from gold and that there were no interferences that may impede the reliability of the results, a gamma spectrum was measured (see Figure 3). The peak at 411 keV is for ^{198}Au; the absence of other peaks shows that the gold used in the experiments did not have impurities, and only gold was quantified in the work.

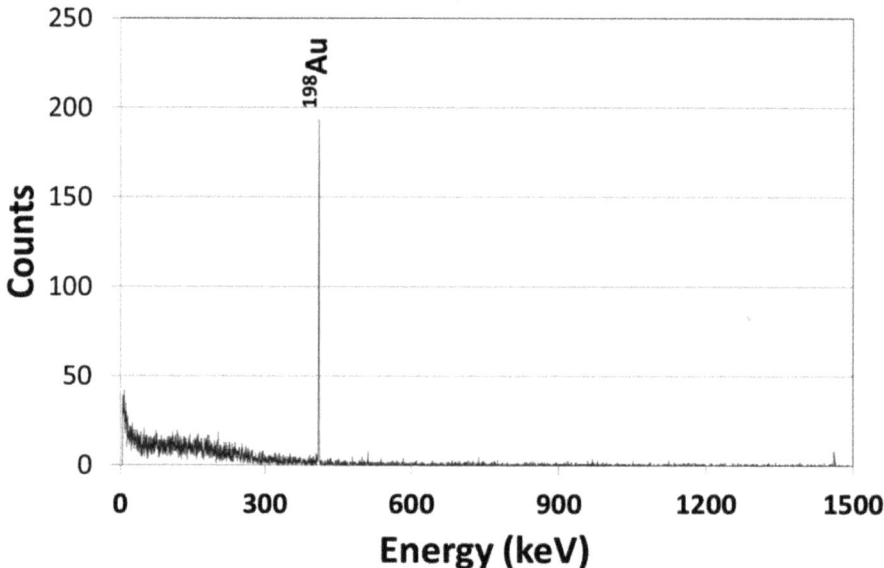

Figure 3. Gamma spectrum of the radioactive gold nanoparticles used in the animal study.

The biodistribution profiles of both the nanoparticle core and surface coating were investigated, with the route of administration and dose level as variables. The amount of Au and citrate was determined using γ-spectroscopy for ^{198}Au and liquid scintillation for ^{14}C at 24 h post a single dose administered intravenously and orally. The amounts of ^{198}Au and ^{14}C in the liver, spleen, lungs and blood were determined in the analysis. Oral administration of AuNPs resulted in no systemic uptake of ^{198}Au; thus, no ^{14}C was measured in the oral group. Activities of the ^{198}Au were only measured/detected in the stomach and other parts of the GI tract, with no measurable activity in all other organs with less than 0.001% injected dose (ID)/g in the blood, liver, spleen and lungs (results not shown). The inclusion of the oral group was because some measurable systemic uptake was expected, as described in the literature [24]. However, our findings correspond well with the results reported in previous studies and are attributed to the size of the particles used in the study [23,24,31]. Therefore, only results from the two intravenous groups (see Figure 4) are reported here. The results are expressed as the percentage injected dose per gram of organ/tissue (%ID/g) for the Au and citrate. The amounts of ^{198}Au measured in the urine and feces were used to perform a mass balance for Au.

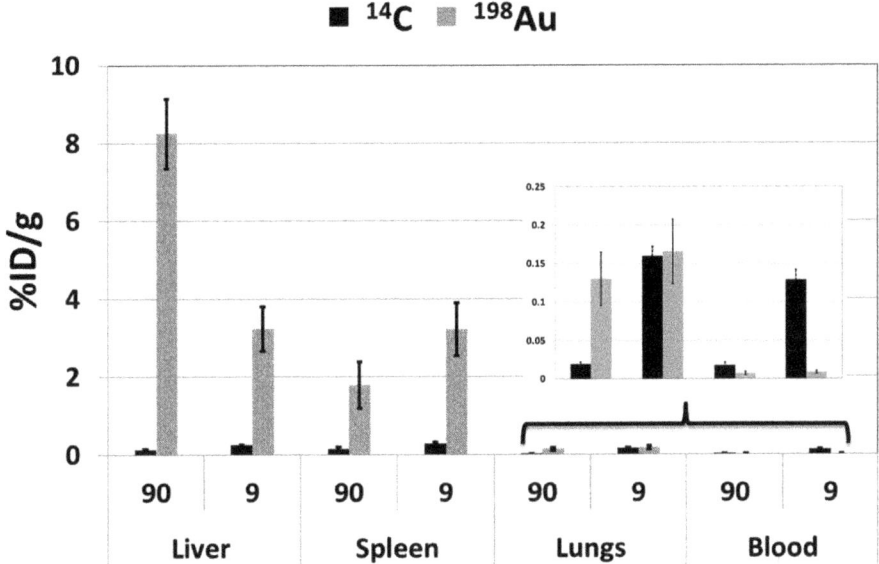

Figure 4. Amounts of gold and citrate expressed as a percentage of the injected dose per gram of organ/tissue (%ID/g) in the liver, spleen, lungs and blood, 24 h after intravenous administration of dual-radiolabeled AuNPs. Results are expressed as the mean ± SD.

Liver

The liver had the highest %ID/g of gold when administered in both 90-µg (high) and 9-µg (low) doses, with 8.2% and 3.2%, respectively. The %ID/g of citrate was less than 0.5% irrespective of the administered dose. The difference in the %ID/g of the Au was statistically significant using both the Mann–Whitney and Student t-tests; the p-values were calculated to be 0.03 and 0.0004, respectively, at the two dose levels used.

Spleen

The spleen had the second highest %ID/g of gold. However, contrary to the liver, the %ID/g was inverted relative to the administered doses, with 1.8% for ID of 90 µg and 3.2% when 9 µg were administered. The differences were not statistically significant. The values for the citrate were also determined to be under 0.5%.

Lungs

The %ID/g of both the Au and citrate were under 0.25% in the lung tissue. The biodistribution pattern was comparable between the Au and citrate only for the administered dose of 9 µg (low dose). The difference in the %ID/g of the Au was significant using both Mann–Whitney and Student t-tests; p-values of 0.03 and 0.002, respectively, at the two dose levels used.

Blood

The %ID/g of citrate was low (≤0.15%) and that of Au even lower (≤0.02%). There was not a statistically significant difference in the %ID/g of citrate at the two dose levels. The biodistribution profiles of the Au were independent of the dose quantity.

Summary of Biodistribution Profiles

In general, Au and citrate had unique biodistribution profiles, as shown by the differences in %ID/g portrayed in Figure 5 with the exception of the 9-µg administered dose in the lungs, where the %ID/g values were comparable.

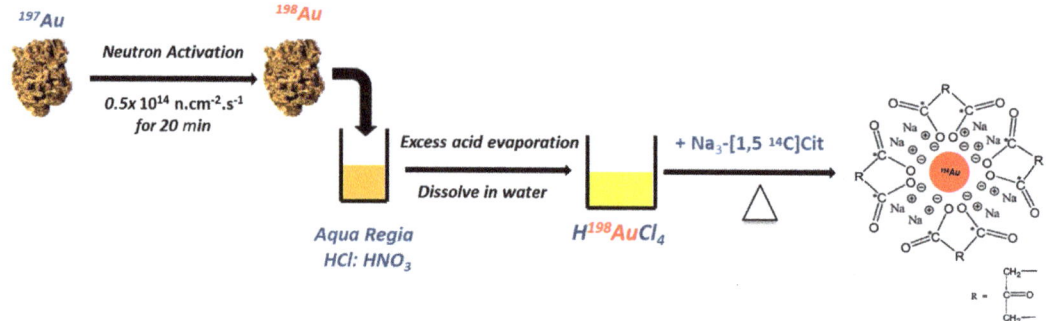

Figure 5. Schematic showing the synthesis of dual-radiolabeled AuNPs starting with the stable ^{197}Au isotope of gold. The neutron activation step is unique to the radioactive synthesis.

The biodistribution profile of the Au varied based on the dosing level, 90 µg vs. 9 µg. The ratios of %ID/g of Au between the 90 µg and 9 µg dose were: liver, 2.6; spleen, 0.6; lungs, 0.8; blood, 0.8; whilst for citrate, the ratios were: liver, 0.5; spleen, 0.5; lungs, 0.1; blood, 0.1.

3. Discussion

The need for extensive biodistribution studies to assess the safety of AuNPs can never be over emphasized, as this will ensure that AuNPs reach the clinic faster. With no consensus on the toxicity profile of AuNPs, a need to understand the biodistribution profile of each component of the AuNP system becomes apparent. It has been generally accepted that surface functionalization is an important determinant of the *in vivo* dynamics and toxicity [11,20,32]. In this study, we synthesized AuNPs, both non-radioactive and radioactive, and compared the two formulations to assess the impact of using radioactive precursors on the physico-chemical properties. The dual-radiolabeled AuNPs were used to determine the biodistribution profile of the Au core using ^{198}Au and surface coating using [^{14}C]citrate. The influences of the dose and route of administration were also investigated.

The use of radioactive precursors had no impact on the quality attributes of the synthesized dual-radiolabeled AuNPs. This was shown when the physico-chemical properties of the non-radioactive and radioactive AuNPs were compared. The UV/Vis spectra of the non-radioactive and radioactive batches were comparable and characteristic of the 14-nm size range, which has a defined plasmon resonance peak maxima around 520 nm [19]. The absence of secondary peaks at wavelengths higher than 600 nm also confirms the absence of agglomerates and/or aggregates in the suspensions [30]. The polydispersity index (PDI), a measure of monodispersity obtained with the Zetasizer Nano ZS, also showed that the suspensions were free of agglomerates/aggregates. The zeta potential was as expected: a high negative charge due to the negative charge of the citrate surface coating. This was comparable for both the non-radioactive and radioactive AuNPs. The molar ratio of the hydrogen chloroauric acid:citrate used in the synthesis of the AuNPs yielded nanoparticles with a core diameter around 14 nm, which is consistent with the sizes obtained by other researchers when similar molar ratios were used [15,16,33].

The radiotracers used in the synthesis of the dual-radiolabeled [^{14}C]citrate-[^{198}Au]AuNPs were well controlled and were adjusted to meet the varying requirements. The photons emitted during the decay of ^{198}Au have energies that can be detected by a gamma camera; thus, a change in the activity concentration of the Au during uptake in the various organs can be imaged. The method used in this work solves some challenges that are normally encountered when other ways of incorporating radiotracers into AuNPs are used. Agglomeration and/or aggregate formation when synthesized AuNPs are irradiated to neutrons activate the Au core [20–23]. The activity of both labels was homogenously distributed in the solution. This was shown experimentally when the doses were measured using both volume and radioactivity. There was a correlation between the expected and determined value for each dose using ^{198}Au.

In this study, the biodistribution profiles observed for the Au core and surface coating were very different. Use of surface attachment as the radiotracer has been done [34] and suffers the disadvantage of misinterpretation of the biodistribution profiles. The radiotracers can be displaced from the core due to the formation of a bio-corona [35,36]. Usually, the biodistribution of the radiotracer is assumed to represent that of the Au core and the whole nanoparticle system. From our results, it is seen that surface attachments will not have to have the same biodistribution profile as that of the core or carrier molecule used to transport it. Caution must therefore be exercised when interpreting the results of biodistribution and toxicity studies of AuNPs with surface attachments that will not be present in those intended for biomedical applications. The addition of different surface attachments will most likely alter the biodistribution and toxicity profile of AuNPs *in vivo*, as surface chemistry plays an integral part in the toxicity and biodistribution of AuNPs and other nanomaterials [37].

The biodistribution profiles of the Au core and citrate surface coating were different in the organs/tissues used in the analysis. This can be explained by the formation of the bio-coronas around the nanoparticle core [38–43], which results in the dissociation of the surface attachment from the core. These results indicate that during the synthesis and design of therapeutic agents, the type of interaction between the Au surfaces and "cargo" should be carefully considered when surface modifications are made to AuNPs. This is especially important for the delivery of drug molecules to ensure that the cargo is not lost before the intended destination. Electrostatic interactions might be desirable, since covalent bonds require energy for the cargo to dissociate from the surface. A similar dissociation of surface attachments that had electrostatic interactions with the nanoparticle surface has been reported for superparamagnetic iron [28,29].

The effect of the dose was more prominent for the Au compared to the citrate surface coating. With the exception of the liver, the %ID/g was higher in the lower dose level in all of the organs/tissues. For citrate, the opposite was observed: the %ID/g was lower in the higher dose level for blood and lung (a blood-rich organ). This can possibly be explained by isotopic exchange between the citrate (which predominates in blood) and its radiolabeled analogue. With higher dose, the amount of ^{14}C citrate will be the same in the blood as that in the lower dose, thus giving a lower %ID/g. The ratios of the %ID/g of the 90 µg:9 µg dosages for Au (liver: 2.6; spleen: 0.6; lungs: 0.8; blood: 0.8) may be an indication of a saturable transport mechanism of the Au into tissues/organs, with the liver taking up excess Au in the case of higher dosing levels. If this can be repeatedly shown, it may be a useful consideration when planning to use AuNPs as a drug delivery vector. To date, there is little evidence that AuNPs lead to histological changes and toxicity [33,44]. Whether this will be the case in an extensive treatment regime, with multiple doses administered over the course of weeks or months remains unknown. It is also not known whether the systemic/tissue concentrations will be maintained by the prolonged exposure of the repeated doses, unlike in this acute study. The subchronic and chronic use of AuNPs presents another variable and so does the level of biopersistence. All of the above scenarios will need to be investigated.

4. Experimental Section

4.1. Preparation of AuNPs and Dual-Radiolabeled AuNPs

Elemental gold (24 carat) was purchased from Cape Precious Metals Holding Pvt. Ltd., Johannesburg, South Africa. 1,5-[^{14}C] citric acid (concentration: 3.7 GBq/mL; specific activity: 2.07 GBq/mmol) was purchased from American Radiolabeled Chemicals, Inc. (St. Louis, MO, USA). Hydrochloric acid (HCl, 37%), nitric acid (HNO$_3$, 68%) (used to prepare aqua regia using an HCl:HNO$_3$ in a 3:1 ratio) and trisodium citrate (Na$_3$C$_6$O$_7$H$_5$·H$_2$O), were all of analytical grade and purchased from Merck (Billerica, MA, USA). Deionized water (resistance >18 MΩ) was prepared by an in-house ultrapure water system (Merck Millipore, Billerica, MA, USA). All chemicals, except for the 1,5-[^{14}C] citric acid (deprotonated using NaOH to make trisodium citrate), were used as received without purification. All radioactive materials were produced and handled at the South African Nuclear Energy Corporation (Necsa, Pelindaba, South Africa) facilities and laboratories.

Two 5-mg samples of natural gold (^{197}Au) metal were weighed using an analytical balance (5-decimal place Mettler Toledo). One sample was used as natural gold, while the other sample (target) was irradiated in the SAFARI 1 20 MW research reactor situated at Necsa in a hydraulic position with a neutron flux of 0.5×10^{14} n·cm^{-2}·s^{-1} for 20 min to obtain ^{198}Au. Both Au samples were dissolved in aqua regia (5 mL) dried down (using heat) and reconstituted in 0.5–1 mL 0.005 N HCl to yield HAuCl$_4$·HAuCl$_4$ and [^{198}Au]HAuCl$_4$·[^{198}Au]HAuCl$_4$ in 0.05 N HCl [45], the starting material in the synthesis of AuNPs. The activity of the [^{198}Au]HAuCl$_4$·[^{198}Au]HAuCl$_4$ was measured using a CRC-15R dose calibrator (Capintec Inc., Ramsey, NJ, USA). The radioactive HAuCl$_4$ sample was used to synthesize the dual-radiolabeled AuNPs. The activity concentration of 1,5-[^{14}C]trisodium citrate was determined by liquid scintillation. Three counting solutions were used to determine the activity concentrations of the 1,5-[^{14}C]trisodium citrate. These solutions were prepared using a standard containing 10 µL (37KBq) of 1,5-[^{14}C]trisodium citrate whose volume was made up to 1 mL (stock solution). Five, then and one hundred microliters of the stock solution were added to 20-mL glass vials containing 15 mL of the liquid scintillation cocktail (Bioscint). The activity measurements in the vials were 8491, 14,420 and 139,818 disintegrations per minute (DPM), respectively.

An adaptation of the method published by Turkevich, *et al.* [46] and Frens [47] was used to synthesize sterile radioactive and natural AuNPs. The volumes of the prepared solution of radioactive [^{198}Au]HAuCl$_4$·[^{198}Au]HAuCl$_4$ in 0.05 N HCl and the non-radioactive HAuCl$_4$·HAuCl$_4$ in 0.05 N HCl were diluted to 25 mL using deionized water to make 1 mM solutions. Solutions of hydrogen chloroauric acid were heated to the boiling point with vigorous stirring, and the reducing agents were added to the solutions and boiled under reflux for a further 30 min. For the non-radioactive synthesis, 2.5 mL of 38.8 mM trisodium citrate were used as the reducing agent. For the dual radiolabel synthesis, 2.5 mL (38.8 mM) of solution containing 1,5-[^{14}C]trisodium citrate (1.52 MBq: 600 µL, 1.07×10^{-3} mmol) and non-labeled trisodium citrate (1.9 mL: 9.743×10^{-2} mM) were used as the reducing agent. Figure 4 shows the adapted method used to synthesize dual-radiolabeled [^{14}C]citrate-[^{198}Au]AuNPs.

4.2. Characterization of Dual-Radiolabeled AuNPs

Both the radioactive and non-radioactive AuNPs were characterized using the same techniques to assess the impact of using radioactive precursors in the quality attributes of AuNPs. With the exception of the UV/Vis spectra, the radioactive sample was analyzed after 10 half-lives (27 days), when the radioactivity of the samples was low enough to be safely cleared from Necsa laboratories and analyzed in non-radiological laboratories.

The hydrodynamic size (Z-average size) and polydispersity index (PDI) of the nanoparticles was acquired by dynamic light scattering with a Zetasizer Nano ZS (Malvern Instruments Ltd., Worcestershire, UK) operated in backscattering mode at 173° with a He–Ne laser beam ($\lambda = 632.8$ nm). For the zeta potential measurements, which were performed at 25 °C with a scattering angle of 90°, the particles were dispersed in aqueous solution with an average pH of 6.2. The experiment was done in triplicate, and the results were averaged.

The morphology and primary size distributions of AuNPs were determined using transmission electron microscopy (TEM) (FEI Tecnai G2, Eindhoven, The Netherlands). Specimens were prepared by drop casting of a 10-µL aliquot of a dilute NP solution on an Athene® grid (Plano GmbH, Wetzlar, Germany). At least 250 particles were used to determine the primary size distributions using ImageJ software (Version 1.48; National Institutes of Health, Bethesda, MD, USA).

UV/Vis spectra were recorded for both the radioactive and non-radioactive AuNP suspensions using a PerkinElmer LAMBDA 1050 UV/Vis/NIR spectrophotometer (Waltham, MA, USA). The spectra were also used to determine the concentration [48].

4.3. In Vivo Study

4.3.1. Animals

The study was conducted in accordance with the South African National Standard for the Care and Use of Animals for Scientific Purpose. Ethical approval was sought and granted by the North-West University (NWU) Ethical Committee. Twelve (12) male Sprague Dawley rats, 8–10 weeks old, weighing 200–250 g, were used in the study. The rats were bred and procured from the Department of Science and Technology (DST)/NWU/Preclinical Drug Development Platform (PCDDP) Vivarium (Potchefstroom, South Africa) and housed in stainless steel cages in groups of 4. The rats were kept under standard environmental conditions (23 ± 1 °C, 55% ± 5% humidity and 12/12 h light/dark cycle) with water and food provided *ad libitum* throughout the study.

4.3.2. Experimental Design

The rats were randomly divided into 3 treatment groups ($n = 4$ per group; see Figure 6). The dual-radiolabeled [^{14}C]citrate-[^{198}Au]AuNPs suspensions were administered as a slow intravenous injection using the tail vein in Groups 1 and 2 and orally via gavage in Group 3. The administered doses were 90 µg (high dose) for Group 1 and 9 µg (low dose) for Group 2. The administered doses were within the ranges found in the literature [13]. The volume of all of the administrations was 500 µL. The accuracy of the dose was controlled using both the volume injected and the radioactivity of ^{198}Au. Any activity remaining in the syringe was measured and used to calculate the exact dose injected.

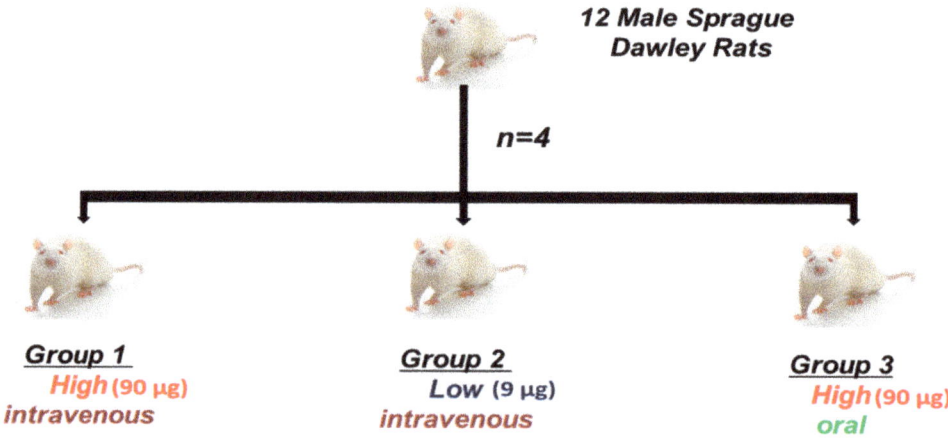

Figure 6. Study design of the animal experiment. Groups 1 and 2 received intravenous doses, while Group 3 got an oral dose.

After each administration, the rats were individually placed in metabolic cages to collect the total amount of urine and feces. All of the administrations were well tolerated with no apparent adverse events being observed during the 24-h study. The 24-h time point was selected based on acute biodistribution studies found in the literature [20,22]. At the termination of the study, the rats were euthanized using an overdose of Euthapent® (sodium pentobarbitone 200 mg/mL; Kyron laboratories, Johannesburg,

South Africa), administered intravenously. Blood was collected using the cardiac puncture technique and stored without further processing before cutting open the chest cavity and abdomen of the rat to collect the liver, spleen and lungs. Together with the blood, these were used to determine whether the gold core of the AuNPs is distributed in a similar pattern as the citrate surface coating, which is an indication of whether the NPs remain intact in physiological conditions. The mass of each sample, including the carcass, was measured and used in calculating the percentage of the injected dose per gram of the organ/tissue.

4.4. Quantification of Gold and Citrate in Samples

The quantification of citrate was done after at least 30 days (10 half-lives) of ^{198}Au, to avoid measuring the *beta* decay of ^{198}Au, as well. The quantities of citrate were determined only in the intravenous groups, since no absorption of Au was seen, thus negligible ($\geq$0.001 %ID/g) amounts in the blood, liver, lungs and spleen (results not shown).

4.4.1. Gold

The ^{198}Au radioactivity of the blood, liver, lungs, spleen and the remainder (total remaining carcass) was measured without further sample preparation by γ-spectroscopy using a CRC-15R dose calibrator (Capintec Inc., Ramsey, NJ, USA) and a lead-shielded well-type NaI (TI) scintillation detector using the winTMCA32 software (FLIR Radiation, GmbH, Solingen, Germany). The counts were corrected for physical decay from the time of injection and any background radiation. A ^{198}Au standard prepared in the laboratory was used to correlate ^{198}Au radioactivities to the masses, numbers and surface areas of the AuNP suspension used in the study. To ensure that the entire administered dose was accounted for, the amounts of ^{198}Au in the total urine and feces and the total remaining carcass was measured. A gamma spectrum of the ^{198}Au was also measured to give evidence that the signal measured was only from gold, and there were no interferences that may impede the reliability of the results.

4.4.2. Citrate

To measure the ^{14}C radioactivity in liver, lungs, spleen and blood, a known mass of approximately 200 mg of the liver, lungs and spleen and 500 μL of whole blood were added to a 20-mL glass scintillation vial. To each sample, 1–2 mL of the solubilizer (Biosol, National Diagnostics, Atlanta, GA, USA) were added, and the samples were incubated between 55 and 60 °C until the samples were completely solubilized or had a brown/green color in the case of the blood. The digestion times varied depending on the tissues (up to 5 h for liver). Two hundred microliters of 30% hydrogen peroxide (H_2O_2) were added in 2 aliquots to discolor the dissolved tissues. The samples were allowed to stand for 24 h. Scintillation cocktail (Bioscint, National Diagnostics, Atlanta, GA, USA) was added to fill up the vial to 20 mL. The samples were stored in a cool dark place and counted for 10 min using a Perkin-Elmer Tri-Carb 3100 TR scintillation spectrophotometer (Waltham, MA, USA). All measurements were done in triplicate.

4.5. Statistics

The statistical significance of the differences between the mean %ID/g values in the different groups was assessed by use of the non-parametric Mann–Whitney test and a Student *t*-test. Statistical probability (*p*) values less than 0.05 were considered significantly different.

5. Conclusions

With the present study, we have shown that the use of radioactive precursors does not have a negative impact on the physico-chemical properties of AuNPs, and dual radiolabeling is a good technique for studying the biodistribution of a multi-component nano-particulate system. The biodistribution profile of the Au core and citrate surface coating are different, and for the Au component, the biodistribution is dose dependent. At both dose levels, the majority of the Au accumulates in the liver and spleen, and an unexpected deposition in the lungs occurs after intravenous administration.

Acknowledgments

We would like to acknowledge the South African Nanotechnology Initiative for funding. The assistance of Liezl-Marie Scholtz in preparation of the ethics application, Cor Bester in the animal study and Deon Kotze with the liquid scintillation counting is gratefully acknowledged. We thank Jesper Knijnenburg, Florentine Hilty-Vancura and Rose Hayeshi for the assistance in the preparation of this manuscript. We thank Frankline Keter and Hendriette van der Walt from Mintek for discussions on the preparation of AuNPs.

Author Contributions

C.R. was involved in all aspects of the experiments and drafted the manuscript. A.G. and J.R.Z. were involved in the design of all of the experiments and reviewed the manuscript. N.B. was involved in the optimization of the transfer of the synthetic method to the radiochemical method and reviewed the manuscript. M.S. and D.J. were involved the handling of the radioisotopes, ^{14}C and ^{198}Au, respectively, during the preparation of dual-radiolabeled AuNPs and reviewed the manuscript. H.B. was involved in the design and conducting of the animal experiments. A.J. did the TEM analysis. H.S. was involved in the statistical considerations during the design of the animal studies and did the statistical analysis.

Conflicts of Interest

The authors declare no conflict of interest.

References

1. Paciotti, G.F.; Myer, L.; Weinreich, D.; Goia, D.; Pavel, N.; McLaughlin, R.E.; Tamarkin, L. Colloidal gold: A novel nanoparticle vector for tumor directed drug delivery. *Drug Deliv.* **2004**, *11*, 169–183.

2. Eustis, S.; El-Sayed, M.A. Why gold nanoparticles are more precious than pretty gold: Noble metal surface plasmon resonance and its enhancement of the radiative and nonradiative properties of nanocrystals of different shapes. *Chem. Soc. Rev.* **2006**, *35*, 209–217.
3. Boisselier, E.; Astruc, D. Gold nanoparticles in nanomedicine: Preparations, imaging, diagnostics, therapies and toxicity. *Chem. Soc. Rev.* **2009**, *38*, 1759–1782.
4. Connor, E.E.; Mwamuka, J.; Gole, A.; Murphy, C.J.; Wyatt, M.D. Gold nanoparticles are taken up by human cells but do not cause acute cytotoxicity. *Small* **2005**, *1*, 325–327.
5. Papasani, M.R.; Wang, G.; Hill, R.A. Gold nanoparticles: The importance of physiological principles to devise strategies for targeted drug delivery. *Nanomedicine* **2012**, *8*, 804–814.
6. Rana, S.; Bajaj, A.; Mout, R.; Rotello, V.M. Monolayer coated gold nanoparticles for delivery applications. *Adv. Drug Deliv. Rev.* **2012**, *64*, 200–216.
7. Kumar, A.; Zhang, X.; Liang, X.-J. Gold nanoparticles: Emerging paradigm for targeted drug delivery system. *Biotechnol. Adv.* **2013**, *31*, 593–606.
8. Vigderman, L.; Zubarev, E.R. Therapeutic platforms based on gold nanoparticles and their covalent conjugates with drug molecules. *Adv. Drug Deliv. Rev.* **2013**, *65*, 663–676.
9. Hwang, J.H.; Kim, S.J.; Kim, Y.H.; Noh, J.R.; Gang, G.T.; Chung, B.H.; Song, N.W.; Lee, C.H. Susceptibility to gold nanoparticle-induced hepatotoxicity is enhanced in a mouse model of nonalcoholic steatohepatitis. *Toxicology* **2012**, *294*, 27–35.
10. Simpson, C.A.; Salleng, K.J.; Cliffel, D.E.; Feldheim, D.L. *In vivo* toxicity, biodistribution, and clearance of glutathione-coated gold nanoparticles. *Nanomedicine* **2013**, *9*, 257–263.
11. Alkilany, A.; Murphy, C. Toxicity and cellular uptake of gold nanoparticles: What we have learned so far? *J. Nanopart. Res.* **2010**, *12*, 2313–2333.
12. Khlebtsov, N.; Dykman, L. Biodistribution and toxicity of engineered gold nanoparticles: A review of *in vitro* and *in vivo* studies. *Chem. Soc. Rev.* **2011**, *40*, 1647–1671.
13. Zhang, X.D.; Wu, D.; Shen, X.; Liu, P.X.; Yang, N.; Zhao, B.; Zhang, H.; Sun, Y.M.; Zhang, L.A.; Fan, F.Y. Size-dependent *in vivo* toxicity of peg-coated gold nanoparticles. *Int. J. Nanomed.* **2011**, *6*, 2071–2081.
14. De Jong, W.H.; Hagens, W.I.; Krystek, P.; Burger, M.C.; Sips, A.N.J.A.M.; Geertsma, R.E. Particle size-dependent organ distribution of gold nanoparticles after intravenous administration. *Biomaterials* **2008**, *29*, 1912–1919.
15. Sonavane, G.; Tomoda, K.; Makino, K. Biodistribution of colloidal gold nanoparticles after intravenous administration: Effect of particle size. *Colloids Surf. B Biointerfaces* **2008**, *66*, 274–280.
16. Cho, W.S.; Cho, M.; Jeong, J.; Choi, M.; Cho, H.Y.; Han, B.S.; Kim, S.H.; Kim, H.O.; Lim, Y.T.; Chung, B.H.; *et al.* Acute toxicity and pharmacokinetics of 13 nm-sized peg-coated gold nanoparticles. *Toxicol. Appl. Pharmacol.* **2009**, *236*, 16–24.
17. Sadauskas, E.; Danscher, G.; Stoltenberg, M.; Vogel, U.; Larsen, A.; Wallin, H. Protracted elimination of gold nanoparticles from mouse liver. *Nanomedicine* **2009**, *5*, 162–169.
18. Balasubramanian, S.K.; Jittiwat, J.; Manikandan, J.; Ong, C.N.; Yu, L.E.; Ong, W.Y. Biodistribution of gold nanoparticles and gene expression changes in the liver and spleen after intravenous administration in rats. *Biomaterials* **2010**, *31*, 2034–2042.

19. Lasagna-Reeves, C.; Gonzalez-Romero, D.; Barria, M.A.; Olmedo, I.; Clos, A.; Urayama, A.; Sadagopa Ramanujam, V.M.; Vergara, L.; Kogan, M.J.; Soto, C. Bioaccumulation and toxicity of gold nanoparticles after repeated administration in mice. *Biochem. Biophys. Res. Commun.* **2010**, *393*, 649–655.
20. Hirn, S.; Semmler-Behnke, M.; Schleh, C.; Wenk, A.; Lipka, J.; Schäffler, M.; Takenaka, S.; Möller, W.; Schmid, G.; Simon, U.; *et al.* Particle size-dependent and surface charge-dependent biodistribution of gold nanoparticles after intravenous administration. *Eur. J. Pharm. Biopharm.* **2011**, *77*, 407–416.
21. Lipka, J.; Semmler-Behnke, M.; Sperling, R.A.; Wenk, A.; Takenaka, S.; Schleh, C.; Kissel, T.; Parak, W.J.; Kreyling, W.G. Biodistribution of peg-modified gold nanoparticles following intratracheal instillation and intravenous injection. *Biomaterials* **2010**, *31*, 6574–6581.
22. Semmler-Behnke, M.; Kreyling, W.G.; Lipka, J.; Fertsch, S.; Wenk, A.; Takenaka, S.; Schmid, G.; Brandau, W. Biodistribution of 1.4- and 18-nm gold particles in rats. *Small* **2008**, *4*, 2108–2111.
23. Schleh, C.; Semmler-Behnke, M.; Lipka, J.; Wenk, A.; Hirn, S.; Schäffler, M.; Schmid, G.; Simon, U.; Kreyling, W.G. Size and surface charge of gold nanoparticles determine absorption across intestinal barriers and accumulation in secondary target organs after oral administration. *Nanotoxicology* **2012**, *6*, 36–46.
24. Hillyer, J.F.; Albrecht, R.M. Gastrointestinal persorption and tissue distribution of differently sized colloidal gold nanoparticles. *J. Pharm. Sci.* **2001**, *90*, 1927–1936.
25. Balogh, L.; Nigavekar, S.S.; Nair, B.M.; Lesniak, W.; Zhang, C.; Sung, L.Y.; Kariapper, M.S.T.; El-Jawahri, A.; Llanes, M.; Bolton, B.; *et al.* Significant effect of size on the *in vivo* biodistribution of gold composite nanodevices in mouse tumor models. *Nanomedicine* **2007**, *3*, 281–296.
26. Henriksen-Lacey, M.; Bramwell, V.; Perrie, Y. Radiolabelling of antigen and liposomes for vaccine biodistribution studies. *Pharmaceutics* **2010**, *2*, 91–104.
27. Black, K.C.L.; Akers, W.J.; Sudlow, G.; Xu, B.; Laforest, R.; Achilefu, S. Dual-radiolabeled nanoparticle spect probes for bioimaging. *Nanoscale* **2015**, *7*, 440–444.
28. Freund, B.; Tromsdorf, U.I.; Bruns, O.T.; Heine, M.; Giemsa, A.; Bartelt, A.; Salmen, S.C.; Raabe, N.; Heeren, J.; Ittrich, H.; *et al.* A simple and widely applicable method to ^{59}Fe-radiolabel monodisperse superparamagnetic iron oxide nanoparticles for *in vivo* quantification studies. *ACS Nano* **2012**, *6*, 7318–7325.
29. Wang, H.; Kumar, R.; Nagesha, D.; Duclos, R.I., Jr.; Sridhar, S.; Gatley, S.J. Integrity of ^{111}In-radiolabeled superparamagnetic iron oxide nanoparticles in the mouse. *Nucl. Med. Biol.* **2015**, *42*, 65–70.
30. Shim, J.Y.; Gupta, V.K. Reversible aggregation of gold nanoparticles induced by pH dependent conformational transitions of a self-assembled polypeptide. *J. Colloid Interface Sci.* **2007**, *316*, 977–983.
31. Smith, C.A.; Simpson, C.A.; Kim, G.; Carter, C.J.; Feldheim, D.L. Gastrointestinal bioavailability of 2.0 nm diameter gold nanoparticles. *ACS Nano* **2013**, *7*, 3991–3996.
32. Harper, S.; Usenko, C.; Hutchison, J.E.; Maddux, B.L.S.; Tanguay, R.L. *In vivo* biodistribution and toxicity depends on nanomaterial composition, size, surface functionalisation and route of exposure. *J. Exp. Nanosci.* **2008**, *3*, 195–206.

33. Terentyuk, G.S.; Maslyakova, G.N.; Suleymanova, L.V.; Khlebtsov, B.N.; Kogan, B.Y.; Tuchin, V.V.; Akchurin, G.G.; Shantrocha, A.V.; Maksimova, I.L.; Khlebtsov, N.G. Circulation and distribution of gold nanoparticles and induced alterations of tissue morphology at intravenous particle delivery. *J. Biophotonics* **2009**, *2*, 292–302.
34. Zhang, G.; Yang, Z.; Lu, W.; Zhang, R.; Huang, Q.; Tian, M.; Li, L.; Liang, D.; Li, C. Influence of anchoring ligands and particle size on the colloidal stability and *in vivo* biodistribution of polyethylene glycol-coated gold nanoparticles in tumor-xenografted mice. *Biomaterials* **2009**, *30*, 1928–1936.
35. Casals, E.; Puntes, V.F. Inorganic nanoparticle biomolecular corona: Formation, evolution and biological impact. *Nanomedicine* **2012**, *7*, 1917–1930.
36. Tenzer, S.; Docter, D.; Kuharev, J.; Musyanovych, A.; Fetz, V.; Hecht, R.; Schlenk, F.; Fischer, D.; Kiouptsi, K.; Reinhardt, C.; *et al*. Rapid formation of plasma protein corona critically affects nanoparticle pathophysiology. *Nat. Nanotechnol.* **2013**, *8*, 772–781.
37. Tay, C.Y.; Setyawati, M.I.; Xie, J.; Parak, W.J.; Leong, D.T. Back to basics: Exploiting the innate physico-chemical characteristics of nanomaterials for biomedical applications. *Adv. Funct. Mater.* **2014**, *24*, 5936–5955.
38. Lynch, I.; Cedervall, T.; Lundqvist, M.; Cabaleiro-Lago, C.; Linse, S.; Dawson, K.A. The nanoparticle-protein complex as a biological entity; a complex fluids and surface science challenge for the 21st century. *Adv. Colloid Interface Sci.* **2007**, *134–135*, 167–174.
39. Dobrovolskaia, M.A.; Aggarwal, P.; Hall, J.B.; McNeil, S.E. Preclinical studies to understand nanoparticle interaction with the immune system and its potential effects on nanoparticle biodistribution. *Mol. Pharm.* **2008**, *5*, 487–495.
40. Dobrovolskaia, M.A.; Patri, A.K.; Zheng, J.; Clogston, J.D.; Ayub, N.; Aggarwal, P.; Neun, B.W.; Hall, J.B.; McNeil, S.E. Interaction of colloidal gold nanoparticles with human blood: Effects on particle size and analysis of plasma protein binding profiles. *Nanomedicine* **2009**, *5*, 106–117.
41. Aggarwal, P.; Hall, J.B.; McLeland, C.B.; Dobrovolskaia, M.A.; McNeil, S.E. Nanoparticle interaction with plasma proteins as it relates to particle biodistribution, biocompatibility and therapeutic efficacy. *Adv. Drug Deliv. Rev.* **2009**, *61*, 428–437.
42. Nel, A.E.; Madler, L.; Velegol, D.; Xia, T.; Hoek, E.M.V.; Somasundaran, P.; Klaessig, F.; Castranova, V.; Thompson, M. Understanding biophysicochemical interactions at the nano-bio interface. *Nat. Mater.* **2009**, *8*, 543–557.
43. Monopoli, M.P.; Walczyk, D.; Campbell, A.; Elia, G.; Lynch, I.; Baldelli Bombelli, F.; Dawson, K.A. Physical-chemical aspects of protein corona: Relevance to *in vitro* and *in vivo* biological impacts of nanoparticles. *J. Am. Chem. Soc.* **2011**, *133*, 2525–2534.
44. Chen, Y.S.; Hung, Y.C.; Liau, I.; Huang, G. Assessment of the *in vivo* toxicity of gold nanoparticles. *Nanoscale Res. Lett.* **2009**, *4*, 858–864.
45. Katti, K.; Kannan, R.; Katti, K.; Kattumori, V.; Pandrapragada, R.; Rahing, V.; Cutler, C.; Boote, E.; Casteel, S.; Smith, C.; *et al*. Hybrid gold nanoparticles in molecular imaging and radiotherapy. *Czechoslov. J. Phys.* **2006**, *56*, D23–D34.
46. Turkevich, J.; Stevenson, P.C.; Hillier, J. The formation of colloidal gold. *J. Phys. Chem.* **1953**, *57*, 670–673.

47. Frens, G. Controlled nucleation for the regulation of the particle size in monodisperse gold suspensions. *Nat. Phys. Sci.* **1973**, *241*, 20–22.
48. Haiss, W.; Thanh, N.T.K.; Aveyard, J.; Fernig, D.G. Determination of size and concentration of gold nanoparticles from UV-Vis spectra. *Anal. Chem.* **2007**, *79*, 4215–4221.

Sample Availability: Not available.

© 2015 by the authors; licensee MDPI, Basel, Switzerland. This article is an open access article distributed under the terms and conditions of the Creative Commons Attribution license (http://creativecommons.org/licenses/by/4.0/).

Molecules **2015**, *20*, 9591-9615; doi:10.3390/molecules20069591

ISSN 1420-3049
www.mdpi.com/journal/molecules

Article

Synthesis, ^{18}F-Radiolabelling and Biological Characterization of Novel Fluoroalkylated Triazine Derivatives for *in Vivo* Imaging of Phosphodiesterase 2A in Brain via Positron Emission Tomography

Susann Schröder [1],*, Barbara Wenzel [1], Winnie Deuther-Conrad [1], Rodrigo Teodoro [1], Ute Egerland [2], Mathias Kranz [1], Matthias Scheunemann [1], Norbert Höfgen [2], Jörg Steinbach [1] and Peter Brust [1]

[1] Department of Neuroradiopharmaceuticals, Institute of Radiopharmaceutical Cancer Research, Helmholtz-Zentrum Dresden-Rossendorf, Permoserstraße 15, Leipzig 04318, Germany; E-Mails: b.wenzel@hzdr.de (B.W.); w.deuther-conrad@hzdr.de (W.D.-C.); r.teodoro@hzdr.de (R.T.); m.kranz@hzdr.de (M.K.); m.scheunemann@hzdr.de (M.S.); j.steinbach@hzdr.de (J.S.); p.brust@hzdr.de (P.B.)

[2] BioCrea GmbH, Meissner Str. 191, Radebeul 01445, Germany; E-Mails: Ute.Egerland@biocrea.com (U.E.); Norbert.Hoefgen@biocrea.com (N.H.)

* Author to whom correspondence should be addressed; E-Mail: s.schroeder@hzdr.de; Tel.: +49-341-234-179-4631.

Academic Editor: Svend Borup Jensen

Received: 17 April 2015 / Accepted: 18 May 2015 / Published: 26 May 2015

Abstract: Phosphodiesterase 2A (PDE2A) is highly and specifically expressed in particular brain regions that are affected by neurological disorders and in certain tumors. Development of a specific PDE2A radioligand would enable molecular imaging of the PDE2A protein via positron emission tomography (PET). Herein we report on the syntheses of three novel fluoroalkylated triazine derivatives (**TA2–4**) and on the evaluation of their effect on the enzymatic activity of human PDE2A. The most potent PDE2A inhibitors were ^{18}F-radiolabelled ([^{18}F]**TA3** and [^{18}F]**TA4**) and investigated regarding their potential as PET radioligands for imaging of PDE2A in mouse brain. *In vitro* autoradiography on rat brain displayed region-specific distribution of [^{18}F]**TA3** and [^{18}F]**TA4**, which is consistent with the expression pattern of PDE2A protein. Metabolism studies of both [^{18}F]**TA3** and [^{18}F]**TA4** in mice showed a significant accumulation of two major radiometabolites of each radioligand in brain as investigated by micellar radio-chromatography. Small-animal PET/MR studies in mice using [^{18}F]**TA3** revealed a constantly increasing uptake of activity in the non-target region cerebellum, which may be caused by the accumulation of brain penetrating radiometabolites. Hence, [^{18}F]**TA3** and [^{18}F]**TA4** are exclusively suitable for *in vitro* investigation of PDE2A. Nevertheless, further structural modification of these promising radioligands might result in metabolically stable derivatives.

Keywords: PDE2A; Alzheimer's disease; PET imaging in brain; micellar HPLC

1. Introduction

Phosphodiesterases (PDEs) are a class of intracellular enzymes consisting of 11 families and 21 isoforms [1–4]. The PDE family subtypes differ in their three-dimensional structure, kinetic and regulatory properties, cellular expression, intracellular location, inhibitor sensitivities, substrate selectivity and distribution within the organism. Due to hydrolysis of the cyclic nucleotides, PDEs affect the second messenger signaling cascades of cyclic adenosine monophosphate (cAMP) and/or cyclic guanosine monophosphate (cGMP) [1,2]. Therefore, pharmacological inhibition of PDEs has the potential to enhance cyclic nucleotide signaling and can provide a strategy for the prevention or treatment of various diseases [2,5–8].

Phosphodiesterase 2A (PDE2A) is a dual-substrate specific enzyme degrading both nucleotides cAMP and cGMP. The PDE2A protein is highly and specifically expressed in particular brain regions such as striatum, cortex, hippocampus, substantia nigra, amygdala, and in olfactory neurons [9–12].

The specific distribution of the PDE2A protein in brain indicates a modulation of important neuronal functions associated to learning and memory [13–15]. Inhibition of PDE2A activity in brain is suggested to improve neuronal plasticity and memory formation due to increased cGMP levels in active synapses [16,17]. Hence, PDE2A inhibitors might be promising compounds regarding drug development for treatment of neurodegenerative disorders such as Alzheimer's disease (AD) [3,12]. In addition, this enzyme is highly expressed in certain tumors such as malignant melanoma and mammary carcinoma [18,19], and it is assumed that PDE2A activity is related to highly proliferative processes [20,21].

The most common PDE2A inhibitors, shown in Figure 1, are *erythro*-9-(2-hydroxy-3-nonyl)-adenine (EHNA) [2,16,22,23] and BAY 60–7550 [2,16,17,24–28].

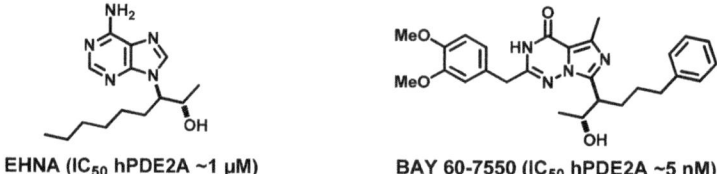

Figure 1. Most common PDE2A inhibitors: EHNA and BAY 60-7550 [2,16,17,22–28].

Besides further PDE2A inhibitors developed for treatment of neurological disorders [3,29–35], two PDE2A radioligands for molecular imaging of this protein in the brain via positron emission tomography (PET) have been reported to date (Figure 2).

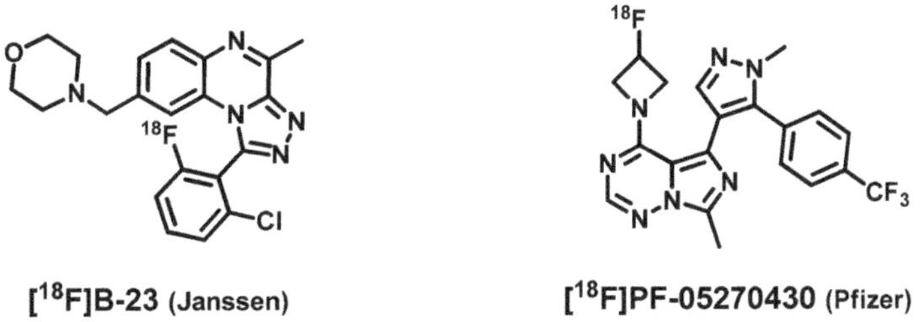

[¹⁸F]B-23 (Janssen)   **[¹⁸F]PF-05270430 (Pfizer)**

Figure 2. PDE2A radioligands developed by Janssen ([¹⁸F]B-23) and Pfizer ([¹⁸F]PF-05270430) [3,36,37].

The PET ligand [¹⁸F]B-23 developed by Janssen [3,36] is highly affine towards the PDE2A protein but also for PDE10A (IC$_{50}$ hPDE2A = 1 nM; IC$_{50}$ rPDE10A = 11 nM). In microPET imaging studies with rats the highest accumulation of [¹⁸F]B-23 has been observed in the striatum, however, radio-metabolites have been detected in the brain (at 2 min p.i.: 4%; at 10 min p.i.: 10% of total activity). The distribution pattern of PDE2A and PDE10A is comparable: e.g., both proteins are highly expressed in the caudate nucleus [9]. Thus, the reported uptake in the striatum may be caused by binding of [¹⁸F]B-23 to both enzymes in this brain region. Hence, to develop appropriate radioligands for PET imaging of PDE2A, high selectivity toward this protein is of major importance. The radioligand [¹⁸F]PF-05270430 published by Pfizer [3,37] is a highly affine and selective PDE2A inhibitor (IC$_{50}$ hPDE2A = 0.5 nM; IC$_{50}$ hPDE10A = 3.0 µM) with good brain uptake. In PET studies on monkeys a rapid and high uptake of [¹⁸F]PF-05270430 in the striatum has been reported. Supplementary data of [¹⁸F]PF-05270430 concerning dosimetry and radiometabolite analysis have not been published yet.

The goal of our work is the imaging of the PDE2A protein in brain via PET that may enable early diagnostics of related diseases. Furthermore, specific PDE2A radioligands may also be used for the pharmacological characterization and evaluation of novel PDE2A inhibitors as therapeutics. Herein, we present the development of three novel fluoroalkylated derivatives (**TA2–4**) as PDE2A ligands starting from a triazine lead compound (**TA1**) [38] (Figure 3).

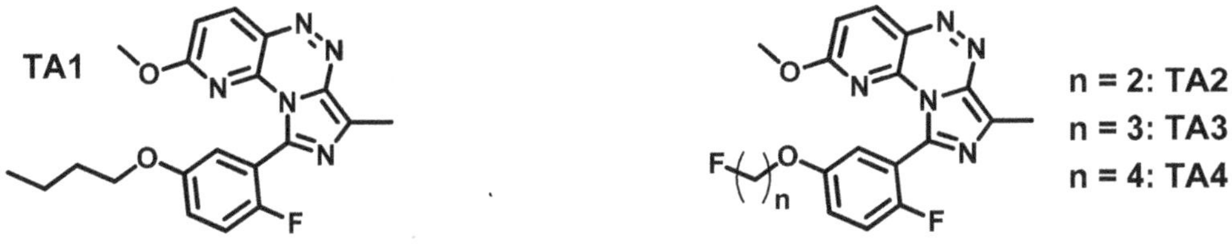

TA1   n = 2: **TA2**
n = 3: **TA3**
n = 4: **TA4**

Figure 3. Triazine lead compound **TA1** and the novel fluoroalkylated derivatives **TA2–4** as PDE2A ligands.

Out of this series of derivatives, the two most suitable PDE2A inhibitors were ¹⁸F-radiolabelled and investigated regarding their (i) *in vitro* properties using rat brain slices; (ii) metabolic stability in mice; and (iii) *in vivo* imaging potential using small-animal PET in mice.

2. Results and Discussion

2.1. Synthesis and in Vitro Binding

The novel fluoroalkylated derivatives **TA2–4** presented as PDE2A ligands were developed starting from the triazine lead compound **TA1** [38]. The structure of **TA1** (Figure 3) contains a fluorine atom on the benzene ring. In terms of a planned ^{18}F-radiolabelling, this position is not activated for a nucleophilic aromatic substitution of a leaving group by [^{18}F]fluoride due to the enhanced electron density [39]. Therefore, we favored the introduction of a second fluorine atom in the phenolic ether group enabling a nucleophilic ^{18}F-radiolabelling at the alkyl side chain. The five steps synthesis of the lead compound **TA1** has already been reported [38] and was partly optimized in this study (Scheme 1).

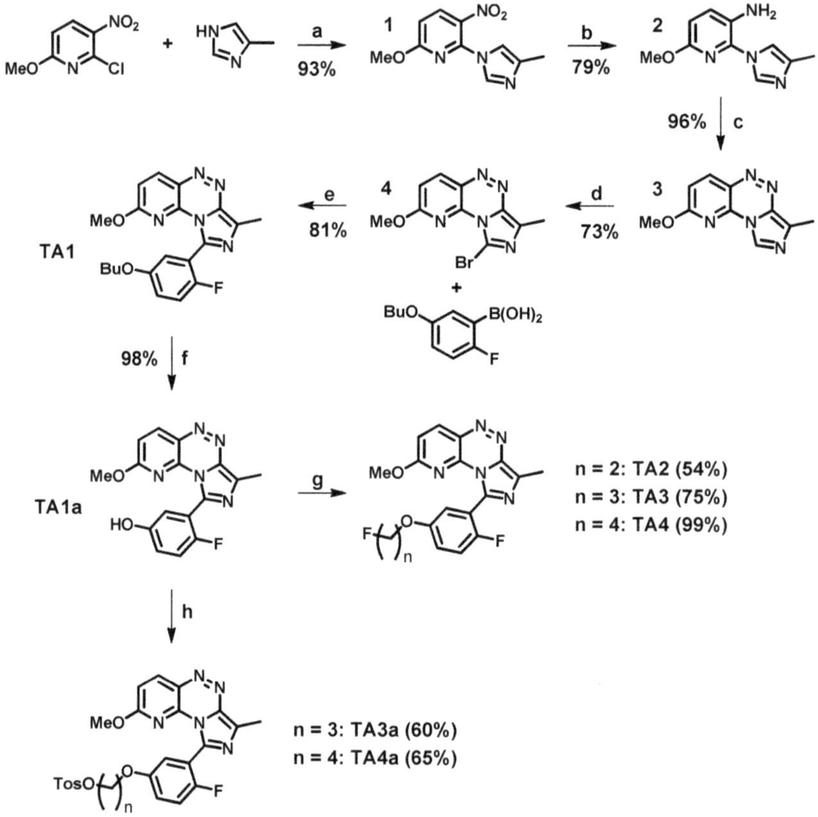

Reagents and Conditions: (**a**) 3 eq. TEA, 10 mol % DMAP, CHCl$_3$, 0 °C to RT, overnight; (**b**) Pd(C)/H$_2$, EtOH, RT, overnight; (**c**) 1.5 eq NaNO$_2$, H$_2$O/CH$_3$COOH, ≤5 °C, 30 min; (**d**) 1.5 eq *N*-bromosuccinimide (NBS), CH$_2$Cl$_2$, ≤5 °C to RT, overnight; (**e**) 1 eq 5-butoxy-2-fluorophenyl boronic acid, 0.05 eq [(Ph$_3$)P]$_4$Pd(0), 3 eq K$_2$CO$_3$, dioxan/H$_2$O, 90 °C, 5 h; (**f**) 3.05 eq BBr$_3$ (1 M in CH$_2$Cl$_2$), CH$_2$Cl$_2$, ≤5 °C, 2 h; (**g**) *n* = 2: F-(CH$_2$)$_2$-I, *n* = 3: F-(CH$_2$)$_3$-I, *n* = 4: F-(CH$_2$)$_4$-Br, 3 eq K$_2$CO$_3$, MeCN, 70–80 °C, 5 h; (**h**) *n* = 3: TosO-(CH$_2$)$_3$-OTos, *n* = 4: TosO-(CH$_2$)$_4$-OTos, 4 eq K$_2$CO$_3$, MeCN, 60–70 °C, 5–10 h.

Scheme 1. Syntheses of the triazine lead compound **TA1**, the phenolic intermediate **TA1a**, the novel fluoroalkylated PDE2A ligands **TA2–4** and the tosylate precursors **TA3a** and **TA4a**.

The first step relates to the coupling reaction between the substituted 2-chloropyridine and the 4-methylimidazole component to afford **1**. By using triethylamine (TEA) as base in the presence of 4-(dimethylamino)pyridine (DMAP) as catalyst instead of potassium carbonate, and replacing *N,N*-dimethylformamide (DMF) by chloroform as solvent, **1** was obtained isomerically pure and with a 20% higher yield (93% *vs.* 71% [38]). Notably, a synthesis of compound **1** [38] resulting in an inseparable 4:1 mixture of the imidazole regioisomers has also been published by Malamas *et al.* [40]. For the reduction of the nitro group to the corresponding amine, milder reaction conditions (e.g., temperature and pressure) were used, affording **2** in similar yields (79% *vs.* 81% [38]). Afterwards, a diazotization was performed followed by an intramolecular cyclisation (azo coupling) to get the triazine basic structure **3**. By washing the precipitate instead of recrystallization or column chromatography, **3** was obtained in comparable yields (96% *vs.* 93% [38]). The bromination at the imidazole site has been reported for the corresponding 4-methoxy compound [38]. The 2-methoxy-bromo derivative **4** was obtained in similar yields according to the literature (73% *vs.* 76% [38]). Finally, the Suzuki coupling with the 5-butoxy-2-fluorophenyl boronic acid was performed as previously reported [38] affording the lead compound **TA1** in 81% yield.

The subsequent cleavage of the butoxy group by boron tribromide resulted in the phenol compound **TA1a** in 98% yield. Notably, the butoxy group was selectively cleaved while the 2-methoxy function remained stable, even in the presence of a large excess of boron tribromide (up to 10 eq.). The novel derivatives **TA2**, **TA3** and **TA4** were successfully synthesized in 54%, 75% and 99% yield, respectively, using the phenolic intermediate **TA1a** and appropriate fluoroalkyl halides (Scheme 1).

The novel fluoroalkylated derivatives **TA2–4** were evaluated in an enzyme assay [38] to determine their inhibitory potencies for the human recombinant PDE2A and PDE10A proteins. The IC_{50} values obtained by this assay represent relative measures of the respective target affinity of the compounds. We have previously shown for a specific PDE10A radioligand that the target affinity is within the same order of magnitude as the inhibitory potency of the corresponding non-radioactive reference compound [41].

As mentioned, only ligands with high affinity and selectivity are suitable for PET imaging of PDE2A due to the comparable distribution pattern of PDE2A and PDE10A in the brain [9]. Table 1 summarizes the IC_{50} values of the lead compound **TA1** and the novel fluoroalkylated derivatives **TA2–4** for the inhibition of human PDE2A and human PDE10A.

Table 1. IC_{50} values of the lead compound **TA1** and the novel fluoroalkylated derivatives **TA2–4** for the inhibition of human PDE2A and human PDE10A.

Ligand	IC_{50} hPDE2A	IC_{50} hPDE10A	Selectivity Ratio PDE10A/PDE2A
TA1 (lead)	4.5 nM	670 nM	148.9
TA2 (2-fluoroethyl)	10.4 nM	77 nM	7.4
TA3 (3-fluoropropyl)	**11.4 nM**	**318 nM**	27.9
TA4 (4-fluorobutyl)	**7.3 nM**	913 nM	125.1

Compared to the lead compound **TA1** the affinity and selectivity of **TA2** and **TA3** are slightly lower. Besides its high potency, **TA4** shows the highest selectivity of all tested fluoroalkylated derivatives. These findings can be explained by the recently reported binding-induced hydrophobic pocket (H-pocket) in the active center of the PDE2A protein [3,42]. This mechanism suggests that hydrophobic interactions between the propylphenyl group of BAY 60–7550 and the H-pocket might be responsible for the high PDE2A selectivity of this compound [42]. Accordingly, the strength of hydrophobic interactions with the H-pocket may be related to the increased chain lengths within the novel derivatives **TA2–4**.

Thus, the ligand **TA3** and the very promising derivative **TA4** were selected as candidates for ^{18}F-radiolabelling. The corresponding tosylate precursors **TA3a** and **TA4a**, necessary for a one-step nucleophilic ^{18}F-radiolabelling strategy, were synthesized with yields of 60%–65% by reaction of propane-1,3-diyl bis(4-methyl-benzenesulfonate) or butane-1,4-diyl bis(4-methyl-benzenesulfonate) with the phenolic intermediate **TA1a** (see Scheme 1).

2.2. Radiochemistry, Lipophilicity and in Vitro Stability

The novel PDE2A radioligands [^{18}F]**TA3** and [^{18}F]**TA4** were prepared in a one-step radiosynthesis by nucleophilic substitution of the tosylate group of the precursors **TA3a** and **TA4a** with the anhydrous K$^+$/[^{18}F]F$^-$/K$_{2.2.2}$-carbonate complex in acetonitrile (for the drying procedure see Section 3.4.1.).

Optimization of the aliphatic radiolabelling was performed by varying the amount of precursor (1–3 mg) and reaction time (up to 20 min) under conventional heating at 80 °C in acetonitrile. As shown in Figure 4, the highest labelling yields were achieved by using 1 mg of the corresponding tosylate precursor **TA3a** or **TA4a**. Accordingly, an increase of the amount of tosylate precursor **TA3a** resulted in an unexpected decrease of the labelling yield, which is more pronounced at the early time points. Independent of the amount of precursor **TA3a**, only a single ^{18}F-side product was detected with 4% of total activity at 15 min reaction time. Both precursors were stable under the reaction conditions over 20 min, proved by HPLC.

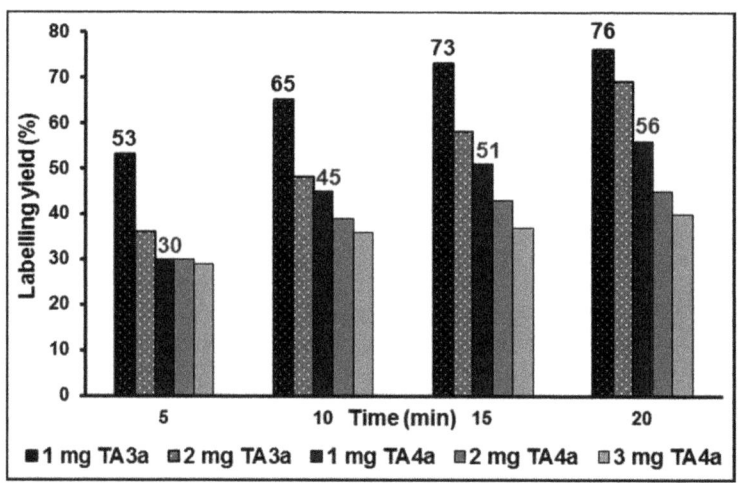

Figure 4. Optimization of the labelling yields for aliphatic ^{18}F-radiolabelling of the tosylate precursors **TA3a** and **TA4a** by varying the amount of precursor and the reaction time (in MeCN at 80 °C, conventional heating).

A time-dependent increase of the labelling yields was observed, however, 73% for [^{18}F]**TA3** and 51% for [^{18}F]**TA4** after 15 min reaction time are sufficiently high for further experiments. Thus, manual radiosyntheses of [^{18}F]**TA3** and [^{18}F]**TA4** were performed in acetonitrile at 80 °C for 15 min under no-carrier-added (n.c.a.) conditions using 1 mg of each tosylate precursor **TA3a** or **TA4a** (Scheme 2).

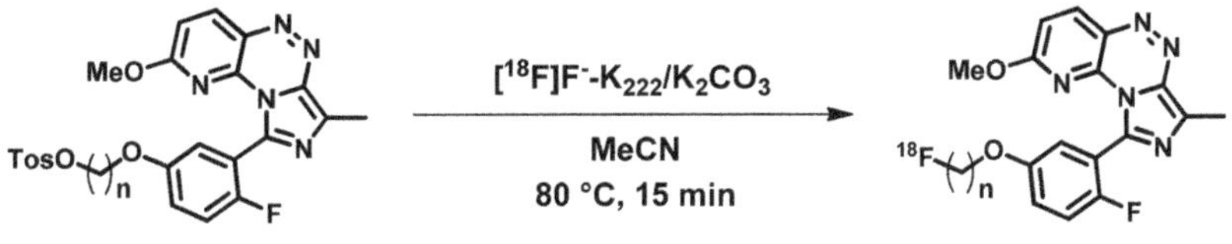

Scheme 2. Nucleophilic ^{18}F-radiolabelling of tosylate precursors **TA3a** and **TA4a** to generate the novel PDE2A radioligands [^{18}F]**TA3** and [^{18}F]**TA4**.

[^{18}F]**TA3** and [^{18}F]**TA4** were isolated by semi-preparative HPLC (system **A**; t_R = 27–33 min, see Figure 5), purified using solid phase extraction on a pre-conditioned RP cartridge and eluted with absolute ethanol. The solvent was evaporated at 70 °C and the radioligands were formulated in sterile isotonic saline containing 10% of ethanol (v/v) for better solubility. Aliquots of the final products were spiked with the related non-radioactive reference compounds **TA3** and **TA4** to verify the identities of [^{18}F]**TA3** and [^{18}F]**TA4** (Figure 5) by analytical HPLC (system **B**).

The radioligands were synthesized with moderate to high labelling yields of 72.5% ± 6.4% for [^{18}F]**TA3** (n = 8) and 41.5% ± 9.6% for [^{18}F]**TA4** (n = 4), radiochemical yields of 48.7% ± 8.5% for [^{18}F]**TA3** (n = 4) and 25.4% ± 3.9% for [^{18}F]**TA4** (n = 4), specific activities (EOS) of 60.4 ± 11.6 GBq/µmol for [^{18}F]**TA3** (n = 3) and 77.1 ± 23.8 GBq/µmol for [^{18}F]**TA4** (n = 3), and high radiochemical purities of ≥99%. The decreased labelling yield of [^{18}F]**TA4** in comparison with [^{18}F]**TA3** could be a result of the additional CH$_2$ group. The electron withdrawing effect of the phenolic oxygen on the alkyl side chain might be slightly reduced due to the longer distance to the carbon atom that is attacked by the nucleophilic [^{18}F]F$^-$. For example, a similar effect has been observed for [^{18}F]fluoroalkoxy derivatives of harmine (7-methoxy-1-methyl-9*H*-β-carboline) where a decreased radiochemical yield of the 7-(3-[^{18}F]fluoropropoxy) analogue in comparison to the 7-(2-[^{18}F]fluoroethoxy) derivative has been reported [43].

To estimate the lipophilicity of [^{18}F]**TA3** and [^{18}F]**TA4**, the distribution coefficients were determined by partitioning between *n*-octanol and phosphate buffered saline (PBS, pH 7.4) at ambient temperature using the conventional shake-flask method. LogD values of 3.37 ± 0.14 for [^{18}F]**TA3** and 2.99 ± 0.15 for [^{18}F]**TA4** were obtained, indicating a high to moderate lipophilicity regarding passive transport across the blood brain barrier. *In vitro* stability of each radioligand was investigated in phosphate buffered saline (PBS, pH 7.4), *n*-octanol and pig plasma after 60 min incubation at 37 °C. [^{18}F]**TA3** and [^{18}F]**TA4** proved to be stable in all media tested *in vitro*, and no defluorination or degradation was observed by radio-TLC or analytical radio-HPLC.

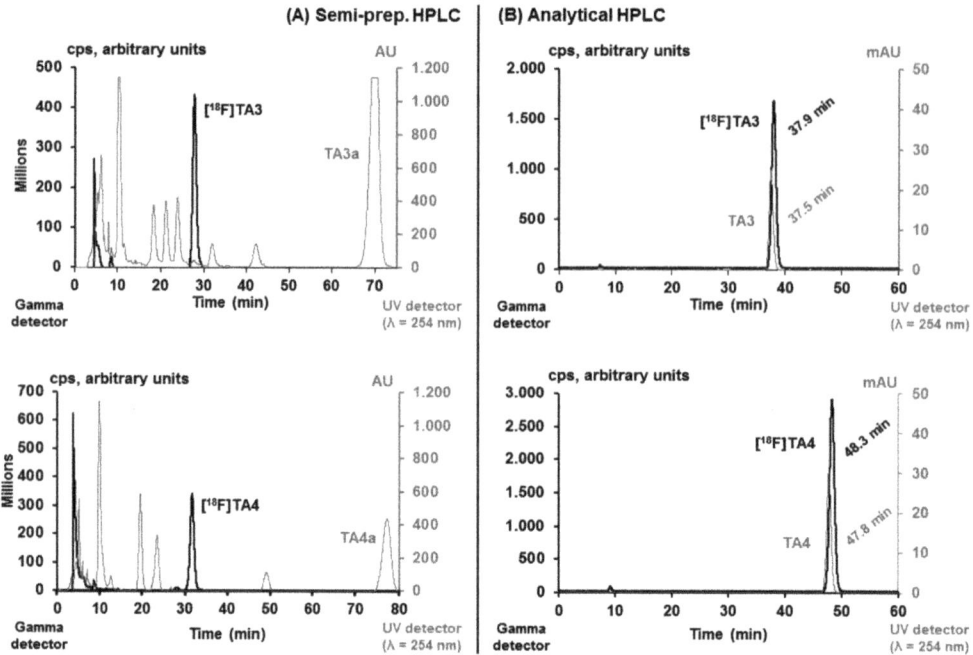

Figure 5. (**A**) HPLC profile of the crude reaction mixtures for semi-preparative isolation of [^{18}F]**TA3** and [^{18}F]**TA4** (column: Reprosil-Pur C18-AQ, 250 × 10 mm, particle size: 10 μm; eluent: 50% MeCN/20 mM NH$_4$OAc$_{aq.}$; flow: 3 mL/min for [^{18}F]**TA3**, 3.5 mL/min for [^{18}F]**TA4**); (**B**) Analytical HPLC profile of the formulated radioligands [^{18}F]**TA3** and [^{18}F]**TA4** spiked with the non-radioactive reference compounds **TA3** and **TA4** (column: Reprosil-Pur C18-AQ, 250 × 4.6 mm, particle size: 5 μm; eluent: 44% MeCN/20 mM NH$_4$OAc$_{aq.}$; flow: 1 mL/min).

Due to the higher metabolic stability of [^{18}F]**TA3** *in vivo* compared to that of [^{18}F]**TA4** as described below in Section 2.4., the automated radiosynthesis was performed only of [^{18}F]**TA3**. The conditions for the manual radiosynthesis of [^{18}F]**TA3** (^{18}F-labelling: 1 mg of tosylate precursor **TA3a** in MeCN, 80 °C, 15 min) were transferred to an automated process using a TRACERlab™ FX F-N synthesis module. After automated isolation of [^{18}F]**TA3** by semi-preparative HPLC (system **A**), purification by solid phase extraction and elution of the RP cartridge, the radioligand was formulated manually in sterile isotonic saline containing 10% of ethanol (*v*/*v*) as described above. Analytical HPLC (system **B**) of the final product spiked with the non-radioactive reference compound **TA3** confirmed the identity of [^{18}F]**TA3** (see Figure 5). The radioligand was obtained with a radiochemical purity higher than 99%, a radiochemical yield of 41.7% ± 6.3% (*n* = 3), and a specific activity of 142.6 ± 35.2 GBq/μmol (EOS; *n* = 3) in a total synthesis time of 75 min.

Compared to the manual procedure, there were no significant differences regarding the radiochemical yield of [^{18}F]**TA3** achieved in the automated process. Besides the high reproducibility of the automated radiosyntheses, higher starting activities could be used (5–7 GBq *vs.* max. 3 GBq) and thus a significantly increased specific activity of [^{18}F]**TA3** was obtained (143 GBq/μmol *vs.* 60 GBq/μmol).

2.3. In Vitro Autoradiographic Studies in Rat Brain

In vitro autoradiographic studies were accomplished by incubating sagittal sections of rat brain with [^{18}F]**TA3** or [^{18}F]**TA4**. Non-specific binding of each radioligand was assessed by co-incubation with an excess of lead compound **TA1**. The images shown in Figure 6 indicate region-specific accumulation of both radioligands [^{18}F]**TA3** and [^{18}F]**TA4**, which is consistent with the distribution pattern of PDE2A protein in rat brain [9,11]. Therefore, we assume that the novel radioligands might be appropriate for *in vitro* imaging of PDE2A. Notably, a specific PDE2A-radioligand (e.g., ^{3}H-labelled) is not commercially available and *in vitro* autoradiographic images of the PDE2A distribution in brain have not been published yet.

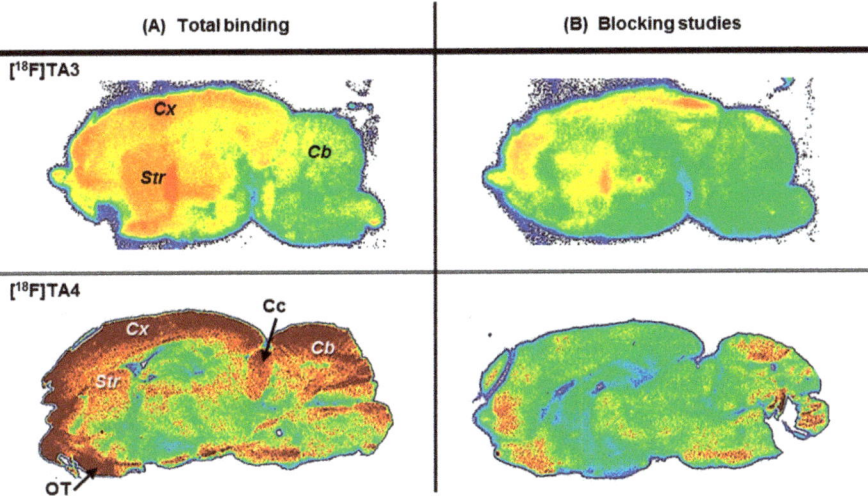

Figure 6. Representative autoradiographic images of sagittal rat brain slices: (**A**) *In vitro* distribution of activity after incubation with ~1 MBq/mL of [^{18}F]**TA3** or [^{18}F]**TA4**; (**B**) Non-specific binding of [^{18}F]**TA3** or [^{18}F]**TA4** determined in the presence of 1 µM of **TA1** as blocking compound. *Abbreviations*: Cb—cerebellum, Cc—colliculi, Cx—cortex, OT—olfactory tubercle, Str—striatum.

The radioligand [^{18}F]**TA3** shows higher binding densities in cortex and striatum than in cerebellum, while the radioligand [^{18}F]**TA4** also binds to the olfactory tubercle, colliculi and partly to cerebellum.

2.4. In Vivo Metabolism of [^{18}F]TA3 and [^{18}F]TA4 in Mice

In vivo metabolism of [^{18}F]**TA3** and [^{18}F]**TA4** was investigated in plasma and brain samples obtained from CD-1 mice at 30 min post injection of 150 MBq of each radioligand. Analysis of the samples was performed after protein precipitation and twofold extraction using an organic solvent. For both radioligands, a high fraction of radiometabolites was detected in plasma with only 8% and 6% of total activity representing non-metabolized [^{18}F]**TA3** (Figure 7; recovery of total activity: 76%) and [^{18}F]**TA4** (recovery: 83%), respectively. In brain samples, 50% of total activity were represented by intact [^{18}F]**TA3** (Figure 7; recovery: 65%). No intact [^{18}F]**TA4** was detectable in related brain samples (recovery: 85%).

The conventional extraction procedure is very work-intensive, time-consuming and—most important—an absolute quantification of the real composition in the samples is not possible due to 65%–85% recovery of total activity. In order to understand the impact of polar radiometabolites on the recovery yields, we used [^{18}F]fluoride to investigate its extractability from denatured proteins *in vitro* taking into account that (i) defluorination is a common process in the *in vivo* metabolism of fluorine-bearing molecules; and (ii) [^{18}F]fluoride is the most polar radiometabolite resulting from the metabolic degradation of ^{18}F-compounds. Thus, pig plasma samples were incubated *in vitro* with [^{18}F]fluoride and subsequent extraction was performed, applying the same protocol as for the *in vivo* metabolism studies, revealing that only 34% of the total activity could be recovered.

These findings indicate that the binding of polar radiometabolites, especially with an ionic character, to the denatured proteins is stronger than interactions of the proteins with the intact radioligand and thus, these radiometabolites are not completely extractable. Accordingly, the real percentage ratio of intact radioligand to its radiometabolites cannot be quantified by RP-HPLC based on samples obtained from an extraction procedure.

Therefore we used micellar HPLC (MLC), because this method allows the direct injection of plasma samples without deproteination. MLC was recently investigated by Nakao *et al.* [44] regarding their suitability for direct plasma metabolite analysis of PET radioligands. As eluent a mixture of an aqueous solution of sodium dodecyl sulphate (SDS), phosphate buffer and 1-propanol was used. According to the published procedure [44], a gradient mode was applied starting under micellar conditions to completely elute the protein fraction and release the protein bound ^{18}F-compounds by the SDS micelles. Subsequent increase of the amount of 1-propanol as organic modifier leads to high submicellar conditions [44,45] resulting in the elution of the intact radioligand and its radio-metabolites.

Analysis of the MLC chromatograms revealed that only 5% of total activity in plasma are represented by intact radioligand [^{18}F]**TA3** (Figure 7) or [^{18}F]**TA4**. For the first time, the micellar HPLC method was also applied to evaluate mouse brain radiometabolites. In the analyzed brain samples, 29% and 4% of total activity were represented by intact [^{18}F]**TA3** (Figure 7) and [^{18}F]**TA4**, respectively.

The radio-chromatograms of the RP-HPLC (conventional extraction) and the micellar HPLC method displayed slightly different elution profiles (for [^{18}F]**TA3** see Figure 7) which are probably caused by the various retention mechanisms in these two systems. RP-HPLC revealed the presence of two major radiometabolites ([^{18}F]**M1** and [^{18}F]**M2**) in plasma and brain for each radioligand. It is supposed that the second radiometabolite [^{18}F]**M2** elutes within the first fraction under micellar conditions.

Furthermore, comparing the results of the RP-HPLC (for [^{18}F]**TA3**: t_R = 33 min, see Figure 7) and the MLC (for [^{18}F]**TA3**: t_R = 39 min, see Figure 7), slightly different concentrations of the detected radioligands in the mouse plasma samples (e.g., for [^{18}F]**TA3**: RP: 8% *vs.* MLC: 5%) and brain homogenates (e.g., for [^{18}F]**TA3**: RP: 50% *vs.* MLC: 29%) can be observed. These findings might be caused by insufficient extraction of radioactive compounds (recovery: 65%–85%) from the precipitated proteins, mainly of the more polar radiometabolites detected by RP-HPLC (for [^{18}F]**TA3**: [^{18}F]**M1**: t_R = 3 min; [^{18}F]**M2**: t_R = 22 min, see Figure 7). Therefore we conclude that only MLC provides the detection of 100% of the real composition in the analyzed samples due to direct injection into the micellar HPLC system.

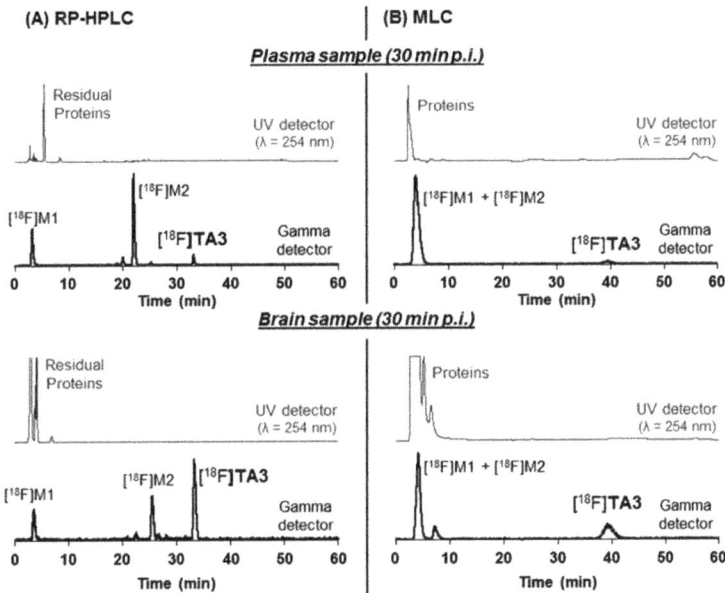

Figure 7. Representative *in vivo* metabolism study of mouse plasma and brain samples at 30 min p.i. of [^{18}F]**TA3** (150 MBq): (**A**) RP-HPLC chromatograms of extracted samples (column: Reprosil-Pur C18-AQ, 250 × 4.6 mm, particle size: 5 µm; gradient: 10% → 90% → 10% MeCN/20 mM NH$_4$OAc$_{aq.}$; flow: 1 mL/min); (**B**) MLC chromatograms of samples directly injected into the MLC system (column: Reprosil-Pur C18-AQ, 250 × 4.6 mm, particle size: 10 µm; gradient: 3% → 30% → 3% 1-PrOH/100 mM SDS$_{aq.}$, 10 mM Na$_2$HPO$_{4aq.}$; flow: 1 mL/min).

In comparison with the intact radioligands, the shorter retention times of [^{18}F]M1 and [^{18}F]M2 observed in the radio-chromatograms of the RP-HPLC (for [^{18}F]**TA3** see Figure 7) point to a higher polarity of these radiometabolites. The similar elution profiles of extracted plasma and brain samples (for [^{18}F]**TA3** see Figure 7) indicate that both radiometabolites may cross the blood-brain barrier. The significant accumulation of the highly polar radiometabolite [^{18}F]M1 in brain could be explained by cytochrome P450 enzyme induced metabolic degradation of the ^{18}F-bearing alkyl side chains in [^{18}F]**TA3** or [^{18}F]**TA4** and formation of the corresponding brain penetrating ^{18}F-alkyl alcohols, aldehydes or carboxylic acids [46,47]. Regarding the formation of the radiometabolite [^{18}F]M2, the related mechanism of metabolic degradation and thus the molecular structure of [^{18}F]M2 remain unclear.

2.5. PET/MR Studies of [^{18}F]TA3 in Mice

The metabolism studies indicated that the metabolic stability of [^{18}F]**TA3** in mouse is significantly higher than that of [^{18}F]**TA4**. Thus, further *in vivo* studies using dynamic PET imaging were performed only with [^{18}F]**TA3**.

For PET/MR baseline studies of [^{18}F]**TA3** in anaesthetized CD-1 mice, the radioligand was injected intravenously (9.7 ± 1.3 MBq, SA$_{EOS}$ ~70 GBq/µmol, n = 4), and whole body scans were performed for

60 min in listmode with a Mediso nanoScan PET/MR scanner followed by dynamic reconstruction. Time-activity curves (TACs) were generated for regions of interest such as whole brain, striatum, and cerebellum (Figure 8).

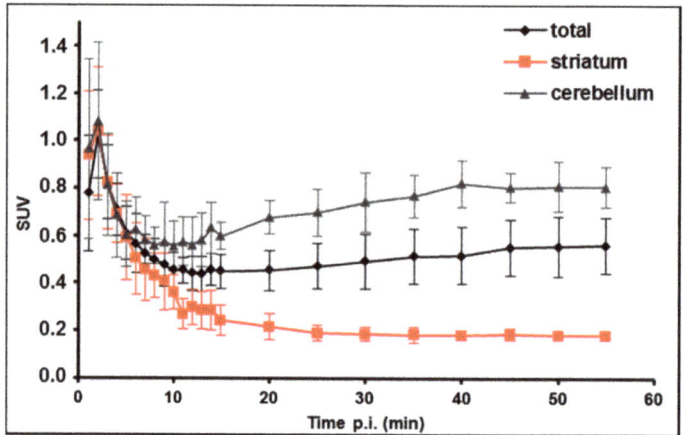

Figure 8. Averaged time-activity curves of [^{18}F]**TA3** in CD-1 mice ($n = 4$) with standard uptake values (SUV) in whole brain (total), striatum, and cerebellum.

As presented in Figure 8, the TACs between 0 and 55 min p.i. show a fast wash out of activity from the striatum while a constantly increased uptake in the cerebellum was observed. This finding is not consistent with the distribution of PDE2A protein in murine brain which corresponds the *in vitro* autoradiography of [^{18}F]**TA3** in rat brain slices (see Figure 6), but likely reflects the accumulation of the radiometabolite [^{18}F]**M2** capable of penetrating the blood-brain barrier as detected by RP-HPLC analysis of extracted brain samples (see Figure 7). With regard to that, we assume that the constantly increasing uptake of activity in the cerebellum as a non-target region of a PDE2A radioligand, which is in contrast to the fast wash out of activity from the target region striatum, indicates specific binding of the structurally not analyzed radiometabolite [^{18}F]**M2** of [^{18}F]**TA3** to an unknown target in this brain region.

3. Experimental Section

3.1. General Information

Chemicals were purchased from standard commercial sources in analytical grade and were used without further purification. Radio-/TLCs were performed on pre-coated silica gel plates (Alugram® Xtra SIL G/UV$_{254}$; Polygram® SIL G/UV$_{254}$, Roth, Karlsruhe, Germany). The compounds were localized at 254 nm (UV lamp) and/or by staining with aqueous KMnO$_4$ solution or ninhydrin solution. Radio-TLC was recorded using a bioimaging analyzer system (BAS-1800 II, Fuji Photo Film, Co. Ltd., Tokyo, Japan) and images were evaluated with Aida 2.31 software (raytest Isotopenmessgeräte GmbH, Straubenhardt, Germany). Column chromatography was conducted on silica gel (0.06–0.20 mm, Roth). HPLC separations were performed on JASCO systems equipped with UV detectors from JASCO and activity detectors from raytest Isotopenmessgeräte GmbH (GABI Star, Straubenhardt, Germany).

HPLC columns and conditions were: System **A**, semi-preparative HPLC (column: Reprosil-Pur C18-AQ, 250 × 10 mm, particle size: 10 μm; eluent: 50% MeCN/20 mM NH$_4$OAc$_{aq.}$; flow: 3 mL/min for [^{18}F]**TA3** or 3.5 mL/min for [^{18}F]**TA4**; ambient temperature; UV detection at 254 nm); system **B**, analytical HPLC (column: Reprosil-Pur C18-AQ, 250 × 4.6 mm, particle size: 5 μm; gradient: 0–10 min: 10% MeCN, 10–35 min: 10% → 90% MeCN, 35–45 min: 90% MeCN, 45–50 min: 90% → 10% MeCN, 50–60 min: 10% MeCN/20 mM NH$_4$OAc$_{aq.}$; isocratic: 44% MeCN/20 mM NH$_4$OAc$_{aq.}$; flow: 1 mL/min; ambient temperature; UV detection at 254 nm). The NH$_4$OAc concentration stated as 20 mM NH$_4$OAc$_{aq.}$ corresponds to the concentration in the aqueous component of an eluent mixture.

Specific activity was determined on the base of a calibration curve carried out under isocratic HPLC conditions (44% MeCN/20 mM NH$_4$OAc$_{aq.}$; system **B**) using chromatograms obtained at 270 nm as an appropriate maximum of UV absorbance.

NMR spectra (^{1}H, ^{13}C, ^{19}F) were recorded on Mercury 300/Mercury 400 (Varian, Palo Alto, CA, USA) or Fourier 300/Avance DRX 400 Bruker (Billerica, MA, USA) instruments. The hydrogenated residue of deuteriated solvents and/or tetramethylsilane (TMS) were used as internal standards for ^{1}H-NMR (CDCl$_3$, δ = 7.26; DMSO-d_6, δ = 2.50) and ^{13}C-NMR (CDCl$_3$, δ = 77.2; DMSO-d_6, δ = 39.5). The chemical shifts (δ) are reported in ppm (s, singlet; d, doublet; t, triplet; qui, quintet; m, multiplet) and the corresponding coupling constants (*J*) are reported in Hz. High resolution mass spectra (ESI +/−) were recorded on an Esquire 300Plus instrument (Bruker; equipped with ion trap). No-carrier-added (n.c.a.) [^{18}F]fluoride ($t_{1/2}$ = 109.8 min) was produced via the [^{18}O(p,n)^{18}F] nuclear reaction by irradiation of [^{18}O]H$_2$O (Hyox 18 enriched water, Rotem Industries Ltd, Beer-Sheba, Israel) on a Cyclone®18/9 (iba RadioPharma Solutions, Louvain-la-Neuve, Belgium) with fixed energy proton beam using Nirta® [^{18}F]fluoride XL target.

3.2. Syntheses

The synthesis of the lead compound **TA1** is already reported [38] and involves five steps, which were partly optimized. Furthermore, the patent does not provide NMR data of the published compounds **1**–**4** or **TA1**, which are reported herein.

6-Methoxy-2-(4-methyl-1H-imidazol-1-yl)-3-nitropyridine (**1**) [38]. To 4.02 g (1.6 eq.) 4-methylimidazole dissolved in CHCl$_3$ (8 mL, instead of DMF [38]) a solution of 4-(dimethylamino) pyridine (DMAP, 0.37 g, 10 mol %) in CHCl$_3$ (2 mL) was added dropwise. The reaction mixture was stirred at 0 °C for 10 min followed by the addition of triethylamine (TEA, 12.73 mL, 3 eq.; instead of solid KOH [38]) and stirring at 0 °C for further 10 min. To this reaction mixture was added dropwise 2-chloro-6-methoxy-3-nitropyridine (5.77 g, 1 eq.) in CHCl$_3$ (14 mL). After 30 min at 0 °C the mixture was stirred at ambient temperature overnight. The mixture was washed twice with water and aq. NaCl saturated solution (20 mL). The aqueous phase was extracted with CHCl$_3$ (20 mL). The organic phase was dried over Na$_2$SO$_4$ and filtered. Evaporation of the solvent and subsequent purification by column chromatography (EtOAc/ *n*-hexane, 1:2, *v/v*) afforded **1** as a yellow solid (6.65 g, 93%). ^{1}H-NMR (400 MHz, CDCl$_3$): δ (ppm) = 2.27 (s, 3H); 4.03 (s, 3H); 4.75 (s, 2H); 6.78 (d, ^3J = 8.8, 1H); 6.90 (dd, ^4J = 2.0, ^4J = 1.2, 1H); 7.94 (d, ^4J = 1.2, 1H); 8.23 (d, ^3J = 8.8, 1H).

6-Methoxy-2-(4-methyl-1H-imidazol-1-yl)pyridin-3-amine (**2**) [38]. Compound **1** (6.62 g, 1 eq.) was dissolved in absolute EtOH (180 mL) and after addition of palladium on charcoal (0.20 g, 10%) in absolute EtOH (20 mL) the reaction mixture was hydrogenated under pressure (1.5–2 bar instead of 10 to 15 bar [38]) at ambient temperature (instead of 40 °C [38]). Afterwards the mixture was stirred overnight and filtered over kieselguhr (suspended in EtOH). The red brown filtrate was concentrated under reduced pressure and the residue was treated with cold methyl *tert*-butyl ether (MTBE, 20 mL). The mixture was stirred at 0 °C for 30 min and stored at 4 °C (fridge) overnight to precipitate the product. The light brown solid was filtered and dried in a desiccator under vacuum for two days. This procedure was repeated twice with the filtrate and the dry precipitates were combined to afford 4.54 g (79%) of **2**. ^{1}H-NMR (300 MHz, DMSO-d_6): δ (ppm) = 2.17 (s, 3H); 3.74 (s, 3H); 4.75 (s, 2H); 6.67 (d, ^3J = 8.4, 1H); 7.29 (m, 1H); 7.35 (d, ^3J = 8.4, 1H); 7.96 (d, ^4J = 1.2, 1H).

2-Methoxy-7-methylimidazo[5,1-c]pyrido[2,3-e][1,2,4]triazine (**3**) [38]. Under ice bath cooling, compound **2** (100 mg, 1 eq.) was dissolved in CH$_3$COOH (30 mL) followed by addition of aq. NaNO$_2$ solution (51 mg, 1.5 eq. in 500 μL water) and water (2 mL). Precipitation of a yellow solid started immediately. The reaction mixture was stirred for 30 min (instead of 1–2 h [38]) and the precipitate was filtered, washed with ethyl acetate and water, and dried in a desiccator under vacuum for three days (instead of re-crystallization from *iso*-propanol or column chromatography [38]). 101 mg (96%) of **3** as a yellow solid were obtained. ^{1}H-NMR (400 MHz, DMSO-d_6): δ (ppm) = 2.77 (s, 3H); 4.13 (s, 3H); 7.21 (d, ^3J = 8.8, 1H); 8.72 (d, ^3J = 8.8, 1H); 8.97 (s, 1H).

9-Bromo-2-methoxy-7-methylimidazo[5,1-c]pyrido[2,3-e][1,2,4]triazine (**4**) [38]. A solution of compound **3** (0.74 g, 1 eq.; instead of 4-methoxy-7-methyl-imidazo[5,1-*c*]pyrido[2,3-e][1,2,4]triazine [38]) in CH$_2$Cl$_2$ (50 mL; instead of MeCN [38]) was stirred for 10 min under ice bath cooling. A suspension of *N*-bromosuccinimide (NBS, 0.92 g, 1.5 eq.) in CH$_2$Cl$_2$ (10 mL) was added dropwise and the round bottom flask was covered using aluminum foil. The reaction was stirred in an ice bath for 5 h and overnight at ambient temperature (instead of stirring at ambient temperature for 16 h [38]). The mixture was washed once with aq. saturated solutions of Na$_2$SO$_3$, NaHCO$_3$, and NaCl and water (20 mL each). The aqueous phase was extracted with CH$_2$Cl$_2$ (20 mL). The organic phase was dried over Na$_2$SO$_4$ and filtered. Evaporation of the solvent and subsequent purification by column chromatography (EtOAc/CH$_2$Cl$_2$, 1:6 to 1:4, *v/v*) afforded a yellow solid of **4** (0.74 g, 73%). ^{1}H-NMR (300 MHz, CDCl$_3$): δ (ppm) = 2.85 (s, 3H); 4.19 (s, 3H); 7.04 (d, ^3J = 8.8, 1H); 8.57 (d, ^3J = 8.8, 1H).

9-(5-Butoxy-2-fluorophenyl)-2-methoxy-7-methylimidazo[5,1-c]pyrido[2,3-e][1,2,4]triazine (**TA1**) [38]. The synthesis of the lead compound **TA1** was performed as described in the General Suzuki Coupling Procedure [38]. Subsequent evaporation of the solvent at 60 °C and purification by column chromatography (EtOAc/CH$_2$Cl$_2$, 1:6, *v/v*) afforded a yellow solid of **TA1** (0.34 g, 81%). ^{1}H-NMR (300 MHz, DMSO-d_6): δ (ppm) = 0.90 (t, ^3J = 7.4, 3H); 1.35–1.50 (m, 2H); 1.60–1.75 (m, 2H); 2.79 (s, 3H); 3.64 (s, 3H); 3.98 (t, ^3J = 6.4, 2H); 7.07–7.14 (m, 1H); 7.10 (d, ^3J = 8.7, 1H); 7.21 (dd, ^4J = 5.7, ^5J = 3.1, 1H); 7.29 (t, ^3J = 9.2, 1H); 8.64 (d, ^3J = 8.7, 1H). ^{13}C-NMR (75 MHz, DMSO-d_6): δ (ppm) = 12.3 (s); 13.6 (s); 18.7 (s); 30.7 (s); 54.2 (s); 68.0 (s); 112.1 (s); 115.9 (d, ^2J = 22.5); 117.3 (d, ^3J = 8.3); 117.5 (d, ^3J = 1.5); 120.2 (d, ^2J = 16.5); 127.8 (s); 131.6 (s); 133.5 (s); 136.6 (s); 138.9 (s); 140.6 (s); 154.2 (d, ^4J = 2.3);

154.8 (d, ^1J = 238.5); 163.8 (s). ^{19}F-NMR (282 MHz, DMSO-d_6): δ (ppm) = −121.7 (ddd, ^1J = 9.2, ^2J = 5.7, ^2J = 4.3). HR-MS (ESI): 382.17 [M+H]$^+$(Lit.: 382 [38]).

4-Fluoro-3-(2-methoxy-7-methylimidazo[5,1-c]pyrido[2,3-e][1,2,4]triazin-9-yl)phenol (**TA1a**) A solution of compound **TA1** (1.00 g, 1 eq.) in dry CH$_2$Cl$_2$ (50 mL) was placed in a 100 mL twin-neck flask under ice bath cooling and argon atmosphere. Afterwards boron tribromide (8 mL, 3.05 eq.) in CH$_2$Cl$_2$ (1 M solution) was added dropwise over a period of 45 min whereby the color of the mixture changed from yellow to dark red. The reaction mixture was stirred at 0 °C (ice bath) for 1 h and then warmed up to ambient temperature. The mixture was quenched by adding it dropwise to ice water (100 mL) and the organic phase was washed twice with water (10 mL each). The aqueous phase was extracted twice with CH$_2$Cl$_2$ (5 mL each). The combined organic phases were dried over Na$_2$SO$_4$ and filtered. Evaporation of the solvent and subsequent purification by column chromatography (EtOAc/CH$_2$Cl$_2$, 1:6 to 100% EtOAc, *v/v*) afforded a yellow solid of **TA1a** (0.84 g, 98%). ^{1}H-NMR (300 MHz, DMSO-d_6): δ (ppm) = 2.82 (s, 3H); 3.52 (s, 3H); 6.92 (ddd, ^3J = 9.0, ^4J = 4.2, ^4J = 3.1, 1H); 7.04 (dd, ^4J = 5.7, ^4J = 3.1, 1H); 7.16 (d, ^3J = 9.0, 1H); 7.19 (t, ^3J = 9.0, 1H); 8.71 (d, ^3J = 9.0, 1H); 9.65 (s, 1H).

3.2.1. General Procedure for the Preparation of Fluoroalkylated Triazine Derivatives **TA2–4**

To a solution of compound **TA1a** (1 eq.) in MeCN (10 mL), K$_2$CO$_3$ (3 eq.) and the fluoroalkylating agent (1.5 eq.) were added. The yellow suspension was stirred at 70–80 °C for 5 h and at ambient temperature overnight. The mixture was filtered and the solvent was evaporated. The residue was dissolved in CH$_2$Cl$_2$ (10 mL), washed with water (5 mL) and then with citric acid (5 mL, 25%). The aqueous phase was extracted with CH$_2$Cl$_2$ (2 mL). The combined organic phases were dried over Na$_2$SO$_4$, filtered and evaporation of the solvent yielded the crude fluoroalkylated triazine derivative.

9-(2-Fluoro-5-(2-fluoroethoxy)phenyl)-2-methoxy-7-methylimidazo[5,1-c]pyrido[2,3-e] [1,2,4]triazine (**TA2**) The fluoroalkylation procedure described above was used with 100 mg of **TA1a** and 38 µL of 1-fluoro-2-iodoethane and afforded after column chromatography (EtOAc/CH$_2$Cl$_2$, 1:4 to 1:1, *v/v*) a yellow solid of **TA2** (60 mg, 54%). ^{1}H-NMR (300 MHz, DMSO-d_6): δ (ppm) = 2.82 (s, 3H); 3.48 (s, 3H); 4.23 (t, ^2J = 7.5, ^3J = 30.3, ^3J = 3.9, 1H); 4.33 (t, ^2J = 7.8, ^3J = 30.3, ^3J = 3.9, 1H) 4.66 (t, ^2J = 7.8, ^2J = 48.0, ^3J = 3.9, 1H), 4.82 (t, ^2J = 7.5, ^2J = 48.0, ^3J = 3.9, 1H); 7.15 (d, ^3J = 9.0, 1H); 7.18–7.22 (m, 1H); 7.26–7.31 (m, 1H); 7.35 (t, ^3J = 9.3, 1H); 8.69 (d, ^3J = 9.0, 1H). ^{13}C-NMR (75 MHz, DMSO-d_6): δ (ppm) = 12.4 (s); 54.2 (s); 54.2 (s); 67.9 (d, ^3J = 18.8); 82.1 (d, ^1J = 165.8); 112.2 (s); 116.1 (d, ^2J = 23.0); 117.57 (s); 117.7 (d, ^3J = 2.0); 120.4 (d, ^2J = 16.4); 127.9 (s); 131.4 (s); 133.6 (s); 136.6 (s); 139.0 (s); 140.7 (s); 153.7 (d, ^4J = 1.9); 155.1 (d, ^1J = 239.3); 163.9 (s). ^{19}F-NMR (282 MHz, DMSO-d_6): δ (ppm) = −121.3 to −121.5 (m); -222.6 (tt, ^2J = 48.0, ^3J = 30.3). HR-MS (ESI): 372.13 [M+H]$^+$.

9-(2-Fluoro-5-(2-fluoropropoxy)phenyl)-2-methoxy-7-methylimidazo[5,1-c]pyrido[2,3-e][1,2,4]- triazine (**TA3**) The fluoroalkylation procedure described above was performed with 140 mg of **TA1a** and 66 µL of 1-fluoro-3-iodopropane and afforded after column chromatography (EtOAc/CH$_2$Cl$_2$, 1:4, *v/v*) a yellow solid of **TA3** (120 mg, 75%). ^{1}H-NMR (400 MHz, DMSO-d_6): δ (ppm) = 2.10 (dqui, ^2J = 12.2, ^3J = 25.6, ^3J = 6.0, 2H); 2.81 (s, 3H); 3.47 (s, 3H); 4.10 (t, ^2J = 12.4, ^3J = 6.0, 2H); 4.55 (t, ^2J = 12.0, ^2J = 47.2, ^3J = 6.0, 1H), 4.67 (t, ^2J = 11.6, ^2J = 47.2, ^3J = 6.0, 1H); 7.13 (d, ^3J = 8.8, 1H);

7.13–7.18 (m, 1H);7.26 (dd, ^4J = 5.7, ^5J = 3.2, 1H); 7.32 (t, ^3J = 9.2, 1H); 8.67 (d, ^3J = 8.8, 1H). ^{13}C-NMR (101 MHz, DMSO-d_6): δ (ppm) = 12.4 (s); 29.7 (d, ^2J = 19.7); 54.2 (s); 64.4 (s); 80.8 (d, ^1J = 162.3); 112.1 (s); 116.0 (d, ^2J = 23.1); 117.4 (d, ^3J = 8.1); 117.7 (d, ^3J = 1.9); 120.3 (d, ^2J = 16.5); 127.8 (s); 131.5 (s); 133.6 (s); 136.6 (s); 138.9 (s); 140.6 (s); 154.0 (d, ^4J = 2.0); 155.0 (d, ^1J = 241.3); 163.8 (s). ^{19}F-NMR (376 MHz, DMSO-d_6): δ (ppm) = −121.5 to −121.6 (m); -220.9 (tt, ^2J = 47.2, ^3J = 25.6). HR-MS (ESI): 386.14 [M+H]$^+$.

9-(2-Fluoro-5-(2-fluorobutoxy)phenyl)-2-methoxy-7-methylimidazo[5,1-c]pyrido[2,3-e] [1,2,4]triazine (**TA4**) The fluoroalkylation procedure described above was performed with 100 mg of **TA1a** and 50 μL of 1-bromo-4-fluorobutane and afforded after column chromatography (EtOAc/CH$_2$Cl$_2$, 1:4, v/v) a yellow solid of **TA4** (120 mg, 99%). ^{1}H-NMR (400 MHz, DMSO-d_6): δ (ppm) = 1.70–1.86 (m, 4H); 2.80 (s, 3H); 3.46 (s, 3H); 4.01 (t, ^2J = 10.4, ^3J = 6.0, 2H); 4.42 (t, ^2J = 11.6, ^2J = 47.6, ^3J = 6.0, 1H), 4.54 (t, ^2J = 11.6, ^2J = 47.6, ^3J = 5.6, 1H); 7.08–7.16 (m, 1H); 7.11 (d, ^3J = 8.8, 1H); 7.22 (dd, ^4J = 5.6, ^5J = 3.2, 1H); 7.29 (t, ^3J = 9.2, 1H); 8.66 (d, ^3J = 8.8, 1H).

3.2.2. General Procedure for the Preparation of alkyl bis(4-methylbenzenesulfonates)

The synthesis of various alkyl bis(4-methylbenzenesulfonates) is already reported [48] and was performed according to the literature. A solution of propane-1,3-diol or butane-1,4-diol (1 eq.) in CH$_2$Cl$_2$ (5 mL) was stirred at 0 °C (ice bath) for 10 min. The round bottom flask was covered using aluminium foil and 4 eq. of triethylamine (TEA) were added dropwise. Afterwards a solution of 4-methylbenzenesulfonyl chloride (2 eq.) in CH$_2$Cl$_2$ (10 mL) was added over a period of 15–30 min followed by the addition of 10 mol % of DMAP. The reaction mixture was stirred each for 1 h at 0 °C and at ambient temperature. The white precipitate (triethylamine hydrochloride) was filtered off. The light yellow filtrate was washed once with each 1 M HCl, aq. saturated solution of NaHCO$_3$, water, and aq. saturated solution of NaCl. The aqueous phase was extracted once with CH$_2$Cl$_2$. The organic phase was dried over Na$_2$SO$_4$ and filtered. Evaporation of the solvent and subsequent purification by column chromatography (EtOAc/n-hexanes, 1:3 to 1:1, v/v) afforded a white solid (≥60% of each: propane-1,3-diyl bis(4-methylbenzenesulfonate) and butane-1,4-diyl bis(4-methyl-benzenesulfonate)), identified by ^{1}H-NMR.

3.2.3. General Procedure for the Preparation of Tosylate Precursors **TA3a** and **TA4a**

To a solution of compound **TA1a** (1 eq.) in MeCN (10 mL), K$_2$CO$_3$ (4 eq.) and the alkyl bis(4-methylbenzenesulfonate) (2 eq.) were added. The yellow suspension was stirred at 60–70 °C for 5 h and at ambient temperature overnight. In the case of an incomplete reaction (TLC) the mixture was heated to 60–70 °C for 5 h again and stirred at ambient temperature overnight. The mixture was filtered and the solvent was evaporated. The residue was resolved in CH$_2$Cl$_2$ (10 mL), washed with water (5 mL) and then with citric acid (5 mL, 25%). The aqueous phase was extracted with CH$_2$Cl$_2$ (2 mL). The combined organic phases were dried over Na$_2$SO$_4$, filtered and evaporation of the solvent yielded the crude tosylate precursors.

3-(4-Fluoro-3-(2-methoxy-7-methylimidazo[5,1-c]pyrido[2,3-e][1,2,4]triazin-9-yl)phenoxy)propyl 4-methylbenzenesulfonate (**TA3a**) The procedure for tosylate precursors described above was performed with 170 mg of **TA1a** and 410 mg of propane-1,3-diyl bis(4-methylbenzenesulfonate), and afforded after column chromatography (EtOAc/CH$_2$Cl$_2$, 1:6 to 1:4, v/v) a yellow solid of **TA3a** (170 mg, 60%). ^{1}H-NMR (400 MHz, DMSO-d_6): δ (ppm) = 2.03 (qui, ^2J = 24.4, ^3J = 6.0, 2H); 2.34 (s, 3H); 2.83 (s, 3H); 3.46 (s, 3H); 3.97 (t, ^2J = 12.0, ^3J = 6.0, 2H); 4.19 (t, ^2J = 12.0, ^3J = 6.0, 1H); 7.02–7.07 (m, 1H); 7.13–7.17 (m, 1H); 7.16 (d, ^3J = 8.8, 1H); 7.31 (t, ^3J = 9.2, 1H); 7.41 (d, ^3J = 8.4, 2H); 7.76 (dd, ^3J = 8.4, ^4J = 1.6, 2H); 8.71 (d, ^3J = 8.8, 1H). ^{13}C-NMR (101 MHz, DMSO-d_6): δ (ppm) = 12.4 (s); 21.0 (s); 28.1 (s); 54.2 (s); 64.1 (s); 67.6 (s); 112.2 (s); 116.0 (d, ^2J = 23.3); 117.3 (d, ^3J = 8.3); 117.6 (d, ^3J = 1.7); 120.3 (d, ^2J = 16.4); 127.5 (s); 127.9 (s); 130.1 (s); 131.5 (s); 132.2 (s); 133.6 (s); 136.6 (s); 139.0 (s); 140.7 (s); 153.8 (d, ^4J = 2.0); 155.0 (d, ^1J = 241.4); 163.9 (s). ^{19}F-NMR (376 MHz, DMSO-d_6): δ (ppm) = −121.5 to −121.6 (m). HR-MS (ESI): 538.16 [M+H]$^+$.

4-(4-Fluoro-3-(2-methoxy-7-methylimidazo[5,1-c]pyrido[2,3-e][1,2,4]triazin-9-yl)phenoxy)butyl 4-methylbenzenesulfonate (**TA4a**) The procedure for tosylate precursors described above was performed with 200 mg of **TA1a** and 490 mg of butane-1,4-diyl bis(4-methylbenzenesulfonate), and afforded after column chromatography (EtOAc/CH$_2$Cl$_2$, 1:6 to 1:4, v/v) an orange to red solid of **TA4a** (220 mg, 65%). ^{1}H-NMR (300 MHz, DMSO-d_6): δ (ppm) = 1.62–1.80 (m, 4H); 2.38 (s, 3H); 2.82 (s, 3H); 3.46 (s, 3H); 3.93 (t, ^2J = 11.7, ^3J = 5.8, 2H); 4.09 (t, ^2J = 11.7, ^3J = 5.9, 2H); 7.04–7.11 (m, 1H); 7.15 (d, ^3J = 9.0, 1H); 7.19 (dd, ^4J = 5.7, ^5J = 3.1, 1H); 7.31 (t, ^3J = 9.2, 1H); 7.45 (dd, ^3J = 8.4, ^4J = 0.6, 2H); 7.78 (dd, ^3J = 8.4, ^4J = 0.6, 2H); 8.69 (d, ^3J = 9.0, 1H). ^{13}C-NMR (75 MHz, DMSO-d_6): δ (ppm) = 13.0 (s); 21.72 (s); 25.3 (s); 25.8 (s); 54.9 (s); 68.2 (s); 71.3 (s); 112.9 (s); 116.7 (d, ^2J = 23.0); 118.0–118.4 (m); 121.0 (d, ^2J = 16.4); 128.2 (s); 128.6 (s); 130.8 (s); 132.2 (s); 133.2 (s); 134.3 (s); 137.3 (s); 139.7 (s); 141.4 (s); 145.5 (s); 155.6 (d, ^1J = 238.6); 154.7 (d, ^4J = 1.9); 164.6 (s). ^{19}F-NMR (282 MHz, DMSO-d_6): δ (ppm) = −121.7 (ddd, ^1J = 9.2, ^2J = 5.7, ^2J = 4.3). HR-MS (ESI): 574.15 [M+Na]$^+$.

3.3. In Vitro PDE2A Affinity Assay

The inhibitory potencies of **TA1–4** for human recombinant PDE2A and PDE10A proteins were determined by BioCrea GmbH (Radebeul, Germany) [38].

3.4. Radiochemistry

3.4.1. Manual Radiosyntheses of [^{18}F]**TA3** and [^{18}F]**TA4**

No-carrier-added [^{18}F]fluoride was trapped on a Chromafix® 30 PS-HCO$_3^-$ cartridge (MACHEREY-NAGEL GmbH & Co. KG, Düren, Germany). The activity was eluted with 300 μL of an aqueous solution of K$_2$CO$_3$ (1.78 mg, 12.9 μmol) into a 4 mL V-vial and Kryptofix 2.2.2 (K$_{2.2.2}$, 11.2 mg, 29.7 μmol) in 1 mL MeCN was added. The aqueous [^{18}F]fluoride was azeotropically dried under vacuum and nitrogen flow within 7–10 min using a Discover PETwave Microwave CEM® (75 W, 50–60 °C, power cycling mode). Two aliquots of MeCN (2 × 1.0 mL) were added during the drying procedure and the final complex was dissolved in 500 μL MeCN ready for radiolabelling. The reactivity of the anhydrous K$^+$/[^{18}F]F$^-$/K$_{2.2.2}$-carbonate complex as well as the reproducibility of the drying procedure were checked via the standard reaction with 2 mg (5.4 µmol) of ethylene glycol ditosylate (Sigma-Aldrich, Munich, Germany) at 80 °C for 10 min in MeCN.

Optimization of the aliphatic radiolabelling procedures of the tosylates **TA3a** or **TA4a** was performed by varying the amount of precursor (1–3 mg in 500 µL MeCN) and reaction time (up to 20 min) under conventional heating at 80 °C. After 5, 10, 15 and 20 min, aliquots of the reaction mixtures were analyzed by radio-TLC (EtOAc/DCM, 1:1, v/v). The crude reaction mixtures of [^{18}F]**TA3** or [^{18}F]**TA4** were diluted with water (1:1, v/v) and applied to an isocratic semi-preparative HPLC (system **A**) for isolation of the desired radioligands ([^{18}F]**TA3**: t_R = 27–29 min; [^{18}F]**TA4**: t_R = 31–33 min). The collected fractions were analyzed by radio-TLC, diluted with water, passed through a Sep-Pak® C18 Plus light cartridge (Waters, Milford, MA, USA; pre-conditioned with 20 mL of absolute EtOH and 60 mL water), and eluted with 0.75 mL of absolute EtOH. For biological investigations, the solvent was evaporated at 70 °C under a gentle nitrogen stream. Then, the radioligands were formulated in sterile isotonic saline containing 10% EtOH (v/v).

The identities of each radioligand were verified by analytical radio-HPLC (system **B**) of samples of [^{18}F]**TA3** or [^{18}F]**TA4** spiked with the corresponding non-radioactive reference compounds using a gradient and an isocratic method.

3.4.2. Automated Radiosynthesis of [^{18}F]**TA3**

Automated radiosyntheses were performed in a TRACERlab™ FX F-N synthesis module (GE Healthcare, Waukesha, WI, USA) equipped with a PU-980 pump (JASCO, Gross-Umstadt, Germany), a WellChrom K-2001 UV detector (KNAUER GmbH, Berlin, Germany), a NaI(Tl)-counter and automated data acquisition (NINA software version 4.8 rev. 4, Nuclear Interface GmbH, Dortmund, Germany). The conditions for the optimized manual ^{18}F-labelling of the tosylate precursor **TA3a** as well as for the isolation and purification of the radioligand were used for the automated radiosynthesis of [^{18}F]**TA3** and set up as depicted in Scheme 3. No-carrier-added [^{18}F]fluoride was trapped on a Chromafix® 30 PS-HCO$_3^-$ cartridge (entry **1**) in the remotely controlled synthesis module. The activity was eluted with an aqueous solution of K_2CO_3 (1.78 mg/0.4 mL water; entry **2**), mixed with $K_{2.2.2}$ (11.2 mg/1 mL MeCN; entry **3**) into the reaction vessel (entry **6**) and azeotropically dried for approximately 10 min. Thereafter, 1 mg of the tosylate **TA3a** dissolved in 1 mL MeCN (entry **4**) was added, and the reaction mixture was stirred at 80 °C for 15 min. After cooling, the reaction mixture was diluted with 2 mL water (entry **5**) and transferred into the injection vial. Semi-preparative HPLC (system **A**, entry **7**) was performed for isolation of the desired radioligand [^{18}F]**TA3** (t_R = 27–29 min). [^{18}F]**TA3** was collected in entry **8** previously loaded with 40 mL water. Final purification step took place after passing the solution through a Sep-Pak® C18 Plus light cartridge (Waters; pre-conditioned with 20 mL of absolute EtOH and 60 mL water; entry **11**), followed by washing with 2 mL water (entry **9**) and elution of [^{18}F]**TA3** with 1 mL of absolute EtOH (entry **10**) into the product vial (entry **12**).

To obtain an injectable solution, the solvent was reduced under a gentle nitrogen stream at 70 °C and the radioligand was formulated in sterile isotonic saline containing 10% EtOH (v/v). Analytical radio-HPLC (system **B**, gradient and isocratic mode) of the final product spiked with the non-radioactive reference compound **TA3** confirmed the identity of [^{18}F]**TA3**.

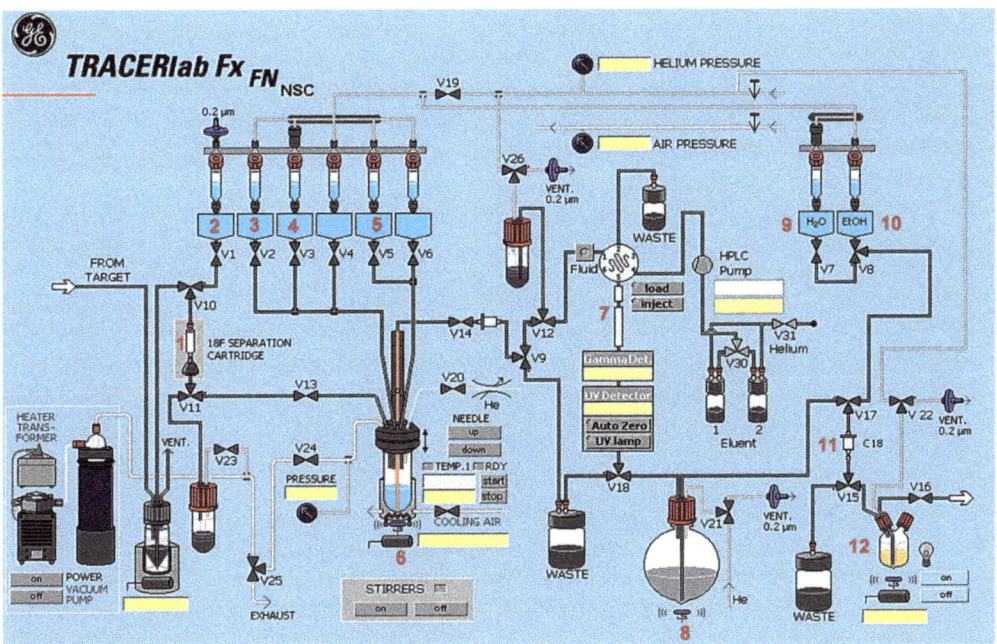

Scheme 3. Tracer Lab™ FX-FN synthesis module for the automated radiosynthesis of [^{18}F]**TA3**: (**1**) Chromafix® 30 PS-HCO$_3^-$ cartridge; (**2**) K$_2$CO$_3$ (1.78 mg in 0.4 mL H$_2$O); (**3**) K$_{2.2.2}$ (11.2 mg in 1 mL MeCN); (**4**) 1 mg **TA3a** in 1 mL MeCN; (**5**) 2 mL H$_2$O; (**6**) reaction vessel; (**7**) Reprosil-Pur C18-AQ (250 × 10 mm; particle size: 10 µm; eluent: 50% MeCN/20 mM NH$_4$OAc$_{aq.}$; flow: 3 mL/min; ambient temperature; UV detection at 254 nm); (**8**) 40 mL H$_2$O; (**9**) 2 mL H$_2$O; (**10**) 1 mL absolute EtOH; (**11**) Sep Pak® C18 light cartridge; (**12**) product vial.

3.5. Determination of Lipophilicity (logD$_{7.4}$) and in Vitro Stability

The lipophilicity of [^{18}F]**TA3** or [^{18}F]**TA4** was determined by partitioning between *n*-octanol and phosphate buffered saline (PBS, pH 7.4) at ambient temperature using the conventional shake-flask method. An aliquot of 10 µL of the formulated solutions of each [^{18}F]**TA3** or [^{18}F]**TA4** with approximately 1 MBq of the radioligand was added to a tube containing 6 mL of the *n*-octanol/PBS-mixture (1:1, *v*/*v*, fourfold determination). The tubes were shaken for 20 min using a mechanical shaker (HS250 basic, IKA Labortechnik GmbH & Co. KG, Staufen, Germany) followed by centrifugation (5000 rpm for 5 min) and separation of the phases. Aliquots of each 1 mL were taken from the organic and the aqueous phase and activity was measured using an automated gamma counter (1480 WIZARD, Fa. Perkin Elmer, Waltham, MA, USA). The distribution coefficient (D) was calculated as [activity (cpm/mL) in *n*-octanol]/[activity (cpm/mL) in PBS, pH 7.4] specified as the decade logarithm (logD$_{7.4}$).

In vitro stabilities of [^{18}F]**TA3** or [^{18}F]**TA4** were investigated by incubation of the radioligands in phosphate buffered saline (PBS, pH 7.4), *n*-octanol and pig plasma at 37 °C for 60 min (~5 MBq of the radioligand added to 500 µL of each medium). Samples were taken at 15, 30 and 60 min after incubation and analyzed by radio-TLC and radio-HPLC (system **B**).

3.6. Animal Studies

All animal procedures were approved by the Animal Care and Use Committee of Saxony (TVV 08/13).

3.6.1. *In Vitro* Autoradiographic Studies in Rat Brain

Frozen sagittal brain sections obtained from female SPRD rats (10–12 weeks old) were thawed, dried in a stream of cold air, and preincubated for 20 min with incubation buffer (50 mM TRIS-HCl, pH 7.4, 120 mM NaCl, 5 mM KCl, 2 mM $CaCl_2$, 5 mM $MgCl_2$) at ambient temperature. Brain sections were incubated with ~1 MBq/mL of [^{18}F]**TA3** or [^{18}F]**TA4** in incubation buffer for 60 min at ambient temperature. Afterwards sections were washed twice with 50 mM TRIS-HCl (pH 7.4) for 2 min at 4 °C, dipped briefly in ice-cold deionized water, dried in a stream of cold air and exposed for 60 min to an ^{18}F-sensitive image plate that was analyzed afterwards using an image plate scanner (HD-CR 35; Duerr NDT GmbH, Bietigheim Bissingen, Germany). Nonspecific binding of the radioligand under investigation was determined by co-incubation with 10 μM **TA1**.

3.6.2. Small-Animal PET/MR Studies in Mice

Female CD-1 mice (n = 4; age: 8 weeks; weight: 28.8 ± 1.2 g; supplier: Medizinisch-Experimentelles Zentrum Leipzig, Leipzig, Germany) were housed under a 12 h:12 h light-dark cycle at 26 °C in a vented animal cabinet. The animals received an injection of 9.7 ± 1.3 MBq of [^{18}F]**TA3** into the tail vein followed by a 60 min PET/MR scan (Mediso nanoScan®, Budapest, Hungary). Each PET image was corrected for random coincidences, dead time, scatter and attenuation (AC), based on a whole body (WB) MR scan.

The reconstruction parameters for the list mode data are: 3D-ordered subset expectation maximization (OSEM), 4 iterations, 6 subsets, energy window: 400–600 keV, coincidence mode: 1–5, ring difference: 81. The mice were positioned prone in a special mouse bed (heated up to 37 °C), with the head fixed to a mouth piece for the anesthetic gas supply with isoflurane in 40% air and 60% oxygen (anesthesia unit: U-410, agnthos, Lidingö, Sweden; Gas blender: MCQ, Rome, Italy). The PET data was collected by a continuous WB scan during the entire investigation. Following the 60 min PET scan a T1 weighted WB gradient echo sequence (GRE, T_R = 20 ms; T_E = 6.4 ms) was performed for AC and anatomical orientation. Image registration and evaluation of the region of interest (ROI) was done with ROVER (ABX advanced biochemical compounds, Radeberg, Germany).

The respective brain regions were identified using the MR information from the GRE scan. The activity data is expressed as mean standardized uptake value (SUV) of the overall ROI.

3.6.3. *In Vivo* Metabolism Studies in Mice

The radioligands [^{18}F]**TA3** or [^{18}F]**TA4** (~60 MBq in 150 μL isotonic saline) were injected via the tail vein in female CD-1 mice (10–12 weeks old). Brain and blood samples were obtained at 30 min p.i., plasma separated by centrifugation (14,000× g, 1 min), and brain homogenized in ~1 mL isotonic saline on ice (10 strokes of a PTFE plunge at 1000 rpm in a borosilicate glass cylinder; Potter S Homogenizer, B. Braun Melsungen AG, Melsungen, Germany).

3.7. Conventional Extraction Procedure

Twofold extractions of plasma (2 × 50 µL, 30 min p.i.) and brain samples (2 × 250 µL, 30 min p.i.) were performed using a mixture of ice-cold acetone/water (8:2, *v/v*; plasma or brain sample/organic solvent, 1:4, *v/v*). The samples were vortexed for 1 min, incubated on ice for 10 min (first extraction) or 5 min (second extraction) and centrifuged at 10,000 rpm for 5 min. Supernatants were collected and the precipitates were re-dissolved in ice-cold acetone/water for the second extraction. Aliquots from supernatants of each extraction step were taken. Activity thereof and of the precipitates was quantified using an automated gamma counter (1480 WIZARD, Fa. Perkin Elmer). The supernatants from both extractions were combined, concentrated under nitrogen stream at 70 °C and analyzed by radio-HPLC (system **B**).

3.8. Micellar Chromatography (MLC)

For preparation of the MLC injection samples, mouse plasma (20–50 µL, 30 min p.i.) was dissolved in 100–300 µL of 200 mM aqueous SDS and injected directly into the MLC system (500 µL sample loop; column: Reprosil-Pur C18-AQ, 250 × 4.6 mm, particle size: 10 µm; gradient: 0–15 min: 3% 1-PrOH, 15–40 min: 3% → 30% 1-PrOH; 40–49 min: 30% 1-PrOH, 49–50 min: 30% → 3% 1-PrOH; 50–60 min: 3% 1-PrOH/100 mM SDS, 10 mM Na_2HPO_4; flow: 1 mL/min; ambient temperature). Notably, a pre-column with 10 mm length was used and frequently exchanged to expand the life time of the RP-column. Homogenized brain material (100–200 µL, 30 min p.i.) was dissolved in 500 µL of 200 mM aqueous SDS, stirred at 75 °C for 5 min and after cooling to ambient temperature injected into the MLC system.

4. Conclusions

In conclusion, the novel radioligands [^{18}F]**TA3** and [^{18}F]**TA4** are well suitable for *in vitro* imaging of PDE2A but further structural modification might prevent the formation of brain penetrating radiometabolites. Thus, these modifications could lead to ^{18}F-radioligands enabling *in vivo* PET imaging of PDE2A in brain.

Furthermore, compared to the conventional extraction procedure combined with analytical RP-HPLC, it could be shown that micellar HPLC is a more reliable method to quantify the amount of non-metabolized [^{18}F]**TA3** and [^{18}F]**TA4** not only in plasma samples but also in brain homogenates.

Acknowledgments

The inhibitory potencies of the triazine lead compound (**TA1** [38]) and the herein presented novel fluoroalkylated derivatives (**TA2–4**) for human recombinant PDE2A and PDE10A proteins were determined by BioCrea GmbH (Radebeul, Germany).

Author Contributions

SS, BW, MS and JS designed and performed organic syntheses; SS, BW, RT, JS and MS designed and performed radiosyntheses; SS, BW, RT, UE, NH, WD-C and PB designed and performed *in vitro* and *in vivo* studies; MK, WD-C and PB designed and performed PET/MR studies; SS, BW, WD-C, MK and PB analyzed the data; SS, BW, RT, WD-C, MK, MS, JS and PB wrote the paper. All authors read and approved the final manuscript.

Conflicts of Interest

The authors declare no conflict of interest.

References

1. Kelly, M.P.; Adamowicz, W.; Bove, S.; Hartman, A.J.; Mariga, A.; Pathak, G.; Reinhart, V.; Romegialli, A.; Kleiman, R.J. Select 3′,5′-cyclic nucleotide phosphodiesterases exhibit altered expression in the aged rodent brain. *Cell Signal.* **2014**, *26*, 383–397.
2. Bender, A.T.; Beavo, J.A. Cyclic nucleotide phosphodiesterases: Molecular regulation to clinical use. *Pharmacol. Rev.* **2006**, *58*, 488–520.
3. Gomez, L.; Breitenbucher, J.G. PDE2 inhibition: Potential for the treatment of cognitive disorders. *Bioorg. Med. Chem. Lett.* **2013**, *23*, 6522–6527.
4. Maurice, D.H.; Ke, H.; Ahmad, F.; Wang, Y.; Chung, J.; Manganiello, V.C. Advances in targeting cyclic nucleotide phosphodiesterases. *Nat. Rev. Drug Discov.* **2014**, *13*, 290–314.
5. Fajardo, A.; Piazza, G.; Tinsley, H. The role of cyclic nucleotide signaling pathways in cancer: Targets for prevention and treatment. *Cancers* **2014**, *6*, 436–458.
6. Lugnier, C. Cyclic nucleotide phosphodiesterase (PDE) superfamily: A new target for the development of specific therapeutic agents. *Pharmacol. Ther.* **2006**, *109*, 366–398.
7. Keravis, T.; Lugnier, C. Cyclic nucleotide phosphodiesterase (PDE) isozymes as targets of the intracellular signalling network: Benefits of PDE inhibitors in various diseases and perspectives for future therapeutic developments. *Br. J. Pharmacol.* **2012**, *165*, 1288–1305.
8. Wang, Z.Z.; Zhang, Y.; Zhang, H.T.; Li, Y.F. Phosphodiesterase: An interface connecting cognitive deficits to neuropsychiatric and neurodegenerative diseases. *Curr. Pharm. Des.* **2015**, *21*, 303–316.
9. Lakics, V.; Karran, E.H.; Boess, F.G. Quantitative comparison of phosphodiesterase mRNA distribution in human brain and peripheral tissues. *Neuropharmacology* **2010**, *59*, 367–374.
10. Van Staveren, W.C.G.; Steinbusch, H.W.M.; Markerink-van Ittersum, M.; Repaske, D.R.; Goy, M.F.; Kotera, J.; Omori, K.; Beavo, J.A.; de Vente, J. MRNA expression patterns of the cGMP-hydrolyzing phosphodiesterases types 2, 5, and 9 during development of the rat brain. *J. Comp. Neurol.* **2003**, *467*, 566–580.
11. Stephenson, D.T.; Coskran, T.M.; Wilhelms, M.B.; Adamowicz, W.O.; O'Donnell, M.M.; Muravnick, K.B.; Menniti, F.S.; Kleiman, R.J.; Morton, D. Immunohistochemical localization of phosphodiesterase 2A in multiple mammalian species. *J. Histochem. Cytochem.* **2009**, *57*, 933–949.
12. Zhang, C.; Yu, Y.; Ruan, L.; Wang, C.; Pan, J.; Klabnik, J.; Lueptow, L.; Zhang, H.T.; O'Donnell, J.M.; Xu, Y. The roles of phosphodiesterase 2 in the central nervous and peripheral systems. *Curr. Pharm. Des.* **2015**, *21*, 274–290.
13. Van Staveren, W.C.G.; Markerink-van Ittersum, M.; Steinbusch, H.W.M.; de Vente, J. The effects of phosphodiesterase inhibition on cyclic GMP and cyclic AMP accumulation in the hippocampus of the rat. *Brain Res.* **2001**, *888*, 275–286.

14. Suvarna, N.U.; O'Donnell, J.M. Hydrolysis of N-methyl-D-aspartate receptor-stimulated cAMP and cGMP by PDE4 and PDE2 phosphodiesterases in primary neuronal cultures of rat cerebral cortex and hippocampus. *J. Pharmacol. Exp. Ther.* **2002**, *302*, 249–256.
15. Blokland, A.; Schreiber, R.; Prickaerts, J. Improving memory: A role for phosphodiesterases. *Curr. Pharm. Des.* **2006**, *12*, 2511–2523.
16. Boess, F.G.; Hendrix, M.; van der Staay, F.J.; Erb, C.; Schreiber, R.; van Staveren, W.; de Vente, J.; Prickaerts, J.; Blokland, A.; Koenig, G. Inhibition of phosphodiesterase 2 increases neuronal cGMP, synaptic plasticity and memory performance. *Neuropharmacology* **2004**, *47*, 1081–1092.
17. Reneerkens, O.A.; Rutten, K.; Bollen, E.; Hage, T.; Blokland, A.; Steinbusch, H.W.; Prickaerts, J. Inhibition of phoshodiesterase type 2 or type 10 reverses object memory deficits induced by scopolamine or MK-801. *Behav. Brain Res.* **2013**, *236*, 16–22.
18. Morita, H.; Murata, T.; Shimizu, K.; Okumura, K.; Inui, M.; Tagawa, T. Characterization of phosphodiesterase 2A in human malignant melanoma PMP cells. *Oncol. Rep.* **2013**, *29*, 1275–1284.
19. Drees, M.; Zimmermann, R.; Eisenbrand, G. 3′,5′-Cyclic nucleotide phosphodiesterase in tumor cells as potential target for tumor growth inhibition. *Cancer Res.* **1993**, *53*, 3058–3061.
20. Abusnina, A.; Keravis, T.; Zhou, Q.; Justiniano, H.; Lobstein, A.; Lugnier, C. Tumour growth inhibition and anti-angiogenic effects using curcumin correspond to combined PDE2 and PDE4 inhibition. *Thromb. Haemost.* **2015**, *113*, 319–328.
21. Bernard, J.J.; Lou, Y.R.; Peng, Q.Y.; Li, T.; Lu, Y.P. PDE2 is a novel target for attenuating tumor formation in a mouse model of UVB-induced skin carcinogenesis. *PLoS ONE* **2014**, *9*, e109862.
22. Podzuweit, T.; Nennstiel, P.; Müller, A. Isozyme selective inhibition of cGMP-stimulated cyclic nucleotide phosphodiesterases by erythro-9-(2-hydroxy-3-nonyl)adenine. *Cell Signal.* **1995**, *7*, 733–738.
23. Biagi, G.; Giorgi, I.; Livi, O.; Pacchini, F.; Rum, P.; Scartoni, V.; Costa, B.; Mazzoni, M.R.; Giusti, L. Erythro- and threo-2-hydroxynonyl substituted 2-phenyladenines and 2-phenyl-8-azaadenines: Ligands for A1 adenosine receptors and adenosine deaminase. *Farmaco* **2002**, *57*, 221–233.
24. Boess, F.G.; Grosser, R.; Hendrix, M.; Koenig, G.; Niewoehner, U.; Schauss, D.; Schlemmer, K.H.; Schreiber, R.; van der Staay, F.J. Novel Substituted Imidazotriazines as PDE II Inhibitors. Patent WO 2002/050078 A1, 27 June 2002.
25. Masood, A.; Huang, Y.; Hajjhussein, H.; Xiao, L.; Li, H.; Wang, W.; Hamza, A.; Zhan, C.G.; O'Donnell, J.M. Anxiolytic effects of phosphodiesterase-2 inhibitors associated with increased cGMP signaling. *J. Pharmacol. Exp. Ther.* **2009**, *331*, 690–699.
26. Xu, Y.; Pan, J.; Chen, L.; Zhang, C.; Sun, J.; Li, J.; Nguyen, L.; Nair, N.; Zhang, H.; O'Donnell, J.M. Phosphodiesterase-2 inhibitor reverses corticosterone-induced neurotoxicity and related behavioural changes via cGMP/PKG dependent pathway. *Int. J. Neuropsychopharmacol.* **2013**, *16*, 835–847.
27. Domek-Łopacińska, K.; Strosznajder, J.B. The effect of selective inhibition of cyclic GMP hydrolyzing phosphodiesterases 2 and 5 on learning and memory processes and nitric oxide synthase activity in brain during aging. *Brain Res.* **2008**, *1216*, 68–77.
28. Rutten, K.; Prickaerts, J.; Hendrix, M.; van der Staay, F.J.; Şik, A.; Blokland, A. Time-dependent involvement of cAMP and cGMP in consolidation of object memory: Studies using selective phosphodiesterase type 2, 4 and 5 inhibitors. *Eur. J. Pharmacol.* **2007**, *558*, 107–112.

29. Abarghaz, M.; Biondi, S.; Duranton, J.; Limanton, E.; Mondadori, C.; Wagner, P. Benzo[1,4]diazepin-2-one Derivatives as Phosphodiesterase PDE2 Inhibitors, Preparation and Therapeutic Use Thereof. Patent WO 2005/063723 A1, 14 July 2005.
30. Dost, R.; Egerland, U.; Grunwald, C.; Höfgen, N.; Langen, B.; Lankau, H.J.; Stange, H. (1,2,4)Triazolo[4,3-*a*]quinoxaline Derivatives as Inhibitors of Phosphodiesterases. Patent WO 2012/104293 A1, 09 August 2012.
31. Andres, J.I.; Buijnsters, P.; de Angelis, M.; Langlois, X.; Rombouts, F.; Trabanco, A.A.; Vanhoof, G. Discovery of a new series of [1,2,4]triazolo[4,3-*a*]quinoxalines as dual phosphodiesterase 2/phosphodiesterase 10 (PDE2/PDE10) inhibitors. *Bioorg. Med. Chem. Lett.* **2013**, *23*, 785–790.
32. Helal, C.J.; Chappie, T.A.; Humphrey, J.M.; Verhoest, P.R.; Yang, E. Imidazo[5,1-*f*][1,2,4]triazines for the Treatment of Neurological Disorders. U.S. Patent WO 2012/114222 A1, 23 August 2012.
33. Schmidt, B.; Weinbrenner, S.; Flockerzi, D.; Kuelzer, R.; Tenor, H.; Kley, H.P. Triazolophthalazines. Patent WO 2006/024640 A2, 09 March 2006.
34. Flockerzi, D.; Kley, H.P.; Kuelzer, R.; Schmidt, B.; Tenor, H.; Weinbrenner, S. Triazolophthalazines as PDE2-inhibitors. Patent WO 2006/072612 A2, 13 July 2006.
35. De Leon, P.; Egbertson, M.; Hills, I.D.; Johnson, A.W.; Machacek, M. Quinolinone PDE2 Inhibitors. U.S. Patent WO 2011/011312 A1, 27 January 2011.
36. Andres, J.I.; Rombouts, F.J.R.; Trabanco, A.A.; Vanhoof, G.C.P.; de Angelis, M.; Buijnsters, P.J.J.A.; Guillemont, J.E.G.; Bormans, G.M.R.; Celen, S.J.L. 1-Aryl-4-methyl-[1,2,4]triazolo[4,3-*a*]quinoxaline Derivatives. Patent WO 2013/000924 A1, 03 January 2013.
37. Zhang, L.; Villalobos, A.; Beck, E.M.; Bocan, T.; Chappie, T.A.; Chen, L.; Grimwood, S.; Heck, S.D.; Helal, C.J.; Hou, X.; *et al.* Design and selection parameters to accelerate the discovery of novel central nervous system positron emission tomography (PET) ligands and their application in the development of a novel phosphodiesterase 2A PET ligand. *J. Med. Chem.* **2013**, *56*, 4568–4579.
38. Stange, H.; Langen, B.; Egerland, U.; Hoefgen, N.; Priebs, M.; Malamas, M.S.; Erdel, J.J.; Ni, Y. Triazine Derivatives as Inhibitors of Phosphodiesterases. Patent WO 2010/054253 A1, 14 May 2010.
39. Brust, P.; van den Hoff, J.; Steinbach, J. Development of ^{18}F-labeled radiotracers for neuroreceptor imaging with positron emission tomography. *Neurosci. Bull.* **2014**, *30*, 777–811.
40. Malamas, M.S.; Ni, Y.; Erdei, J.; Stange, H.; Schindler, R.; Lankau, H.J.; Grunwald, C.; Fan, K.Y.; Parris, K.; Langen, B.; *et al.* Highly potent, selective, and orally active phosphodiesterase 10A inhibitors. *J. Med. Chem.* **2011**, *54*, 7621–7638.
41. Funke, U.; Deuther-Conrad, W.; Schwan, G.; Maisonial, A.; Scheunemann, M.; Fischer, S.; Hiller, A.; Briel, D.; Brust, P. Radiosynthesis and radiotracer properties of a 7-(2-[^{18}F]fluoroethoxy)-6-methoxypyrrolidinylquinazoline for imaging of phosphodiesterase 10A with PET. *Pharmaceuticals* **2012**, *5*, 169–188.
42. Zhu, J.; Yang, Q.; Dai, D.; Huang, Q. X-ray crystal structure of phosphodiesterase 2 in complex with a highly selective, nanomolar inhibitor reveals a binding-induced pocket important for selectivity. *J. Am. Chem. Soc.* **2013**, *135*, 11708–11711.
43. Blom, E.; Karimi, F.; Eriksson, O.; Hall, H.; Långström, B. Synthesis and *in vitro* evaluation of ^{18}F-β-carboline alkaloids as PET ligands. *J. Label. Compd. Rad.* **2008**, *51*, 277–282.

44. Nakao, R.; Schou, M.; Halldin, C. Direct plasma metabolite analysis of positron emission tomography radioligands by micellar liquid chromatography with radiometric detection. *Anal. Chem.* **2012**, *84*, 3222–3230.
45. Rambla-Alegre, M. Basic principles of MLC. *Chromatogr. Res. Int.* **2012**, *2012*, doi:10.1155/2012/898520.
46. Zoghbi, S.S.; Shetty, H.U.; Ichise, M.; Fujita, M.; Imaizumi, M.; Liow, J.S.; Shah, J.; Musachio, J.L.; Pike, V.W.; Innis, R.B. PET imaging of the dopamine transporter with ^{18}F-FECNT: A polar radiometabolite confounds brain radioligand measurements. *J. Nucl. Med.* **2006**, *47*, 520–527.
47. Evens, N.; Vandeputte, C.; Muccioli, G.G.; Lambert, D.M.; Baekelandt, V.; Verbruggen, A.M.; Debyser, Z.; van Laere, K.; Bormans, G.M. Synthesis, *in vitro* and *in vivo* evaluation of fluorine-18 labelled FE-GW405833 as a PET tracer for type 2 cannabinoid receptor imaging. *Bioorg. Med. Chem.* **2011**, *19*, 4499–4505.
48. Burns, D.H.; Chan, H.K.; Miller, J.D.; Jayne, C.L.; Eichhorn, D.M. Synthesis, modification, and characterization of a family of homologues of exo-calix[4]arene: Exo-[*n.m.n.m*]metacyclophanes, *n,m* ≥ 3. *J. Org. Chem.* **2000**, *65*, 5185–5196.

Sample Availability: Samples of the compounds **TA1–4** are available from the authors.

© 2015 by the authors; licensee MDPI, Basel, Switzerland. This article is an open access article distributed under the terms and conditions of the Creative Commons Attribution license (http://creativecommons.org/licenses/by/4.0/).

Molecules **2015**, *20*, 9550-9559; doi:10.3390/molecules20069550

ISSN 1420-3049
www.mdpi.com/journal/molecules

Communication

Practical Radiosynthesis and Preclinical Neuroimaging of [^{11}C]isradipine, a Calcium Channel Antagonist

Benjamin H. Rotstein [1], Steven H. Liang [1], Vasily V. Belov [1,2], Eli Livni [1], Dylan B. Levine [1,2], Ali A. Bonab [1,2], Mikhail I. Papisov [1,2], Roy H. Perlis [3] and Neil Vasdev [1,*]

[1] Department of Radiology, Harvard Medical School, Division of Nuclear Medicine and Molecular Imaging and Center for Advanced Medical Imaging Sciences, Massachusetts General Hospital, 55 Fruit Street, Boston, MA 02114, USA; E-Mails: rotstein.benjamin@mgh.harvard.edu (B.H.R); liang.steven@mgh.harvard.edu (S.H.L.); vbelov@mgh.harvard.edu (V.V.B.); elivni@mgh.harvard.edu (E.L.); dblevine@mgh.harvard.edu (D.B.L.); bonab@pet.mgh.harvard.edu (A.A.B.); papisov@helix.mgh.harvard.edu (M.I.P.)

[2] Department of Research, Shriners Hospitals for Children—Boston, 51 Blossom Street, Boston, MA 02114, USA

[3] Department of Psychiatry and Center for Experimental Drugs and Diagnostics, Massachusetts General Hospital, 185 Cambridge Street, Boston, MA 02114, USA;
E-Mail: rperlis@mgh.harvard.edu

* Author to whom correspondence should be addressed; E-Mail: vasdev.neil@mgh.harvard.edu; Tel.: +1-617-643-4736; Fax: +1-617-726-6165.

Academic Editor: Svend Borup Jensen

Received: 28 April 2015 / Accepted: 20 May 2015 / Published: 26 May 2015

Abstract: In the interest of developing *in vivo* positron emission tomography (PET) probes for neuroimaging of calcium channels, we have prepared a carbon-11 isotopologue of a dihydropyridine Ca^{2+}-channel antagonist, isradipine. Desmethyl isradipine (4-(benzo[*c*][1,2,5]oxadiazol-4-yl)-5-(isopropoxycarbonyl)-2,6-dimethyl-1,4-dihydropyridine-3-carboxylic acid) was reacted with [^{11}C]CH$_3$I in the presence of tetrabutylammonium hydroxide in DMF in an HPLC injector loop to produce the radiotracer in a good yield (6 ± 3% uncorrected radiochemical yield) and high specific activity (143 ± 90 GBq·µmol^{-1} at end-of-synthesis). PET imaging of normal rats revealed rapid brain uptake at baseline (0.37 ± 0.08% ID/cc (percent of injected dose per cubic centimeter) at peak, 15–60 s), which was followed by fast washout. After pretreatment with isradipine (2 mg·kg^{-1}, i.p.), whole brain radioactivity uptake was diminished by 25%–40%. This preliminary study confirms that [^{11}C]isradipine can be synthesized routinely for research studies and is brain penetrating. Further work on Ca^{2+}-channel radiotracer development is planned.

Keywords: carbon-11; radiosynthesis; isradipine; positron emission tomography; neuroimaging; calcium channel blocker

1. Introduction

L-type calcium channels (LTCCs) are cell membrane proteins expressed in most electrically-excitable cells and involved in assorted cellular functions, including neurotransmitter and hormone secretion, Ca^{2+} homeostasis and gene expression [1,2]. A subunit of the LTCC was among the first to be associated with neuropsychiatric diseases, including schizophrenia and bipolar disorder [3,4]. While Ca^{2+}-channel antagonists have a long history in the treatment of hypertension and cardiac disease [5], the promise these pharmaceuticals hold for treatment of neurological and psychiatric disorders has yet to come to fruition [6–14]. For example, it is believed that LTCC antagonists protect dopamine neurons in the substantia nigra from degeneration associated with Parkinson's disease by dopamine D$_2$ receptor desensitization [15]. The 1,4-dihydropyridine (DHP) scaffold has emerged as a privileged structure for LTCC blockers (Figure 1), and several of these compounds, including nimodipine, amlodipine and isradipine (PN 200-110), are among the first-line antihypertensive drugs [16]. Isradipine is currently in clinical trials for the treatment of Parkinson's disease [17] and has demonstrated efficacy for bipolar depression in a proof-of-concept study [18].

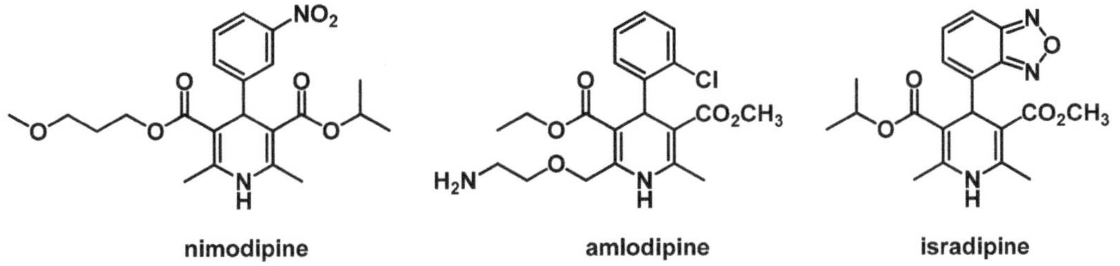

Figure 1. Selected 1,4-dihydropyridine (DHP) Ca^{2+}-channel antagonists.

Efforts to develop neuroimaging probes for ion channels, such as the γ-aminobutyric acid GABA$_A$ receptor [19], nicotinic acetylcholine receptors (nAChRs) [20,21] and the *N*-methyl-D-aspartate (NMDA) receptor [22], have supplied a number of practical tools for clinical research and identified opportunities for subtype-selective ion channel imaging. In contrast, there remains an unmet need for an effective *in vivo* neuroimaging agent for LTCCs [23]. Several DHP-based drugs and derivatives have been radiolabeled with carbon-11 (^{11}C, $t_{1/2}$ = 20.4 min) or fluorine-18 (^{18}F, $t_{1/2}$ = 109.7 min) for positron emission tomography (PET) (Figure 2) [23–28]. Studies using these radiotracers were primarily directed toward imaging of cardiac Ca^{2+}-channels, and an amlodipine-derived radiotracer, [^{11}C]S12968, showed up to 80% specific binding in the myocardium [29–33]. [^{11}C]S12968 was used for *in vivo* measurement of myocardial DHP binding site density in beagles, with low doses of Ca^{2+}-channel antagonists.

Unfortunately, this radiotracer does not cross the blood-brain barrier, and attempts to circumvent this limitation by manipulation of its lipophilicity were ineffective [23].

Isradipine is a highly potent DHP that serves as a reference molecule for *in vitro* studies [34] and has been shown to cross the blood-brain barrier in mice [35]. Furthermore, [O-^{11}C-*methyl*]isradipine ([^{11}C]isradipine) has previously been prepared, though this radiosynthesis demanded the production of an esoteric reagent, [^{11}C]diazomethane [36]. Our goal was to evaluate [^{11}C]isradipine as a potential neuroimaging agent for Ca^{2+}-channels and, in the process, to develop a more convenient radiosynthesis of this molecule from commonly-used [^{11}C]CH$_3$I.

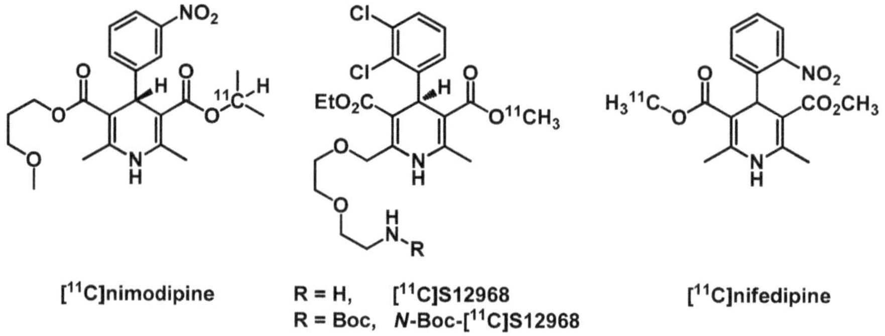

Figure 2. Selected DHP-based PET radiotracers.

2. Results and Discussion

2.1. Radiosynthesis of [^{11}C]Isradipine

Similar to the previous radiosynthesis, the methyl ester was identified as the most convenient site for radiolabeling isradipine with carbon-11. Using the same carboxylic acid precursor (**1**), we sought conditions for selective ^{11}C-methylation using [^{11}C]CH$_3$I (Figure 3). Radiomethylation was conducted using the captive solvent ("loop") method [37,38]. The HPLC injector of a commercial radiosynthesis unit was loaded with a solution of 1 mg of **1** and 0.9 equivalents of tetrabutylammonium hydroxide in 80 μL of anhydrous DMF, and [^{11}C]CH$_3$I was passed through the loop in a stream of helium gas. After 5 min of reaction time, the loop was flushed with the mobile phase directly onto a semi-preparative HPLC column. The radiotracer was purified and reformulated in ethanolic saline, ready for injection.

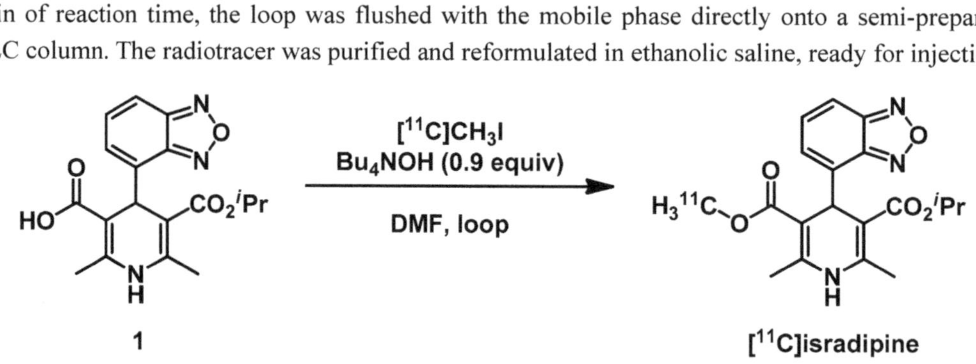

Figure 3. Radiosynthesis of [^{11}C]isradipine.

[^{11}C]Isradipine was prepared in a reasonable overall yield (6 ± 3% non-decay-corrected from starting [^{11}C]CO$_2$) and high specific activity (143 ± 90 GBq·µmol^{-1}; 3.9 ± 2.4 Ci·µmol^{-1} at end-of-synthesis). The radiosynthesis required 40 ± 2 min from end-of-bombardment and yielded the product in >95% radiochemical purity with sufficient quantities for PET imaging (1.5–10.9 GBq, 41–295 mCi). While the radiochemical purity was typically very high, we found that purified [^{11}C]isradipine was susceptible to slow decomposition during the reformulation process if basic aqueous buffers were employed. Using neutral or slightly acidic buffers during reformulation, the radiochemical purity of the final product was consistently >99%. In addition to the improved accessibility of [^{11}C]isradipine on account of widely available [^{11}C]CH$_3$I, compared with the previous report, which used [^{11}C]CH$_2$N$_2$ [36], the radiotracer was isolated in 2–6-fold higher specific activity.

2.2. Partition Coefficient of [^{11}C]isradipine

Octanol/buffer partition coefficients have predictive utility for assessing blood-brain barrier permeability, with an optimum log$D$ range of 2.0–3.5 [39]. Owing to the presence of a basic primary amine, the log$D$ of [^{11}C]S12968 is 1.54, and this tracer does not enter the brain [23]. In contrast, nimodipine possesses a log$D$ of 2.41 and does cross the blood-brain barrier. However, DHPs have been shown to engage in complex interactions with lipid membranes, suggesting that partition coefficients in isotropic solvent systems may hold less predictive value for this class of compounds [40]. For example, N-Boc-[^{11}C]S12968, in which the amine is masked as a carbamate, has an increased log$D$ (2.12), yet still fails to significantly enter the brain (0.04% ID/cc (percent of injected dose per cubic centimeter) at peak). With these factors in mind, we experimentally determined the log$D$ of [^{11}C]isradipine to be 2.15 ($n = 8$), using liquid-liquid partition between 1-octanol and phosphate-buffered saline (PBS, pH 7.4) [41].

2.3. Positron Emission Tomography Neuroimaging of [^{11}C]Isradipine

To determine brain uptake and feasibility for neuroimaging of Ca^{2+}-channels using [^{11}C]isradipine, we conducted preliminary PET brain imaging studies in healthy, large (440–670 g), male Sprague-Dawley rats. Dynamic 30-min PET scans were acquired beginning at time-of-injection (TOI) of a bolus of radiotracer (41–166 MBq) via tail-vein, and time-activity curves were generated after image reconstruction and processing (Figure 4). Whole brain activity uptake was moderate (0.37 ± 0.08% ID/cc, percent of injected dose per cubic centimeter; 1.9 ± 0.0 SUV, standardized uptake value; $n = 3$) and peaked in the first minute after TOI. Washout was rapid, as approximately half of the peak uptake was eliminated from the brain within 20 min.

To determine the fraction of specific binding of [^{11}C]isradipine *in vivo* in rat brain, animals were treated with isradipine at a dose of 2 mg·kg^{-1} i.p., 30 min prior to radiotracer administration and again imaged for 30 min. This pretreatment dose represents the upper level of previously reported i.p. dosing of isradipine [8,9], as well as the solubility threshold in vehicle (5% DMSO, 5% Tween 80, 90% saline). Whole brain uptake peaked at 0.19 ± 0.05% ID/cc (1.1 ± 0.1 SUV) 15–60 s after TOI and cleared much more slowly over the course of the imaging session. It is apparent from the time-activity curves that a significant component of whole brain uptake can be attributed to nonspecific binding. Two observations support this conclusion: (1) the magnitude of uptake after pretreatment is relatively high; and (2) the rate of clearance from 0–5 min is much slower than at baseline. The level of nonspecific binding can be estimated by area-under-the-curve (AUC) analysis. From 0–5 min post-injection of [^{11}C]isradipine, uptake is approximately 60% of that at baseline (paired *t*-test, $p < 0.05$) (Figure 4C).

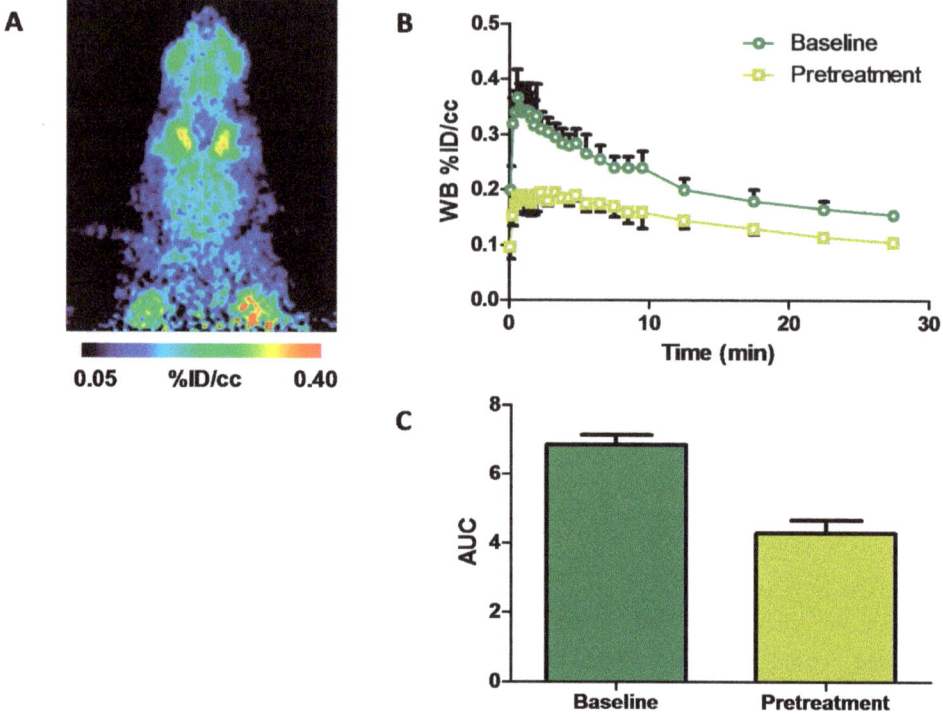

Figure 4. (**A**) Coronal view of summed image 0–7 min post bolus [^{11}C]isradipine injection at baseline; (**B**) time-activity curves in rat whole brain at baseline (–●–) and after pretreatment (–□–) with isradipine (2 mg·kg^{-1} i.p., 30 min prior to time-of-injection (TOI)); (**C**) area-under-the-curve analysis of whole brain uptake 0–5 min after tracer injection at baseline and after pretreatment, paired *t*-test, $p < 0.05$; WB, whole brain; % ID/cc, percent of injected dose per cubic centimeter; AUC, area-under-the-curve.

Given the preclinical and clinical observations of the pharmacology of isradipine and certain other DHPs, such as nimodipine, these drugs appear promising as leads for brain-penetrating Ca^{2+}-channel imaging agents [8,9]. In line with previous studies, which showed that log*D* values in this range may not be strongly predictive of brain uptake for this class of compounds [23], the lipophilicities (as measured by octanol/buffer partitioning) of *N*-Boc-S12968 and isradipine are very similar, and yet, the latter shows much greater levels of brain uptake. Over the whole brain region of rats, a 40% blockade was observed after pretreatment with the nonradioactive drug. The limited effects of pre-treatment on [^{11}C]isradipine uptake could be explained by numerous factors, including a low fraction of specific binding, changes in blood flow as a result of LTCC antagonism or LTCC inhibition by ethanol present in the radiotracer formulation. Further work would be required to evaluate the effects of injection vehicle, blood pressure, anesthesia and species differences on [^{11}C]isradipine brain uptake, radiotracer metabolism and whether the observed specificity is uniform across all brain regions.

3. Experimental Section

Isradipine was purchased from U.S. Pharmacopeial Convention (USP, Rockville, MD, USA). The radiotracer precursor (4-(benzo[c][1,2,5]oxadiazol-4-yl)-5-(isopropoxycarbonyl)-2,6-dimethyl-1,4-dihydropyridine-3-carboxylic acid, **1**) was purchased from Aberjona Laboratories (Beverly, MA, USA). Anhydrous solvents were purchased from Fisher Scientific. Tetrabutylammonium hydroxide (1 M in methanol) was purchased from Sigma-Aldrich. A GE PETtrace 16.5 MeV cyclotron was used for [^{11}C]CO_2 production by the ^{14}N(p,α)^{11}C nuclear reaction using a 50-μA proton beam current to irradiate $^{14}N_2$ containing 1% O_2.

3.1. Radiosynthesis of [^{11}C]Isradipine

Desmethyl isradipine (**1**, 1.2 mg) was dissolved in anhydrous DMF (100 μL) in a glass vial. To this vial, tetrabutylammonium hydroxide (TBAOH, 1 M MeOH, 3.0 μL, 0.9 equiv.) was added and the contents vortexed for 60 s to prepare the precursor solution. An aliquot of the precursor solution (80 μL) was withdrawn and loaded onto the HPLC injection loop of a commercial radiofluorination unit (GE Tracerlab FX N) [38]. The remaining precursor solution (20 μL) was diluted with water (20 μL), and the pH of the aqueous solution was then tested using indicator paper to confirm a pH range of 8–9. [^{11}C]CH_3I, prepared from [^{11}C]CO_2 using a commercial [^{11}C]CH_3I synthesis unit (GE Tracerlab FX MeI), was passed through the loop on a flow of He$_{(g)}$. After transfer of [^{11}C]CH_3I was complete, gas flow was discontinued for 5 min, prior to flushing of the loop contents onto a previously equilibrated semi-preparative C18 Luna HPLC column using 6:4, CH_3CN:0.1 M $NH_4·HCO_2$ mobile phase at 5 mL·min^{-1}. The product peak was collected 10.5–11.0 min after injection into a vessel containing water (24 mL). It is noteworthy that if the bulk collection vessel contained a solution of water (22 mL) and $NaHCO_3$ (8% *w/v*, 2 mL), a radiochemical impurity would be observed in the final product. The diluted collected fraction was passed through an HLB solid-phase extraction cartridge, which was then flushed with sterile water (10 mL). The reformulated product was collected by elution with ethanol (1 mL), followed by saline (9 mL). The product was analyzed by analytical HPLC to determine radiochemical purity and specific activity (stationary phase: C18 Luna, 5 μm, 100 Å, 250 × 4.6 mm or Prodigy ODS-3, 5 μm, 100 Å, 250 × 4.60 mm; mobile phase: 7:3, CH_3CN:0.1 M AMF, 1 mL·min^{-1}). The identity of the product was confirmed by coinjection with the known standard, isradipine.

3.2. PET Imaging

All animal imaging studies were performed in accordance with the National Institutes of Health Guide for the Care and Use of Laboratory Animals and were approved by the Massachusetts General Hospital Institutional Animal Care and Use Committee.

Four male rats Sprague-Dawley rats (440–670 g, Charles River Laboratories) were included in this study. Two animals were studied under both baseline and pretreatment conditions, and an additional two age- and weight-matched animals were studied under either baseline or pretreatment conditions. Animals were pair-housed on a diurnal 12:12 light/dark cycle with *ad libitum* access to food and water. Animals were weighed immediately before or immediately after imaging studies. For pretreatment studies, isradipine was reformulated in 5% DMSO, 5% Tween 80 and 90% saline and administered by intraperitoneal injection at a dose of 2 mg·kg^{-1}, 30 min prior to radiotracer administration. Animals were anesthetized using isoflurane/oxygen (2%–3%, 1.5 L·min^{-1}) and positioned with a custom-fabricated head holder for the duration of the imaging study (~45 min). [^{11}C]Isradipine, reformulated in 0.2–1.0 mL of 10% ethanolic saline, was administered by tail-vein injection at a dose of 41–166 MBq (1.1–4.5 millicurie). Using a Siemens Focus 220 μPET (Siemens Medical Solutions, Knoxville, TN, USA) in line with a CereTom NL 3000 CT scanner (NeuroLogica, Danvers, MA, USA), list mode PET data for brain imaging were acquired for 30 min beginning at the time of radiotracer administration (TOI). Data from a second bed position, with the tail in the field-of-view of the PET camera, were then acquired in list mode for 5 min and used for correction for the activity residing in the tail. Subsequently, CT data were acquired for attenuation correction and anatomic coregistration. PET data from the 30-min scan were reconstructed using the 3D ordered-subset expectation maximization followed by maximum a posteriori reconstruction (OSEM3D/MAP) protocol (smoothing resolution of 1.5 mm, 9 OSEM3D subsets, 2 OSEM3D and 15 MAP iterations) with decay-correction to TOI and the following framing: 12 × 10 s, 6 × 30 s, 5 × 60 s, 4 × 300 s.

Reconstructed PET datasets were analyzed using AMIDE software. Elliptical regions of interest (ROI) were placed around the whole brain, as identified from overlaid CT datasets, anatomical landmarks visible in the 3D PET images and radiotracer uptake intensity. Time-activity curves were generated for each ROI and normalized to injected radiotracer dose (drawn radioactivity, less residual radioactivity in the syringe, less residual radioactivity in the tail, all decay-corrected to TOI) and expressed as the percent of the injected dose per cubic centimeter (%ID/cc) versus time. Time-activity curves were also normalized to animal weight and expressed as standardized uptake value (SUV) versus time. To facilitate comparison between subjects and treatment groups, brain radioactivity uptake was quantified at the peak (15–60 s) and evaluated by area-under-the-curve (AUC) analysis over the entire dataset and the interval showing the highest brain uptake (0–5 min).

4. Conclusions

Carbon-11- and fluorine-18-radiolabeled DHPs have shown utility for myocardial imaging of Ca^{2+}-channels, yet imaging of this important target in the central nervous system has remained elusive. An efficient radiosynthesis of [^{11}C]isradipine can be easily achieved with [^{11}C]CH$_3$I at room temperature using the "loop" method resulting in reasonable radiochemical yields and high specific activity. A preliminary PET neuroimaging study using this radiotracer in rats at baseline and after pretreatment with isradipine demonstrated that *in vivo* binding of this radiotracer showed only 40% blockade in whole brain. Regional analysis of [^{11}C]isradipine in the brains of higher species could reveal specific binding in discrete regions of interest. The facile radiochemistry methodology developed for [^{11}C]isradipine should prove to be widely applicable for O-^{11}CH$_3$ labeling of methyl esters of structurally-related DHPs.

Acknowledgments

B.H.R. was supported by a Natural Sciences and Engineering Research Council of Canada (NSERC) Postdoctoral Fellowship. R.H.P. and E.L. were supported in part by the Stanley Center for Psychiatric Research. We thank Edward Soares for helpful discussions.

Author Contributions

B.H.R. and E.L. conducted the radiochemical experiments. B.H.R., V.V.B., D.B.L. and A.A.B. conducted the PET imaging experiments. B.H.R. analyzed the data and wrote the manuscript. B.H.R., R.H.P. and N.V. conceived of the project. B.H.R., M.I.P., S.H.L. and N.V. supervised the project.

Conflicts of Interest

The authors declare no conflict of interest.

References

1. Catterall, W.A.; Perez-Reyes, E.; Snutch, T.P.; Striessnig, J. International Union of Pharmacology. XLVIII. Nomenclature and Structure-Function Relationships of Voltage-Gated Calcium Channels. *Pharmacol. Rev.* **2005**, *57*, 411–425.
2. Striessnig, J.; Koschak, A. Exploring the function and pharmacotherapeutic potential of voltage-gated Ca^{2+} channels with gene-knockout models. *Channels* **2008**, *2*, 233–251.
3. Ferreira, M.A.R.; O'Donovan, M.C.; Meng, Y.A.; Jones, I.R.; Ruderfer, D.M.; Jones, L.; Fan, J.; Kirov, G.; Perlis, R.H.; Green, E.K.; *et al.* Collaborative genome-wide association analysis supports a role for *ANK3* and *CACNA1C* in bipolar disorder. *Nat. Genet.* **2008**, *40*, 1056–1058.
4. Sklar, P.; Ripke, S.; Scott, L.J.; Andreassen, O.A.; Cichon, S.; Craddock, N.; Mahon, P.B. Large-scale genome-wide association analysis of bipolar disorder identifies a new susceptibility locus near *ODZ4*. *Nat. Genet.* **2011**, *43*, 977–983.
5. Triggle, D.J.; Hawthorn, M.; Gopalakrishnan, M.; Minarini, A.; Avery, S.; Rutledge, A.; Bangalore, R.; Zheng, W. Synthetic Organic Ligands active at Voltage-Gated Calcium Channels. *Ann. N. Y. Acad. Sci.* **1991**, *635*, 123–138.
6. Deyo, R.A.; Straube, K.T.; Disterhoft, J.F. Nimodipine facilitates associative learning in aging rabbits. *Science* **1989**, *243*, 809–811.
7. Scriabine, A.; Schuurman, T.; Traber, J. Pharmacological basis for the use of nimodipine in central nervous system disorders. *FASEB J.* **1989**, *3*, 1799–1806.
8. Pucilowski, O.; Plaźnik, A.; Overstreet, D.H. Isradipine Suppresses Amphetamine-Induced Conditioned Place Preference and Locomotor Stimulation in the Rat. *Neuropsychopharmacology* **1995**, *12*, 239–244.
9. Chan, C.S.; Guzman, J.N.; Ilijic, E.; Mercer, J.N.; Rick, C.; Tkatch, T.; Meredith, G.E.; Surmeier, D.J. "Rejuvenation" protects neurons in mouse models of Parkinson's disease. *Nature* **2007**, *447*, 1081–1086.
10. Link, M.C.; Wiemann, M.; Bingmann, D. Organic and inorganic calcium antagonists inhibit veratridine-induced epileptiform activity in CA3 neurons of the guinea pig. *Epilepsy Res.* **2008**, *78*, 147–154.
11. Casamassima, F.; Huang, J.; Fava, M.; Sachs, G.S.; Smoller, J.W.; Cassano, G.B.; Lattanzi, L.; Fagerness, J.; Stange, J.P.; Perlis, R.H. Phenotypic effects of a bipolar liability gene among individuals with major depressive disorder. *Am. J. Med. Genet.* **2010**, *153B*, 303–309.
12. Casamassima, F.; Hay, A.C.; Benedetti, A.; Lattanzi, L.; Cassano, G.B.; Perlis, R.H. L-type calcium channels and psychiatric disorders: A brief review. *Am. J. Med. Genet.* **2010**, *153B*, 1373–1390.

13. Kang, S.; Cooper, G.; Dunne, S.F.; Dusel, B.; Luan, C.H.; Surmeier, D.J.; Silverman, R.B. Cav1.3-selective L-type calcium channel antagonists as potential new therapeutics for Parkinson's disease. *Nat. Commun.* **2012**, *3*, 1146, doi:10.1038/ncomms2149.
14. Pourbadie, H.G.; Naderi, N.; Mehranfard, N.; Janahmadi, M.; Khodagholi, F.; Motamedi, F. Preventing Effect of L-Type Calcium Channel Blockade on Electrophysiological Alterations in Dentate Gyrus Granule Cells Induced by Entorhinal Amyloid Pathology. *PLoS ONE* **2015**, *10*, e0117555.
15. Dragicevic, E.; Poetschke, C.; Duda, J.; Schlaudraff, F.; Lammel, S.; Schiemann, J.; Fauler, M.; Hetzel, A.; Watanabe, M.; Lujan, R.; *et al.* Cav1.3 channels control D2-autoreceptor responses via NCS-1 in substantia nigra dopamine neurons. *Brain* **2014**, *137*, 2287–2302.
16. Triggle, D.J. 1,4-Dihydropyridines as calcium channel ligands and privileged structures. *Cell. Mol. Neurobiol.* **2003**, *23*, 293–303.
17. Parkinson Study Group. Phase II safety, tolerability, and dose selection study of isradipine as a potential disease-modifying intervention in early Parkinson's disease (STEADY-PD). *Mov. Disord.* **2013**, *28*, 1823–1831.
18. Ostacher, M.J.; Iosifescu, D.V.; Hay, A.; Blumenthal, S.R.; Sklar, P.; Perlis, R.H. Pilot investigation of isradipine in the treatment of bipolar depression motivated by genome-wide association. *Bipolar Disord.* **2014**, *16*, 199–203.
19. Andersson, J.D.; Halldin, C. PET radioligands targeting the brain GABA$_A$/benzodiazepine receptor complex. *J. Label. Compd. Radiopharm.* **2013**, *56*, 196–206.
20. Mo, Y.X.; Yin, Y.F.; Li, Y.M. Neural nAChRs PET imaging probes. *Nucl. Med. Commun.* **2014**, *35*, 135–143.
21. Horti, A.G.; Kuwabara, H.; Holt, D.P.; Dannals, R.F.; Wong, D.F. Recent PET radioligands with optimal brain kinetics for imaging nicotinic acetylcholine receptors. *J. Label. Compd. Radiopharm.* **2013**, *56*, 159–166.
22. Sobrio, F. Radiosynthesis of carbon-11 and fluorine-18 labelled radiotracers to image the ionotropic and metabotropic glutamate receptors: ^{11}C and ^{18}F chemistry to image the glutamate receptors. *J. Label. Compd. Radiopharm.* **2013**, *56*, 180–186.
23. Dollé, F.; Valette, H.; Hinnen, F.; Fuseau, C.; Péglion, J.L.; Crouzel, C. Synthesis and Characterization of a ^{11}C-Labelled Derivative of S12968: An Attempt to Image *in Vivo* Brain Calcium Channels. *Nucl. Med. Biol.* **1998**, *25*, 339–342.
24. Wilson, A.A.; Dannals, R.F.; Ravert, H.T.; Burns, H.D.; Lever, S.Z.; Wagner, H.N. Radiosynthesis of [^{11}C]nifedipine and [^{11}C]nicardipine. *J. Label. Compd. Radiopharm.* **1989**, *27*, 589–598.
25. Stone-Elander, S.; Roland, P.; Schwenner, E.; Halldin, C.; Widén, L. Synthesis of [isopropyl-^{11}C]nimodipine for *in vivo* studies of dihydropyridine binding in man using positron emission tomography. *Int. J. Radiat. Appl. Instrum. Part A* **1991**, *42*, 871–875.
26. Holschbach, M.; Roden, W.; Hamkens, W. Synthesis of carbon-11 labelled calcium channel antagonists. *J. Label. Compd. Radiopharm.* **1991**, *29*, 431–442.
27. Pleiss, U. 1,4-Dihydropyridines (DHPs)-a class of very potent drugs: Syntheses of isotopically labeled DHP derivatives during the last four decades. *J. Label. Compd. Radiopharm.* **2007**, *50*, 818–830.

28. Sadeghpour, H.; Jalilian, A.R.; Shafiee, A.; Akhlaghi, M.; Miri, R.; Mirzaei, M. Radiosynthesis of dimethyl-2-[^{18}F]-(fluoromethyl)-6-methyl-4-(2-nitrophenyl)-1,4-dihydropyridine-3,5-dicarboxylate for L-type calcium channel imaging. *Radiochim. Acta* **2008**, *96*, 849–854.
29. Valette, H.; Crouzel, C.; Syrota, A.; Fuseau, C.; Bourachot, M.L. Canine myocardial dihydropyridine binding sites: A positron emission tomographic study with the calcium channel inhibitor 11C-S11568. *Life Sci.* **1994**, *55*, 1471–1477.
30. Dollé, F.; Hinnen, F.; Valette, H.; Fuseau, C.; Duval, R.; Péglion, J.L.; Crouzel, C. Synthesis of two optically active calcium channel antagonists labelled with carbon-11 for *in vivo* cardiac PET imaging. *Bioorg. Med. Chem.* **1997**, *5*, 749–764.
31. Dollé, F.; Valette, H.; Hinnen, F.; Demphel, S.; Bramoulle, Y.; Peglion, J.L.; Crouzel, C. Highly efficient synthesis of [^{11}C]S12968 and [^{11}C]S12967, for the *in vivo* imaging of the cardiac calcium channels using PET. *J. Label. Compd. Radiopharm.* **2001**, *44*, 481–499.
32. Valette, H.; Dollé, F.; Guenther, I.; Hinnen, F.; Fuseau, C.; Coulon, C.; Péglion, J.L.; Crouzel, C. Myocardial Kinetics of the ^{11}C-Labeled Enantiomers of the Ca^{2+} Channel Inhibitor S11568: An *in Vivo* Study. *J. Nucl. Med.* **2001**, *42*, 932–937.
33. Valette, H.; Dollé, F.; Guenther, I.; Fuseau, C.; Coulon, C.; Hinnen, F.; Péglion, J.L.; Crouzel, C. *In vivo* quantification of myocardial dihydropyridine binding sites: A PET study in dogs. *J. Nucl. Med.* **2002**, *43*, 1227–1233.
34. Maan, A.C.; Ptasienski, J.; Hosey, M.M. Influence of Mg^{++} on the effect of diltiazem to increase dihydropyridine binding to receptors on Ca^{++}-channels in chick cardiac and skeletal muscle membranes. *J. Pharmacol. Exp. Ther.* **1986**, *239*, 768–774.
35. Uchida, S.; Yamada, S.; Nagai, K.; Deguchi, Y.; Kimura, R. Brain pharmacokinetics and *in vivo* receptor binding of 1,4-dihydropyridine calcium channel antagonists. *Life Sci.* **1997**, *61*, 2083–2090.
36. Crouzel, C.; Syrota, A. The use of [^{11}C]diazomethane for labelling a calcium channel antagonist: PN 200-110 (isradipine). *Int. J. Radiat. Appl. Instrum. Part A* **1990**, *41*, 241–242.
37. Wilson, A.A.; Garcia, A.; Jin, L.; Houle, S. Radiotracer synthesis from [^{11}C]-iodomethane: A remarkably simple captive solvent method. *Nucl. Med. Biol.* **2000**, *27*, 529–532.
38. Wilson, A.A.; Garcia, A.; Houle, S.; Vasdev, N. Utility of commercial radiosynthetic modules in captive solvent [^{11}C]-methylation reactions. *J. Label. Compd. Radiopharm.* **2009**, *52*, 490–492.
39. Waterhouse, R. Determination of lipophilicity and its use as a predictor of blood–brain barrier penetration of molecular imaging agents. *Mol. Imaging Biol.* **2003**, *5*, 376–389.
40. Herbette, L.G.; Vant Erve, Y.M.; Rhodes, D.G. Interaction of 1,4 dihydropyridine calcium channel antagonists with biological membranes: Lipid bilayer partitioning could occur before drug binding to receptors. *J. Mol. Cell. Cardiol.* **1989**, *21*, 187–201.
41. Wilson, A.A.; Jin, L.; Garcia, A.; DaSilva, J.N.; Houle, S. An admonition when measuring the lipophilicity of radiotracers using counting techniques. *Appl. Radiat. Isot.* **2001**, *54*, 203–208.

Sample Availability: Samples of the compound **1** are available from the authors.

© 2015 by the authors; licensee MDPI, Basel, Switzerland. This article is an open access article distributed under the terms and conditions of the Creative Commons Attribution license (http://creativecommons.org/licenses/by/4.0/).

Molecules **2015**, *20*, 4902-4914; doi:10.3390/molecules20034902

OPEN ACCESS

molecules
ISSN 1420-3049
www.mdpi.com/journal/molecules

Article

Preliminary Biological Evaluation of ^{18}F-FBEM-Cys-Annexin V a Novel Apoptosis Imaging Agent

Chunxiong Lu [1,*], Quanfu Jiang [1], Minjin Hu [2], Cheng Tan [1], Huixin Yu [1] and Zichun Hua [2,3,*]

[1] Ministry of Health & Jiangsu Key Laboratory of Molecular Nuclear Medicine, Jiangsu Institute of Nuclear Medicine, Wuxi 214063, China;
E-Mails: jiangquanfu@jsinm.org (Q.J.); tangcheng@jsinm.org (C.T.); yuhuixin@jsinm.org (H.Y.)

[2] Jiangsu Target Pharma Laboratories Inc., Changzhou High-Tech Research Institute of Nanjing University, Changzhou 213164, China; E-Mail: huminj98@163.com

[3] The State Key Laboratory of Pharmaceutical Biotechnology, Nanjing University, Nanjing 210093, China

* Authors to whom correspondence should be addressed; E-Mails: luchunxiong@jsinm.org (C.L.); huazc@nju.edu.cn (Z.H.); Tel.: +86-510-85514482-3529 (C.L.); +86-25-8332-4605 (Z.H.); Fax: +86-510-85513113 (C.L.).

Academic Editor: Svend Borup Jensen

Received: 22 December 2014 / Accepted: 6 March 2015 / Published: 17 March 2015

Abstract: A novel annexin V derivative (Cys-Annexin V) with a single cysteine residue at its C-terminal has been developed and successfully labeled site-specifically with ^{18}F-FBEM. ^{18}F-FBEM was synthesized by coupling ^{18}F-fluorobenzoic acid (^{18}F-FBA) with N-(2-aminoethyl)maleimide using optimized reaction conditions. The yield of ^{18}F-FBEM-Cys-Annexin V was 71.5% ± 2.0% ($n = 4$, based on the starting ^{18}F-FBEM, non-decay corrected). The radiochemical purity of ^{18}F-FBEM-Cys-Annexin V was >95%. The specific radioactivities of ^{18}F-FBEM and ^{18}F-FBEM-Cys-Annexin V were >150 and 3.17 GBq/μmol, respectively. Like the 1st generation ^{18}F-SFB-Annexin V, the novel ^{18}F-FBEM-Cys-Annexin V mainly shows renal and to a lesser extent, hepatobiliary excretion in normal mice. In rat hepatic apoptosis models a 3.88 ± 0.05 ($n = 4$, 1 h) and 10.35 ± 0.08 ($n = 4$, 2 h) increase in hepatic uptake of ^{18}F-FBEM-Cys-Annexin V compared to normal rats was observed after injection via the tail vein. The liver uptake ratio (treated/control) at 2 h p.i. as measured via microPET correlated with the ratio of apoptotic nuclei in liver observed using TUNEL histochemistry, indicating that the novel ^{18}F-FBEM-Cys-Annexin V is a potential apoptosis imaging agent.

Keywords: Cys-Annexin V; site-specific labeling; ^{18}F-FBEM; apoptosis imaging

1. Introduction

Apoptosis plays an important role, not only in physiology but also in pathology [1,2]. Dysregulation of apoptosis is associated with many diseases such as cancer, autoimmunity and neurodegenerative disorders. Therefore, it has significant clinical value of the detection and quantification of apoptosis *in vivo* for diagnosis and assessment of therapeutic efficacy. One of the early charateristics of apoptosis is the externalization of the phospholipid phosphatidylserine (PS) at the cell membrane [3,4]. Annexin V, a 36-kDa human protein, shows Ca^{2+}-dependent binding to negatively charged phospholipid surfaces and was discovered as a vascular anticoagulant protein [5,6]. The anticoagulant activity is based on the high-affinity for PS. These characteristics make annexin V derivatives suitable candidates for imaging of apoptosis. Several annexin V tagged with bifunctional chelators (BFC) have been labeled with ^{99m}Tc for single photon emission computed tomography (SPECT) imaging of apoptosis *in vivo* [7–11]. However, conjugation of BFC to annexin V for labeling with ^{99m}Tc is usually done by targeting an amino group of one of the 21 lysine residues using BFC, but this method is rather non-specific as any of the –NH$_2$ groups could be targeted. Recent studies have revealed that after structural modification in the recombinant expression annexin V can be directly marked with ^{99m}Tc, giving derivatives such as ^{99m}Tc(CO)$_3$-HIS-cys-Anx V [12], ^{99m}Tc-annexin V-117 [13] and ^{99m}Tc-His10-annexin V [14]. These new annexin V molecules labeled by site-specific methods greatly improve sensitivity for detecting cell death *in vivo* [15]. Our group has reported a site-specific ^{99m}Tc labeling method of a novel annexin V derivative (Cys-Annexin V) with a single cysteine residue at C-terminal [16]. ^{99m}Tc-Cys-Annexin V is a potential SPECT imaging agent of apoptosis. However, because of its higher sensitivity, better spatial resolution and quantification properties a positron emission tomography (PET) analog would be very desirable. Some groups have reported the labeling of annexin V with N-succinimidy-4-^{18}F-fluorobenzoate (^{18}F-SFB) for PET imaging of apoptosis [17–19], however this labeling method is non-specific, as the ^{18}F-SFB reacts with any available NH$_2$ group in the protein. Thiol-reactive agents such as N-substituted maleimides can be used to modify proteins on the cysteine group [20]. ^{18}F-N-[2-(4-Fluoro-benzamido)ethyl]maleimide (^{18}F-FBEM) was used to label thiol-containing proteins as a novel site-specific labeling prosthetic group [21–23]. We report herein the labeling and preliminary *in vivo* evaluation of the novel ^{18}F-FBEM-Cys-Annexin V in normal mice and in rat models of apoptosis induced by cycloheximide. In mice the tracer uptake was studied by dynamic microPET imaging and microPET in a rat model of hepatic apoptosis. Apoptosis was confirmed *in situ* on liver slices using the terminal deoxynucleotidyl transferase (TdT) dUTP nick end labeling (TUNEL) assay.

2. Results and Discussion

2.1. Radiolabeling

Annexin V has been labeled non-specifically with a number of isotopes, including ^{99m}Tc, ^{124}I, ^{18}F and ^{68}Ga [11,17,18,24,25]. Site-specific labeling of annexin V can help improve its sensitivity for detecting cell death *in vivo*. To take advantage of the higher resolution and more accurate quantification of PET, labeling annexin V with short half-life positron-emitters such as ^{18}F is of particular interest. In this study site-specific labeling of Cys-Annexin V with ^{18}F-FBEM as prosthetic group is presented.

As see in Figure 1, Cys-Annexin V and ^{18}F-FBEM-Cys-Annexin V were eluted at retention times of 8.6 min and 9.1 min, respectively, whereas ^{18}F-FBEM eluted at a retention time of 15.9 min. According to HPLC analysis, the radiochemical purity of ^{18}F-FBEM-Cys-Annexin V was above 95%. The total synthesis time for ^{18}F-FBEM was about 100 min and 428 ± 65 MBq ($n = 4$) pure ^{18}F-FBEM was obtained from 18.5 GBq ^{18}F-fluoride. 81–262 MBq ^{18}F-FBEM-Cys-Annexin V was obtained from 111–370 MBq ^{18}F-FBEM as the radiochemical yield was 71.5% ± 2.0% ($n = 4$, based on the starting ^{18}F-FBEM, non-decay corrected). The specific radioactivities of ^{18}F-FBEM and ^{18}F-FBEM-Cys-Annexin V were above 150 MBq/µmol and 3.17 GBq/µmol, respectively.

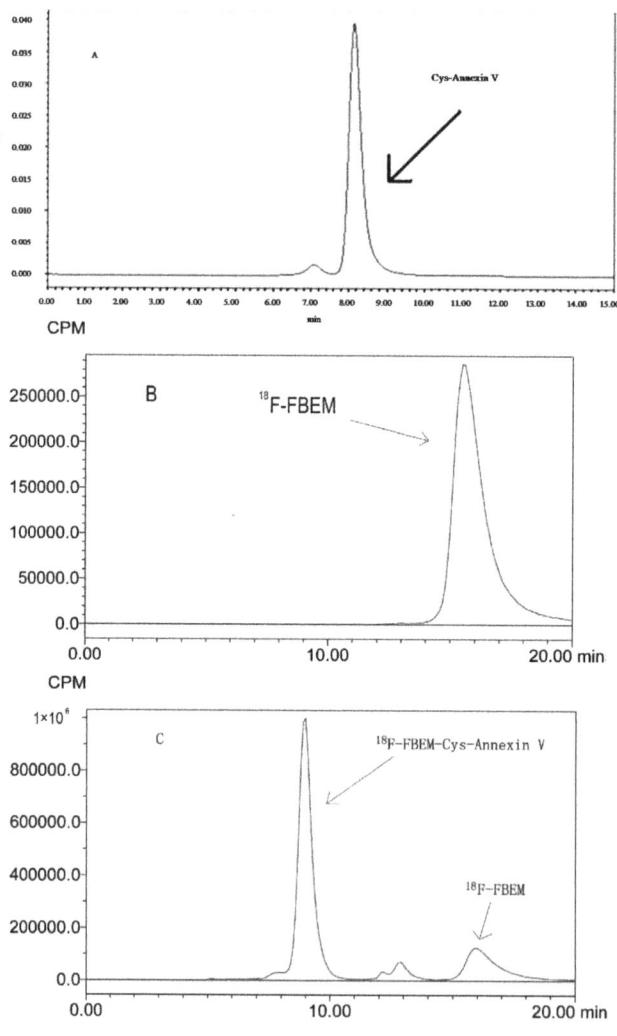

Figure 1. *Cont.*

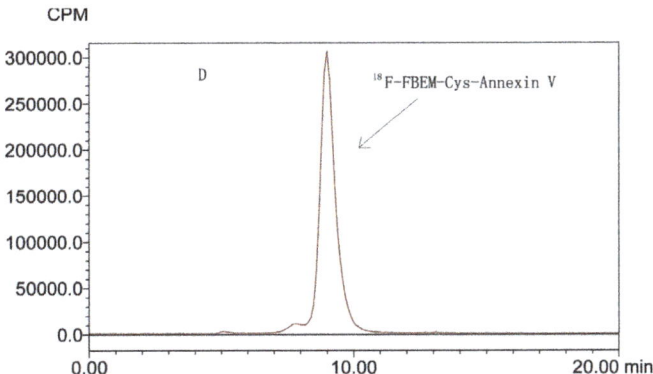

Figure 1. HPLC chromatogram (isocratic, 0.05 mol/L phosphate buffer (pH = 7.0), flow 0.8 mL/min) of: (**A**) Cys-Annexin V, t_R = 8.6 min (UV); HPLC radiochromatograms of (**B**) ^{18}F-FBEM, t_R = 15.9 min, (**C**) reaction mixture (^{18}F-FBEM-Cys-Annexin V, t_R = 9.1 min, ^{18}F-FBEM, t_R = 15.9 min) and (**D**) ^{18}F-FEBM-Cys-Annexin V, t_R = 9.1 min.

2.2. In Vitro Stability of ^{18}F-FBEM-Cys-Annexin V

To determine the radioactive decomposed side products which may accumulate in non-target organs, the stability of ^{18}F-FBEM-Cys-Annexin V was studied. The results of the stability of ^{18}F-FBEM-Cys-Annexin V in (A) phosphate buffered saline (PBS, 0.1 mol/L, pH 7.2) and (B) human serum, respectively, are presented in Figure 2. The results show that the ^{18}F-FBEM-Cys-Annexin V is more stable during biodistribution and PET imaging studies.

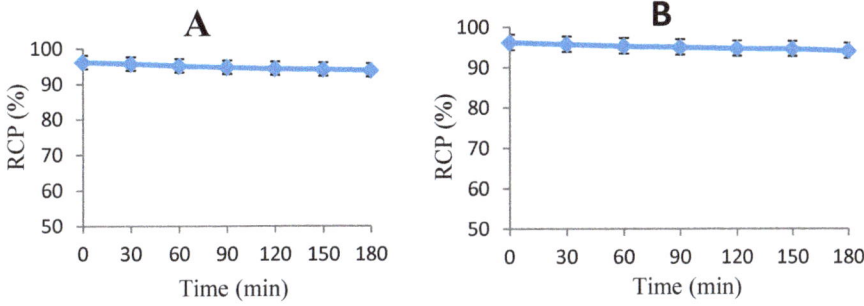

Figure 2. Stability of ^{18}F-FBEM-Cys-Annexin V at different intervals in (**A**) PBS and (**B**) human serum.

2.3. Blood Kinetics Studies

Pharmacokinetic parameters, obtained using the DAS 2.1.1 pharmacokinetic calculation program, are listed in Table 1. Figure 3 shows the blood clearance of ^{18}F-FEBM-Cys-Annexin V in the mice 2 h post-injection. Pharmacokinetics of ^{18}F-FEBM-Cys-Annexin V comply with the two-compartment model with the pharmacokinetic equation of C = $2.359e^{-0.062t} + 3.288e^{-0.005t}$ (where C is radiopharmaceutical activity (%ID/g) in blood, t is the time after injection). The half-life of distribution phase ($t_{1/2\alpha}$) and half-life of elimination phase ($t_{1/2\beta}$) were 11.261 and 134.62 min, respectively, which showed that ^{18}F-FBEM-Cys-Annexin V can be absorbed quickly and eliminated slowly. Biological availability was represented by area under concentration-time curve (AUC) and the values of clearance (CL) and AUC were 0.031 and 643, respectively.

Table 1. Pharmacokinetic parameters of the ^{18}F-FEBM-Cys-Annexin V in mice.

Parameter (units)	^{18}F-FEBM-Cys-Annexin V
K_{12} (min^{-1})	0.02
K_{21} (min^{-1})	0.038
K_e (min^{-1})	0.009
CL (%ID/g/min)	0.031
$T_{1/2\alpha}$ (min)	11.261
$T_{1/2\beta}$ (min)	134.62
AUC (%ID/g·min)	634.123

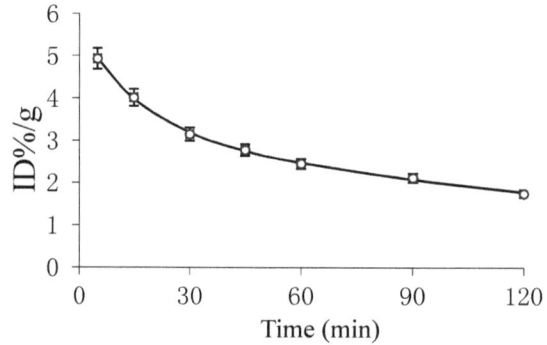

Figure 3. Pharmacokinetic curve for ^{18}F-FEBM-Cys-Annexin V in mice.

In the early phase, the blood clearance of ^{18}F-FEBM-Cys-Annexin V was slow. After 2 h, the radioactivity concentration of the tracer agent in blood reaches an equilibrium which coincides with the pharmacokinetic parameters CL, AUC and the pharmacokinetic curves.

2.4. Dynamic MicroPET Images of Normal ICR Mice

Representative time-activity curves of the major organs (kidneys, liver and heart) were derived from 60-min dynamic microPET scans after intravenous administration of ^{18}F-FBEM-Cys-Annexin V tracers (Figure 4). The radioactivity kinetics were calculated from a region-of-interest analysis of the dynamic small animal PET scans over the heart (squares; mainly representing the cardiac blood pool), kidney (triangles) and liver (diamonds). ^{18}F-FBEM-Cys-Annexin V was excreted mainly through the kidneys, as evidenced by the higher renal uptake at early time points and excretion via the bladder. The kidney uptake reached a peak (11%ID/g) at 13 min after injection and then decreased to 1.43%ID/g at 60 min p.i.

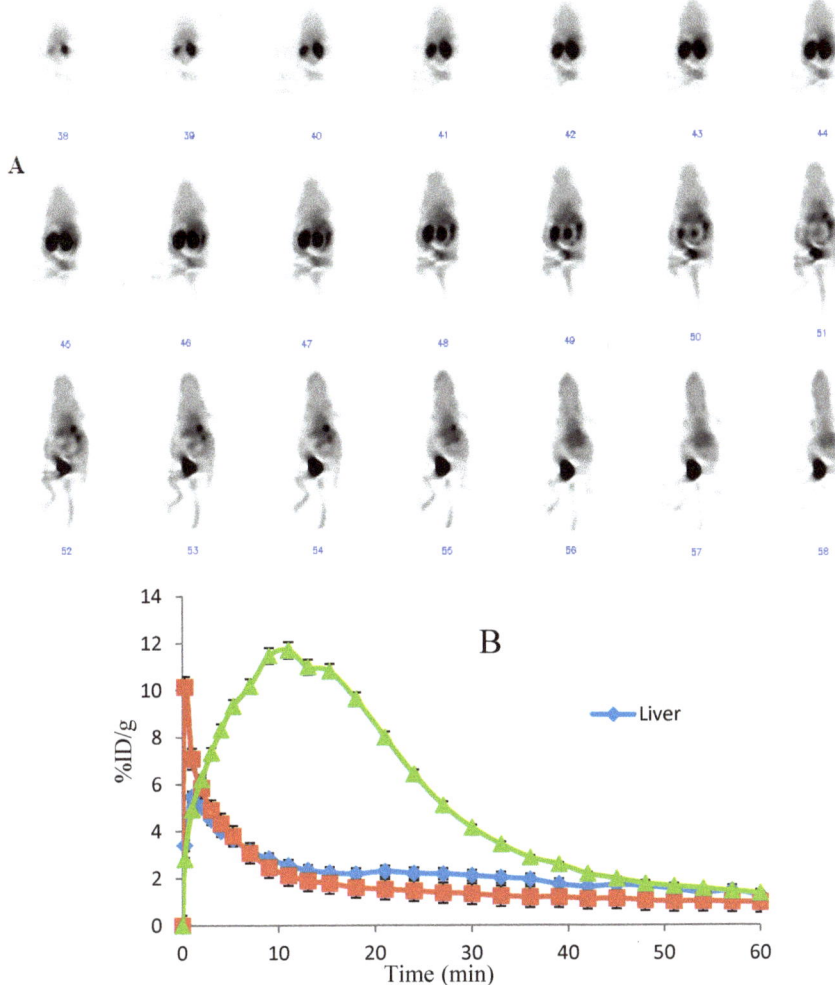

Figure 4. (**A**) Whole body coronal microPET images of ICR mouse from a 60 min dynamic scan after injection of 3.7 MBq ^{18}F-FBEM-Cys-Annexin V. (**B**) Quantified time-activity curves of major organs (liver, heart and kidney) after injection of 3.7 MBq ^{18}F-FBEM-Cys-Annexin V in normal ICR mice (n = 4).

2.5. Imaging of Rat Model of Apoptosis

Four rats were treated with cycloheximide to induce liver apoptosis and two rats were used as the control group. Figure 5 shows the representative coronal microPET images of cycloheximide (CHX)-treated and normal rats at different times after intravenous injection of 8.2 MBq ^{18}F-FBEM-Cys-Annexin V. ^{18}F-FBEM-Cys-Annexin V tracer uptake in the liver (arrow) was increased with CHX treatment. The uptake ratios (treated/control) of liver were 3.88 ± 0.05 (n = 4) and 10.35 ± 0.08 (n = 4), respectively, at 1 h and 2 h p.i.

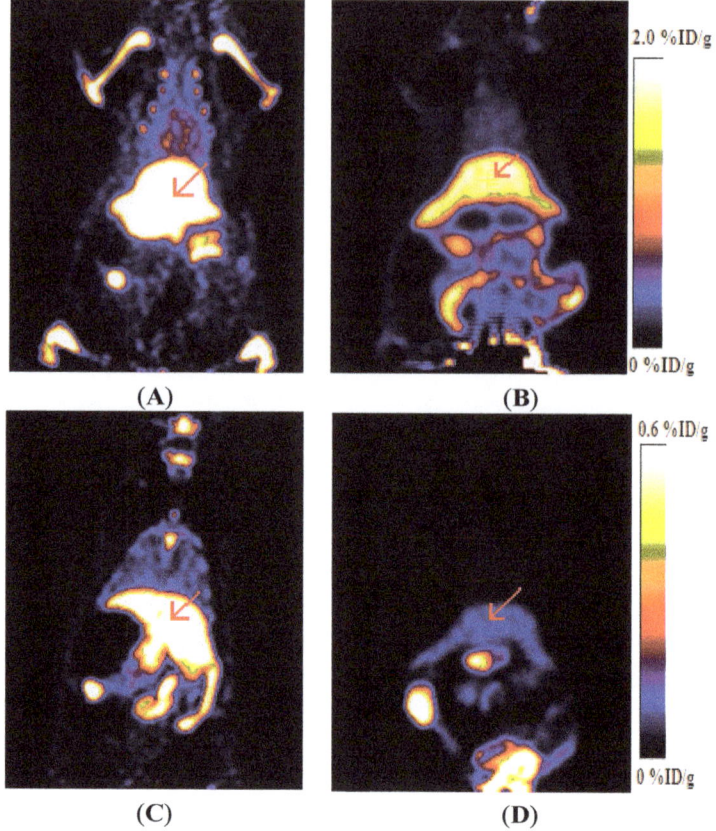

Figure 5. MicroPET images of CHX-treated and normal rats after injection of ^{18}F-FBEM-Cys-Annexin V. (**A**) CHX-treated at 1 h p.i. (**B**) Normal at 1 h p.i. (**C**) CHX-treated at 2 h p.i. (**D**) Normal at 2 h p.i.

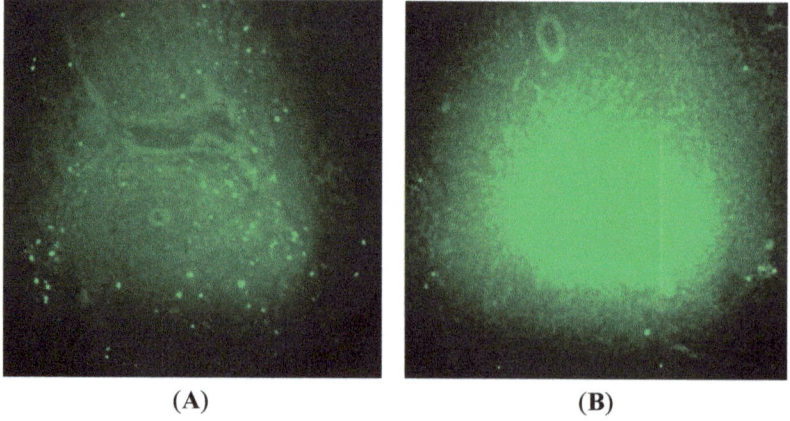

Figure 6. Representative TUNEL-stained images of liver specimen in CHX-treated rats (**A**), control rats (**B**). Green color dot represents positive TUNEL staining.

There were no differences in the blood pool activity between treated and control rats. TUNEL–staining images of liver sections were shown in Figure 6. These images show that the number of apoptotic nuclei in CHX-treated liver was more compared to that in non-treatment control rats. The uptake ratio (treated/control) of liver at 2 h p.i. as measured via microPET correlated with the ratio (treated/control) of apoptotic nuclei in liver observed using TUNEL histochemistry.

3. Experimental Section

3.1. General Information

Unless otherwise specified, all reagents were of analytical grade and were obtained from commercial sources. Cys-Annexin V was supplied by Jiangsu Target Pharma Laboratories Inc. (Changzhou, China). ^{18}F fluoride was obtained from a cyclotron (HM67, Sumitomo Heavy Industries, Ltd, Tokyo, Japan) located at the Jiangsu Institute of Nuclear Medicine by proton irradiation of ^{18}O-enriched water. A Waters high-performance liquid chromatography (HPLC) system (Waters, Milford, MA, USA) equipped with a Waters 2998 photodiode array detector (PDA) and a semi-preparative C18 HPLC column (250 × 10 mm, 5 μm, CHROM-MATRIX BIO-TECH) was used for ^{18}F-FBEM purification. The flow rate was 3 mL/min, and the mobile phase changed from 95% solvent A (0.1% trifluoroacetic acid in water) and 5% solvent B (0.1% trifluoroacetic acid in acetonitrile) (0–2 min). The mobile phase was gradually changed to 35% solvent A and 65% solvent B at 32 min. The UV absorbance was monitored at 218 nm, and the UV spectrum was checked with the PDA detector.

Analytical HPLC was performed on Waters Breeze system with a TSK-GEL column (swG2000SWXL, 300 × 7.8 mm 5 μm, Tosoh Bioscience Co., Ltd, Shanghai, China). The absorbance was measured on the UV detector at 278 nm. Radioanalysis of the labeled compound was conducted using a Cd (Te) detector. The flow rate was adjusted to 0.8 mL/min and the isocratic mobile phase was 0.05 mol/L phosphate buffer (pH = 7.0).

A microPET system (Inveon, Siemens Co. Knoxville, TN, USA) and a fluorescence microscope (X51, Olympus, Tokyo, Japan) were used. The animal experiments in this study were approved by the Animal Care and Ethnics Committee of Jiangsu Institute of Nuclear Medicine.

3.2. Preparation of ^{18}F-N-[2-(4-Fluorobenzamido)ethyl]maleimide

^{18}F-N-[2-(4-Fluorobenzamido)ethyl]maleimide (^{18}F-FBEM) was prepared as previously described using a semi-automatic method with some modifications [22,26]. Briefly, the precursor, ethyl 4-(trimethylammonium) benzoate trifluoromethanesulfonic acid salt (10 mg, 12 μmol) in anhydrous acetonitrile (1.0 mL) was heated at 100 °C for 10 min in a sealed vial with 18.5 GBq [^{18}F]fluoride in the presence of dried Kryptofix2.2.2 (15 mg 6 μmol) and K$_2$CO$_3$ (5 mg, 3.6 mmol). The intermediate was hydrolyzed with NaOH (0.5 M, 0.5 mL) at 90 °C for 5 min. After acidification with 7.5 mL 0.1 M HCl, the solution was loaded onto an activated C18 Sep-Pak column (Waters). The cartridge was then eluted with 3 mL acetonitrile and the eluate was subsequently evaporated at room temperature with a stream of nitrogen to obtain ^{18}F-fluorobenzoic acid (^{18}F-FBA). ^{18}F-FBA was treated with N-(2-aminoethyl)maleimide (MAL, 15 mg, 59 μmol), diethyl cyanophosphonate (20 μL, 99 μmol), and N,N-diisopropylethylamine (40 μL, 240 μmol) in anhydrous acetonitrile (0.5 mL). The resulting solution was heated at 75 °C for 7 min. The reaction was quenched by adding water (8.5 mL) and loaded onto an activated C18 Sep-Pak column. The cartridge was eluted with 1 mL ethanol which was then loaded on to the semi-preparative HPLC. The radioactive peak eluting at ~18 min was collected and passed through a C18 Sep-Pak column which was activated by EtOH/water. The cartridge was washed with 20 mL water and then eluted with 1 mL CH$_2$Cl$_2$. The organic layer was evaporated to dryness at room temperature under a stream of nitrogen and utilized for further Cys-Annexin V labeling. The total synthesis time for ^{18}F-FBEM was about 100 min and 428 ± 65 MBq (n = 4) radiochemically pure ^{18}F-FBEM was obtained from 18.5 GBq ^{18}F-fluoride.

3.3. Preparation of ^{18}F-FBEM-Cys-Annexin V

The isolated ^{18}F-FBEM (111–370 MBq) in 10 µL of ethanol was added to a solution of Cys-Annexin V (50–100 µg in 100 µL, pH = 7.2) PBS (Scheme 1), and the mixture was allowed to react at room temperature for 15–30 min and loaded onto a NAP-5 column (GE Healthcare, Buckinghamshire, UK). The NAP-5 column was eluted with 250 µL portions of PBS. The most concentrated fraction containing the radiolabeled protein (fraction 3, 81–262 MBq) was collected and used for the biological experiments.

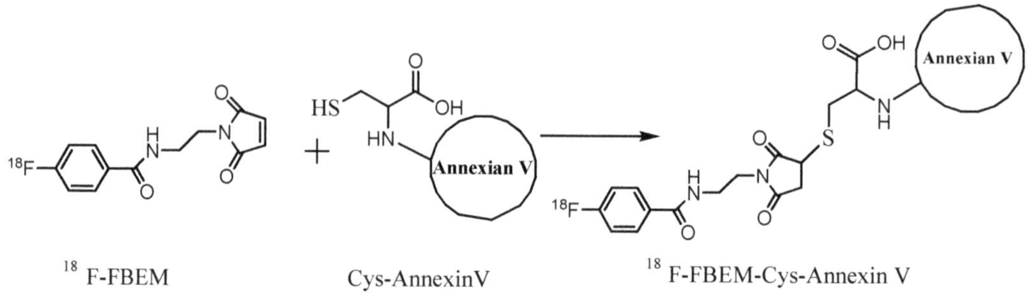

Scheme 1. Syntheses of ^{18}F-FBEM-Cys-Annexin V.

3.4. In Vitro Stability of ^{18}F-FBEM-Cys-Annexin V

The *in vitro* stabilities of freshly prepared ^{18}F-FBEM-Cys-Annexin V were performed in PBS (0.1 mol/L, pH 7.2) and human serum, respectively, for different time intervals (0–6 h) at 37 °C in a water bath.)

3.5. Blood Kinetics Studies of ^{18}F-FBEM-Cys-Annexin V in Normal Mice

Five ICR mice were injected via the tail vein with ^{18}F-FBEM-Cys-Annexin V (0.2 mL) and activity of approximately 3.7 MBq. Ten µL of blood were taken from tails at 5, 15, 30, 45, 60, 90 and 120 min after injection. The activity for each sample was determined by a γ-counter and expressed as percentage of injection dose per gram (%ID/g).

3.6. Dynamical MicroPET Images of Normal Mice

Four ICR mice were anesthetized with 1%–2% isoflurane, positioned prone, immobilized, and injected via the tail vein with 3.7 MBq ^{18}F-FBEM-Cys-Annexin V (0.2 mL) and imaged dynamically for 1 h. The images were reconstructed using a two dimensional ordered-subset expectation maximization (2D OSEM) algorithm without correction for attenuation or scattering. For each scan, regions of interest (ROIs) were drawn over the liver and major organs using the vendor-supplied software (ASI Pro 5.2.4.0) on decay-corrected whole-body coronal images. The radioactivity concentrations (accumulation) within the liver, heart and kidneys were obtained from mean pixel values within the multiple ROI volume and then converted to megabecquerel per milliliter per minute using the calibration factor determined for the Inveon PET system. These values were then divided by the administered activity to obtain (assuming a tissue density of 1 g/mL) an image-ROI-derived percent injected dose per gram (%ID/g).

3.7. MicroPET Images of Rat Model of Apoptosis

Four male SD rats (258 ± 2 g) were treated IV with 10 mg/kg cycloheximide to induce liver apoptosis. Two male SD rats (262 g and 256 g) were treated IV with saline as the control group. 3 h after treatment, the rats were anesthetized with 1%–2% isoflurane and were injected via the tail vein with 8.2 MBq ^{18}F-FBEM-Cys-Annexin V (0.2 mL). Ten-minute static scans were acquired at 1 and 2 h after injection with a MicroPET (Inveon, Siemens), respectively, which was from 1 h to 1 h and 10 min or from 2 h to 2 h and 10 min. Immediately after MicroPET imaging, the livers were dissected. Then, using the livers, formalin-fixed paraffin-embedded specimens were prepared for Terminal deoxynucleotidyl transferase-mediated nick end labeling (TUNEL) staining.

3.8. TUNEL Staining

Because our imaging studies were designed to determine the uptake and biodistribution of ^{18}F-FBEM-Cys-AnnexinV after chemically induced apoptosis, it was important to confirm apoptosis in the livers of treated rats by independent methods that provide quantitative results. A marker of apoptosis was scored by performing a TUNEL assay that measures DNA fragmentation, a characteristic feature of apoptosis. Terminal deoxynucleotide transferase adds labeled nucleotides to the 3' termini at double-stranded breaks in the fragmented DNA. TUNEL assays were performed according to the manufacturer's instructions, using the fluorescein-conjugated Colorimetric TUNEL Apoptosis Assay Kit (Beyotime Institute of Biotechnology, Shanghai, China). Briefly, slices were freed of paraffin through xylene and graded EtOH washes and then incubated with proteinase K (Beyotime Institute of Biotechnology, 2 mg/mL in 10 mmol/L Tris, pH 8.0). After proteinase digestion, the slides were equilibrated in pH 7.4 buffer, the terminal deoxynucleotidetransferase enzyme and Biotin-dUTP labeling mix (Beyotime Institute of Biotechnology) were added, and the slides were incubated at 37 °C for 1 h in a humid chamber. The number of TUNEL-positive cells was counted on 10 randomly selected ×100 fields for each section by use of an Olympus fluorescence microscope.

4. Conclusions

Cys-annexin V, a novel annexinVderivative with a single cysteine residue at the C-terminal, could be site-specifically labeled with ^{18}F-FBEM in high yields and high radiochemical purity. In normal mice, ^{18}F-FBEM-Cys-Annexin V was excreted mainly through the renal pathway. Hepatic uptake of ^{18}F-FBEM-Cys-Annexin V was significantly increased in the rats treated with CHX compared to controls, which correlated well with the increase in cell death observed using TUNEL histochemistry. These results indicate that the novel ^{18}F-FBEM-Cys-Annexin V is a potential apoptosis imaging agent and further study is needed.

Acknowledgments

The authors are very grateful to the Ministry of Health Foundation of China (W201207), Open Fund Project (KF-GN-201304) of State Key Laboratory of Pharmaceutical Biotechnology and Public service platform for Science and technology infrastructure construction project of Jiangsu Province (BM2012066) for their financial support.

Author Contributions

Chunxiong Lu and Zichun Hua conceived and designed the experiments; Chunxiong Lu, Quanfu Jiang and Cheng Tan performed the experiments; Chunxiong Lu and Huixin Yu analyzed the data; Zichun Hua and Minjin Hu contributed Cys-Annexin V; Chunxiong Lu wrote the paper.

Conflicts of Interest

The authors declare no conflict of interest.

References

1. Fink, S.L.; Cookson, B.T. Apoptosis, pyroptosis, and necrosis: Mechanistic description of dead and dying eukaryotic cells. *Infect. Immun.* **2005**, *73*, 1907–1916.
2. Hofstra, L.; Liem, I.H.; Dumont, E.A.; Boersma, H.H.; van Heerde, W.L.; Doevendans, P.A.; DeMuinck, E.; Wellens, H.J.J.; Kemerink, G.J.; Reutelingsperger, C.P.M.; *et al*. Visualisation of cell death *in vivo* in patients with acute myocardial infarction. *Lancet* **2000**, *356*, 209–212.
3. Thiagarajan, P.; Tait, J.F. Binding of Annexin-V Placental Anticoagulant ProteinI to Platelets—Evidence for Phosphatidylserine Exposure in The Procoagulant Response of Activated Platelets. *J. Biol. Chem.* **1990**, *265*, 17420–17423.
4. Tait, J.F.; Gibson, D.; Fujikawa, K. Phospholipid Binding-Properties of Human Placental Anticoagulant Protein-I, A Member of the Lipocortin Family. *J. Biol. Chem.* **1989**, *264*, 7944–7949.
5. Gerke, V.; Moss, S.E. Annexins: From structure to function. *Physiol. Rev.* **2002**, *82*, 331–371.
6. Koopman, G.; Reutelingsperger, C.P.M.; Kuijten, G.A.M.; Keehnen, R.M.J.; Pals, S.T.; Vanoers, M.H.J. Annexin-V For Flow Cytometric Detection of Phosphatidylserine Expression on B-cells Undergoing Apoptosis. *Blood* **1994**, *84*, 1415–1420.

7. Kemerink, G.J.; Boersma, H.H.; Thimister, P.W.; Hofstra, L.; Liem, I.H.; Pakbiers, M.T.; Janssen, D.; Reutelingsperger, C.P.; Heidendal, G.A. Biodistribution and dosimetry of 99mTc-BTAP-annexin-V in humans. *Eur. J. Nucl. Med.* **2001**, *28*, 1373–1378.
8. Boersma, H.H.; Liem, I.H.; Kemerink, G.J.; Thimister, P.W.L.; Hofstra, L.; Stolk, L.M.L.; van Heerde, W.L.; Pakbiers, M.T.W.; Janssen, D.; Beysens, A.J.; *et al.* Comparison between human pharmacokinetics and imaging properties of two conjugation methods for Tc-99m-Annexin A5. *Br. J. Radiol.* **2003**, *76*, 553–560.
9. Vanderheyden, J.L.; Liu, G.; He, J.; Patel, B.; Tait, J.F.; Hnatowich, D.J. Evaluation of 99mTc-MAG3-annexin V: Influence of the chelate on *in vitro* and *in vivo* properties in mice. *Nucl. Med. Biol.* **2006**, *33*, 135–144.
10. Yang, D.J.; Azhdarinia, A.; Wu, P.; Yu, D.F.; Tansey, W.; Kalimi, S.K.; Kim, E.E.; Podoloff, D.A. *In vivo* and *in vitro* measurement of apoptosis in breast cancer cells using 99mTc-EC-annexin V. *Cancer Biother. Radiopharm.* **2001**, *16*, 73–83.
11. Kemerink, G.J.; Liu, X.; Kieffer, D.; Ceyssens, S.; Mortelmans, L.; Verbruggen, A.M.; Steinmetz, N.D.; Vanderheyden, J.L.; Green, A.M.; Verbeke, K. Safety, biodistribution, and dosimetry of ^{99m}Tc-HYNIC-annexin V, a novel human recombinant annexin V for human application. *J. Nucl. Med.* **2003**, *44*, 947–952.
12. De Saint-Hubert, M.; Wang, H.; Devos, E.; Vunckx, K.; Zhou, L.; Reutelingsperger, C.; Verbruggen, A.; Mortelmans, L.; Ni, Y.; Mottaghy, F.M. Preclinical Imaging of Therapy Response Using Metabolic and Apoptosis Molecular Imaging. *Mol. Imaging Biol.* **2011**, *13*, 995–1002.
13. Tait, J.F.; Brown, D.S.; Gibson, D.F.; Blankenberg, F.G.; Strauss, H.W. Development and characterization of annexin V mutants with endogenous chelation sites for (99m)Tc. *Bioconjugate Chem.* **2000**, *11*, 918–925.
14. Ye, F.; Fang, W.; Wang, F.; Hua, Z.-C.; Wang, Z.; Yang, X. Evaluation of adenosine preconditioning with Tc-99m-His(10)-annexin V in a porcine model of myocardium ischemia and reperfusion injury: Preliminary study. *Nucl. Med. Biol.* **2011**, *38*, 567–574.
15. Tait, J.F.; Smith, C.; Levashova, Z.; Patel, B.; Blankenberg, F.G.; Vanderheyden, J.L. Improved detection of cell death *in vivo* with annexin V radiolabeled by site-specific methods. *J. Nucl. Med.* **2006**, *47*, 1546–1553.
16. Lu, C.; Jiang, Q.; Hu, M.; Tan, C.; Ji, Y.; Yu, H.; Hua, Z. Preliminary Biological Evaluation of Novel 99mTc-Cys-Annexin A5 as a Apoptosis Imaging Agent. *Molecules* **2013**, *18*, 6908–6918.
17. Zijlstra, S.; Gunawan, J.; Burchert, W. Synthesis and evaluation of a 18F-labelled recombinant annexin-V derivative, for identification and quantification of apoptotic cells with PET. *Appl. Radiat. Isot.* **2003**, *58*, 201–207.
18. Hu, S.; Kiesewetter, D.O.; Zhu, L.; Guo, N.; Gao, H.; Liu, G.; Hida, N.; Lang, L.; Niu, G.; Chen, X. Longitudinal PET Imaging of Doxorubicin-Induced Cell Death with (18)F-Annexin V. *Mol. Imaging Biol.* **2012**, *14*, 762–770.
19. Zhu, J.C.; Wang, F.; Fang, W.; Hua, Z.C.; Wang, Z.Z. ^{18}F-annexin V apoptosis imaging for detection of myocardium ischemia and reperfusion injury in a rat model. *J. Radioanal. Nucl. Chem.* **2013**, *298*, 1733–1738.

20. De Bruin, B.; Kuhnast, B.; Hinnen, F.; Yaouancq, L.; Amessou, M.; Johannes, L.; Samson, A.; Boisgard, R.; Tavitian, B.; Dolle, F. 1-[3-(2-[^{18}F]-fluoropyridin-3-yloxy)propyl]pyrrole-2,5-dione: Design, synthesis, and radiosynthesis of a new [^{18}F]-fluoropyridine-based maleimide reagent for the labeling of peptides and proteins. *Bioconjugate Chem.* **2005**, *16*, 406–420.
21. Gao, H.; Niu, G.; Yang, M.; Quan, Q.; Ma, Y.; Murage, E.N.; Ahn, J.M.; Kiesewetter, D.O.; Chen, X. PET of Insulinoma Using (18)F-FBEM-EM3106B, a New GLP-1 Analogue. *Mol. Pharm.* **2011**, *8*, 1775–1782.
22. Kiesewetter, D.O.; Jacobson, O.; Lang, L.; Chen, X. Automated radiochemical synthesis of [^{18}F]FBEM: A thiol reactive synthon for radiofluorination of peptides and proteins. *Appl. Radiat. Isot.* **2011**, *69*, 410–414.
23. Wang, H.; Gao, H.; Guo, N.; Niu, G.; Ma, Y.; Kiesewetter, D.O.; Chen, X. Site-Specific Labeling of scVEGF with Fluorine-18 for Positron Emission Tomography Imaging. *Theranostics* **2012**, *2*, 607–617.
24. Keen, H.G.; Dekker, B.A.; Disley, L.; Hastings, D.; Lyons, S.; Reader, A.J.; Ottewell, P.; Watson, A.; Zweit, J. Imaging apoptosis *in vivo* using 124I-annexin V and PET. *Nucl. Med. Biol.* **2005**, *32*, 395–402.
25. Bauwens, M.; De Saint-Hubert, M.; Devos, E.; Deckers, N.; Reutelingsperger, C.; Mortelmans, L.; Himmelreich, U.; Mottaghy, F.M.; Verbruggen, A. Site-specific Ga-68-labeled Annexin A5 as a PET imaging agent for apoptosis. *Nucl. Med. Biol.* **2011**, *38*, 381–392.
26. Li, W.; Niu, G.; Lang, L.; Guo, N.; Ma, Y.; Kiesewetter, D.O.; Backer, J.M.; Shen, B.; Chen, X. PET imaging of EGF receptors using ^{18}F-FBEM-EGF in a head and neck squamous cell carcinoma model. *Eur. J. Nucl. Med. Mol. Imaging* **2012**, *39*, 300–308.

Sample Availability: Sample of the compound Cys-Annexin V is available from the authors.

© 2015 by the authors; licensee MDPI, Basel, Switzerland. This article is an open access article distributed under the terms and conditions of the Creative Commons Attribution license (http://creativecommons.org/licenses/by/4.0/).

Molecules **2015**, *20*, 1712-1730; doi:10.3390/molecules20011712

ISSN 1420-3049
www.mdpi.com/journal/molecules

Article

Synthesis and *in Silico* Evaluation of Novel Compounds for PET-Based Investigations of the Norepinephrine Transporter

Catharina Neudorfer [1,2,*], **Amir Seddik** [3], **Karem Shanab** [1,2], **Andreas Jurik** [3], **Christina Rami-Mark** [1], **Wolfgang Holzer** [2], **Gerhard Ecker** [3], **Markus Mitterhauser** [1], **Wolfgang Wadsak** [1] and **Helmut Spreitzer** [2,*]

[1] Department of Biomedical Imaging and Image-guided Therapy, Division of Nuclear Medicine, Medical University of Vienna, Waehringer Guertel 18-20, 1090 Vienna, Austria; E-Mails: karem.shanab@gmail.com (K.S.); christina.rami-mark@meduniwien.ac.at (C.R.-M.); markus.mitterhauser@meduniwien.ac.at (M.M.); wolfgang.wadsak@meduniwien.ac.at (W.W.)

[2] Division of Drug Synthesis, Department of Pharmaceutical Chemistry, Faculty of Life Sciences, University of Vienna, Althanstraße 14, 1090 Vienna, Austria; E-Mail: wolfgang.holzer@univie.ac.at

[3] Division of Drug Design and Medicinal Chemistry, Department of Pharmaceutical Chemistry, Faculty of Life Sciences, University of Vienna, Althanstraße 14, 1090 Vienna, Austria; E-Mails: amir.seddik@univie.ac.at (A.S.); andreas.jurik@univie.ac.at (A.J.); gerhard.f.ecker@univie.ac.at (G.E.)

* Authors to whom correspondence should be addressed;
E-Mails: catharina.neudorfer@gmail.com (C.N.); helmut.spreitzer@univie.ac.at (H.S.);
Tel.: +43-4277-55629 (C.N.); +43-4277-55621 (H.S.);
Fax: +43-4277-855629 (C.N.); +43-4277-855621 (H.S.).

Academic Editor: Svend Borup Jensen

Received: 20 November 2014 / Accepted: 14 January 2015 / Published: 20 January 2015

Abstract: Since the norepinephrine transporter (NET) is involved in a variety of diseases, the investigation of underlying dysregulation-mechanisms of the norepinephrine (NE) system is of major interest. Based on the previously described highly potent and selective NET ligand 1-(3-(methylamino)-1-phenylpropyl)-3-phenyl-1,3-dihydro-2*H*-benzimidaz- ol-2-one (Me@APPI), this paper aims at the development of several fluorinated methylamine-based analogs of this compound. The newly synthesized compounds were computationally evaluated for their interactions with the monoamine transporters and represent reference compounds for PET-based investigation of the NET.

Keywords: NET; ADHD; cocaine dependence; BAT; PET; FAPPI

1. Introduction

Abnormal regulation of the norepinephrine transporter (NET) or NET dysfunction, respectively, cause either increased or decreased levels of norepinephrine (NE) in the synaptic cleft. Since NE is a fundamental neurochemical messenger, its accurate regulation is of major importance. Thus, the NET, responsible for NE equilibrium in the synaptic cleft, is representing the reuptake site and considered to be involved in a variety of neurological/psychiatric disorders [1,2], but also plays a pivotal role in cardiovascular [1–3] and metabolic diseases [3–5]. Reduced NET levels go along with neurological disorders like major depression [6,7], Parkinson's disease (PD), Alzheimer's disease (AD) [8–18], and cardiovascular diseases such as hypertension, cardiomyopathy, and heart failure [5,13]. Furthermore, a dysfunction of the NE system was reported in Attention Deficit Hyperactivity Disorder (ADHD) [9,17,19], suicide [1,12,20], substance abuse (cocaine dependence) [16], and schizophrenia [10]. A more recent discovery is the involvement of the NET in diseases like diabetes and obesity, due to its presence in brown adipose tissue (BAT) and the proposed activation thereof via NE [4,5,21].

Based on the fact that the NET is involved in such a variety of diseases, the investigation of the underlying dysregulation-mechanism of the NE system is of major interest. For this purpose, information about the transporter abundance and density in healthy and pathological living human brains is required. The most suitable and accurate technique to gain this information is positron emission tomography (PET). As a non-invasive molecular imaging technique, it represents a suitable approach towards the collection of missing data in the living organism and direct quantification of receptor/transporter densities *in vivo*. To fully gain insight in the molecular changes of the noradrenergic system via PET, however, prior development of suitable NET PET radioligands is required.

So far, radiolabeled NET binding reboxetine analogs [^{11}C]MeNER, [^{11}C]MRB, ((S,S)-2-(α-(2-[^{11}C]-methoxyphenoxy)benzyl)morpholine) and [^{18}F]FMeNER-D$_2$ ((S,S)-2-(α-(2-[^{18}F]fluoro[^{2}H$_2$]methoxy-phenoxy)benzyl)morpholine) have been described, which however display certain limitations such as metabolic instability, complex radiosyntheses, or late equilibria [22–26].

Recently, Zhang *et al.* [26] evaluated a series of benzimidazolone-based propanamines with *in vitro* inhibitory activity on the human norepinephrine transporter (hNET). The results of these investigations suggested that compounds containing a phenyl moiety directly attached at the benzimidazolone ring (e.g., **1**, Figure 1) were the most potent, representing a half maximal inhibitory concentration (IC$_{50}$) below 10 nM (IC$_{50}$ < 10 nM). Furthermore, hNET selectivity over human serotonin transporter (hSERT) turned out >300-fold superior to those of reboxetine and atomoxetine (16- and 81-fold) [26]. Fluorination at position 2 of the phenyl moiety attached to the benzimidazolone ring (e.g., **2**, Figure 1), indicated similar hNET potency, comparable to its non-fluorinated analogs (e.g., **1**) and additionally exhibited hNET selectivity over hSERT (80-fold) similar to atomoxetine [26].

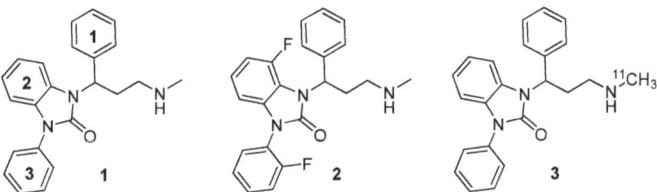

Figure 1. Structures of the highly potent and selective NET ligands 1-(3-(methylamino)- 1-phenylpropyl)-3-phenyl-1,3-dihydro-2*H*-benzimidazol-2-one (Me@APPI, **1**) and 4-fluoro- 1-(2-fluorophenyl)-3-(3-(methylamino)-1-phenylpropyl)-1,3-dihydro-2*H*-benzimidazol-2-one (**2**) as well as radiolabeled analog [^{11}C]Me@APPI (**3**).

Both benzimidazolone derived propanamines with a phenyl moiety (e.g., **1**), as well as a fluorinated phenyl moiety (as in **2**) indicated excellent hNET selectivity over human dopamine transporter (hDAT) with < 50% inhibition of the cocaine analog [^{3}H]WIN-35,428, binding to hDAT at a concentration of 10 μM [26]. Given those findings, both described benzimidazolone-based propanamines **1** and **2** represent excellent candidates for selective and potent NET inhibition with high affinity and low unspecific binding. Thus, on the basis of the results of Zhang *et al.* [26] the methylamino moiety of the core compound **1** has been radiolabeled with ^{11}C and tested by our research group [27]. All investigated preclinical parameters, such as affinity, blood brain barrier penetration, lipophilicity, metabolic degradation, and selectivity showed excellent results, thus suggesting suitability of ^{11}C-radiolabeled 1-(3-(methylamino)-1-phenylpropyl)-3-phenyl-1,3-dihydro-2*H*-benzimidazole-2-one ([^{11}C]Me@APPI) (**3**, Figure 1) as a NET radioligand for use in PET.

Due to successful preclinical testing of [^{11}C]Me@APPI (**3**) and given the excellent *in vitro* results of compound **2**, shown by Zhang *et al.* [26] the aim of this paper is the synthesis and docking studies of several fluorinated analogs **4–6** of compound **1** (Figure 2) as reference compounds for their later prepared radioactive analogs. All methylamine-derived benzimidazolone derivatives **4–6** will be subjected to affinity, selectivity, and lipophilicity studies towards the NET as well as blood brain barrier penetration experiments at the Medical University of Vienna. The most promising NET ligands will then be selected for the development of new, selective PET tracers for the NET and after radiolabeling with both ^{11}C and ^{18}F, they will be the subject of further experiments.

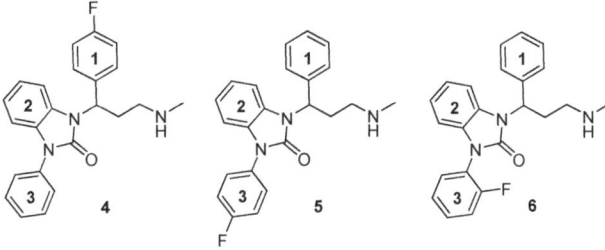

Figure 2. Chemical structure of envisaged reference compounds **4–6** (FAPPI:1-3).

2. Results and Discussion

The synthesis of reference compounds **4–6** first required the preparation of side chains **11** and **12**, as well as core compounds **16–18**. After successful preparation, side chain **11** was merged in a condensation reaction with core compound **16**, whereas side chain **12** was reacted with core compounds **17** and **18**, prior to halogen exchange and substitution with methylamine (Scheme 1).

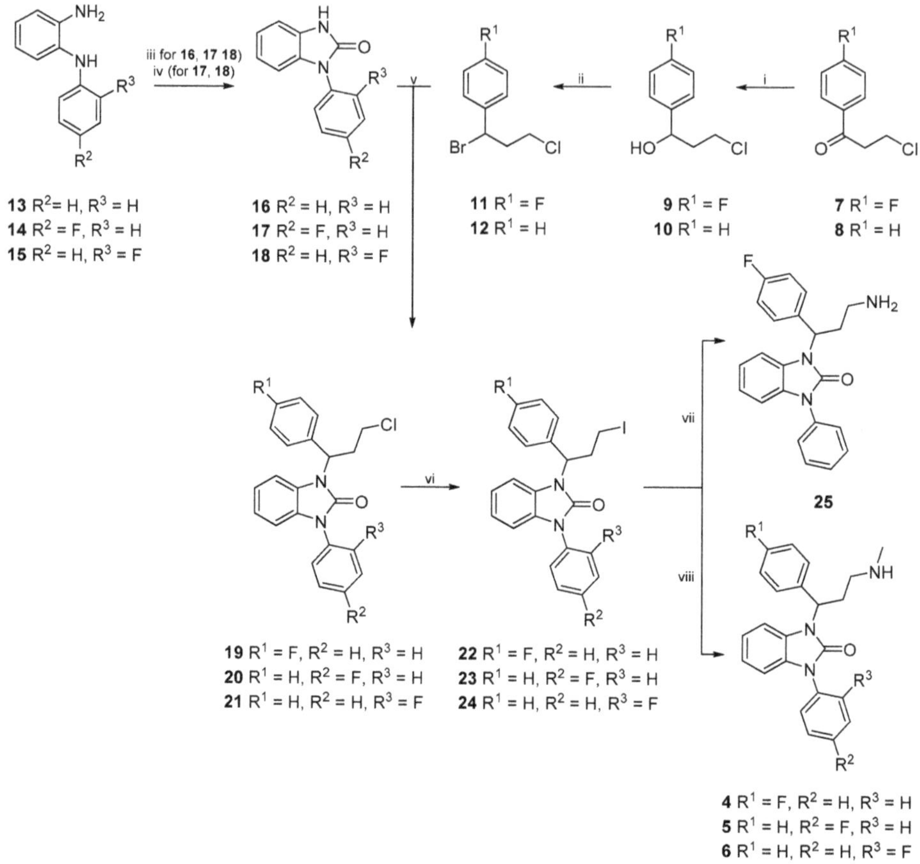

Reagents and conditions: (i) THF, EtOH, NaBH$_4$, −10 °C → −5 °C, 10 min, yields for **9**: 95%, **10**: 100%; (ii) aq. HBr, rt, 3 h, yields for **11**: 64%, **12**: 86%; (iii) 1,1'-carbonyldiimidazole, anhyd. THF, rt, overnight, yields for **16**: 74%, **17**: 61%, **18**: 69%; (iv) 1,1'-carbonyldiimidazole, anhyd. DMF, 90 °C, 2 h, yields for **17**: 80%, **18**: 80%; (v) K$_2$CO$_3$, DMF, rt, 30 min → addition of **11** and **12** → rt, 30 min, yields for **19**: 32%, **20**: 63%, **21**: 55%; (vi) NaI, acetone, reflux, 24 h, yields for **22**: 82%, **23**: 76%, **24**: 53%; (vii) NH$_3$ in isopropanol, 80 °C, 3 h, yield for **25**: 50%; (viii) methylamine in EtOH, 80 °C, 3 h, yields for **4**: 48%, **5**: 29%, **6**: 30%.

Scheme 1. Synthesis of compounds **4–6**.

For the synthesis of side chains **11** and **12**, a protocol of Varney *et al.* [28] was adopted and the keto group of commercially available compounds **7** and **8** was reduced with sodium borohydride in order to obtain intermediate alcohols **9** and **10**. Subsequent bromination of **9** and **10** with aqueous hydrogen bromide led to the formation of products **11** and **12**, respectively.

Core compound **16** was prepared by the reaction of commercially available *N*-phenylbenzene-1,2-diamine (**13**) with 1,1'-carbonyldiimidazole. For the preparation of **17** and **18** however, **14** and **15** first had to be made accessible (Scheme 2). Thus, 1-fluoro-2-nitrobenzene (**26**) reacted with commercially available fluoroanilines **27** and **28**, respectively, to obtain disubstituted amines **29** and **30**. Therefore, two different methods were applied (Scheme 2): The first method (i) was conducted by heating **26** and **27** with anhydrous potassium fluoride and potassium carbonate in a microwave oven [29]. Since the adoption of an alternative method [30]—conventional heating at 180 °C—gave compound **29** in higher yields, this approach (ii) was chosen for the large scale synthesis of **29** as well as for the preparation of **30**.

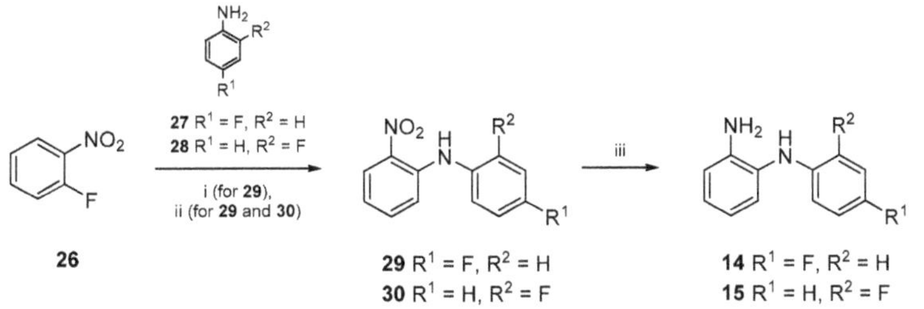

Reagents and conditions: (i) anhyd. KF, K$_2$CO$_3$, microwave oven, 900 W, 10 min, yields for **29**: 58%; (ii) anhyd. KF, K$_2$CO$_3$, 180 °C, 2 d, yields for **29**: 68%, **30**: 52%; (iii) Zn, AcOH, 0 °C → rt, 2 h, yields for **14**: 93%, **15**: n.d.

Scheme 2. Synthesis of compounds **14–15**.

In the next reaction step, the nitro groups of both disubstituted amines **29** and **30** were reduced. For this purpose **29** or **30** were added to a mixture of zinc/acetic acid. The resulting amines **14** or **15** were obtained in excellent yields (Scheme 2) [30].

Freshly prepared intermediates **14** and **15** were then subjected to a cyclization reaction with 1,1'-carbonyldiimidazole in DMF under anhydrous conditions (Scheme 1) by modifying a synthesis protocol according to Zhang *et al.* [26]. DMF was preferred over THF to ensure higher yields and shorter reaction times.

Condensation reactions of benzimidazolones **16**, **17**, and **18** with side chains **11** and **12**, respectively were performed under basic conditions (Scheme 1) by adapting a procedure of Jona *et al.* [31]. After purification, the chloro-substituted derivatives **19–21** were converted into the iodo compounds **22–24** in a Finkelstein reaction. Target compounds **4–6** (FAPPI:1-3) were then obtained by heating derivatives **22–24** in a solution of methylamine in ethanol in a sealed tube for 3 h. In addition to reference compounds **4–6**, free amine **25** was synthesized by dissolving **22** in a solution of ammonia in isopropanol and heating the resulting mixture in a sealed tube for 3 h. As compound **25** features a free amine moiety, it can be considered a precursor for radiosynthesis.

Since compounds **4** and **5** comprise a novel fluoro substitution, a computational docking study was performed to assess if these compounds still would fit in the binding site of the NET. Furthermore, we aimed at creating a binding mode hypothesis which allows gaining insights into the molecular basis of binding and selectivity towards the monoamine transporters. As the basic scaffold has been shown to act in an enantioselective manner, the respective (*R*) enantiomers were used throughout the docking studies [26]. The ligands were docked in the substrate binding site (S1) of the outward-open conformation of the transporter models (see Experimental Section for details), since related inhibitors, such as nortriptyline, sertraline, mazindol, *etc.* were also shown to fit in the S1 of the *Drosophila* DAT (dDAT) and the "SERT"-ized leucine transporter ("LeuBAT") in the same protein conformation [32,33]. Interestingly, the co-crystallized ligand in dDAT, nortriptyline (**31**), has the same ranking of human NET, SERT and DAT activity as reference compound **1**, *i.e.*, 4.4, 18 and 1149 nM K_D *vs.* 9, 2995 and >10,000 nM IC_{50}, respectively [26,34]. Additionally, nortriptyline shares important structural features with the benzimidazolones, *i.e.*, two aromatic moieties and an *N*-methyl-ethylenamine side chain. Therefore their binding mode can be expected to be similar.

Common scaffold clustering [35] revealed two binding hypotheses (see Experimental Section) which indicated that compounds **4–6** fit in the S1 of all three transporters. Hence, additional fluorination does not seem to cause steric clashes. In both hypotheses, the most prominent protein-ligand interaction was the cationic nitrogen atom placed in the A sub pocket [36], located between the central Asp75/98/79 side chain as a salt-bridge and the Phe72/95/76 side chain as a cation-*pi* interaction in NET/SERT/DAT, respectively. This is well in accordance with the X-ray structures of the templates. Additional *pi-pi* stacking interactions with Phe152/176/156 and Phe323/355/341 further promote the binding in both hypotheses obtained:

Hypothesis 1: the benzimidazolone heterocycle (ring 2) is placed in the B sub pocket and ring 3 in the C pocket, whereas ring 1 is solvent exposed (see figure in Experimental Section).

Hypothesis 2: Ligand ring 1 is placed in the B pocket whereas ring 2 is placed at the same height (measured from the membrane-water interface) and overlap with the rings of nortriptyline (**31**, Figure 3). The solvent exposed Tyr151/Tyr175/Phe155 in NET/SERT/DAT, resp. T-stacks with ring 2 whereas ring 3 points extracellularly (Table 1, Figure 4).

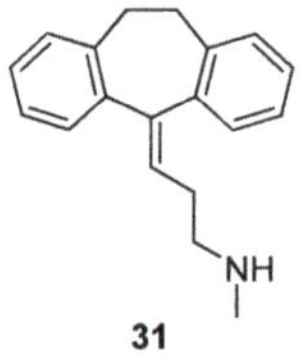

Figure 3. Chemical structure of nortriptyline (co-crystallized ligand from the template, PDB code 4M48) [28].

Table 1. Sequence alignment of all distinct monoamine transporter residues located in sub sites (reference [32]) in the vicinity of the docked compounds. Red: hydrophilic side chain, green: lipophilic side chain, bold: bulkier side chain.

	B Site				C Site			Outer Site	
hNET	S420	M424	G149	V148	A145	F72	D473	Y151	A477
hSERT	T439	L443	A173	I172	A177	Y95	E493	Y175	T497
hDAT	A423	M427	G153	V152	S149	F76	D476	F155	A480

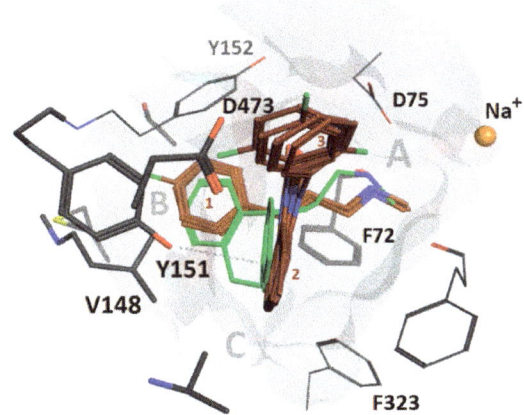

Figure 4. Overlay of compounds **1**, **2**, and **4–6** (maroon) in binding hypothesis 2 in hNET showing agreement with the co-crystal pose of nortriptyline (**31**) (green). Val148 and Asp473 allow more space than Ile172 and Glu439 in hSERT, resp., whereas Tyr151 might induce a more potent stacking interaction as compared to Phe155 in hDAT. The angle between ligand ring 2 and 3 is almost 90° in all poses. The extracellular space is located above.

As binding hypothesis 2 is in close agreement with the co-crystallized ligands in dDAT and LeuBAT, we focus further analysis on this proposed binding mode. Binding hypothesis 2 indicates why the investigated compounds (**4–6**) show weaker affinity to SERT and DAT than to NET: lower SERT affinity may be due to Ile172 and Glu439, allowing less space for the ligand to be accommodated as compared to in NET, that comprises a valine and an aspartate at the homologous positions, respectively.

Lower DAT affinity could be ascribed to weaker T stacking interactions of Phe155 as compared to Tyr151 in NET, based on previous findings that a Tyr-Phe pair has a stronger binding energy than a Phe-Phe pair [37].

Since the docking studies indicate that fluorinated methyl amines **4–6** (FAPPI:1-3) bind in an analogous way to the NET as reference compound **1** [26], compounds **4–6** will be employed in future studies and evaluated for affinity and selectivity towards the NET. Additionally, lipophilicity studies and blood brain barrier penetration experiments are planned for compounds **4–6** at the Medical University of Vienna. The most promising derivatives regarding their suitability as NET ligands will then be selected for the further development of new and selective PET tracers for the NET, which will comprise either a [^{11}C]methylamine, [^{18}F]fluoroalkyl amine or [^{18}F]fluorobenzene radiolabel, respectively. The results of ongoing studies on affinity, selectivity and lipophilicity of the discussed compounds, will be published in a subsequent paper.

3. Experimental Section

3.1. General

The NMR spectra were recorded from CDCl$_3$ or DMSO-d_6 solutions on a Bruker DPX200 spectrometer (200 MHz for ^{1}H, 50 MHz for ^{13}C) or on a Bruker Avance III 400 spectrometer (400 MHz for ^{1}H, 100 MHz for ^{13}C, 40 MHz for ^{15}N, 376 MHz for ^{19}F) at 25 °C. The center of the solvent (residual) signal was used as an internal standard which was related to TMS with δ 7.26 ppm (^{1}H in CDCl$_3$), δ 2.49 ppm (^{1}H in DMSO-d_6), δ 77.0 ppm (^{13}C in CDCl$_3$) and δ 39.5 ppm (^{13}C in DMSO-d_6). ^{15}N NMR spectra (gs-HMBC, gs-HSQC) were referenced against neat, external nitromethane, ^{19}F NMR spectra by absolute referencing via Ξ ratio. Digital resolutions were 0.25 Hz/data point in the ^{1}H and 0.3 Hz/data point in the ^{13}C-NMR spectra. Coupling constants (*J*) are quoted in Hz. The following abbreviations were used to show the multiplicities: s: singlet, d: doublet, t: triplet, q: quadruplet, dd: doublet of doublet, m: multiplet. Mass spectra were obtained on a Shimadzu QP 1000 instrument (EI, 70 eV), high-resolution mass spectrometry (HRMS) was carried out on a Finnigan MAT 8230 (EI, 10 eV) or Finnigan MAT 900 S (ESI, 4 kV, 3 µA, CH$_3$CN/MeOH) electrospray ionization mass spectrometer with a micro-TOF analyzer. Microwave experiments were carried out in a Synthos 3000 microwave oven (SXQ80 rotor, Anton Paar, Graz, Austria) with an internal temperature probe. Compound purity: all compounds synthesized featured a purity of at least 95%.

3.2. Syntheses

3.2.1. General Procedure for the Synthesis of **9** and **10**

Starting materials **7** or **8**, respectively (1 mmol) was dissolved in THF (1 mL) and EtOH (1 mL) was added. The mixture was cooled to −10 °C and NaBH$_4$ (1.05 mmol) was slowly added at this temperature. The solution was stirred at −5 °C for 10 min and thereafter, poured into a mixture of saturated aqueous ammonium chloride (3 mL) in ice (1.5 g). The product was extracted with diethyl ether, dried over Na$_2$SO$_4$ and evaporated to dryness. The crude product was employed directly in the subsequent reaction step without further purification.

3-Chloro-1-(4-fluorophenyl)propan-1-ol (**9**). Yield: 4.78 g (95%), pale yellow oil, analytical data are in complete accordance with literature values [38].

3-Chloro-1-phenylpropan-1-ol (**10**). Yield: 4.61 g (99%), light yellow oil, analytical data are in complete accordance with literature values [28].

3.2.2. General Procedure for the Synthesis of **11** and **12**

To starting material **9** or **10**, respectively (1 mmol) was added 48% aqueous HBr (3 mL) and the mixture was stirred for 3h at room temperature. Thereafter, the solution was poured into a mixture of K$_2$CO$_3$ (1 g) in ice (5.5 g) and additional solid K$_2$CO$_3$ was added for neutralization (pH 7). The crude reaction product was extracted with diethyl ether, the combined organic layers were dried with MgSO$_4$ and evaporated to dryness. The crude product was employed directly in the subsequent reaction step without further purification.

1-(1-Bromo-3-chloropropyl)-4-fluorobenzene (**11**). Yield: 64%, pale yellow oil, analytical data are in complete accordance with literature values [39].

(1-Bromo-3-chloropropyl)benzene (**12**). Yield: 5.64 g (86%), yellow oil, analytical data are in complete accordance with literature values [28].

3.2.3. General Procedure for the Synthesis of **29** and **30**

4-Fluoroaniline or 2-fluoroaniline, respectively (1 mmol), anhydrous KF (1 mmol), and K_2CO_3 (1 mmol) were well powdered with a mortar and a pestle, then 1-fluoro-2-nitrobenzene (1 mmol) was added and the mixture was stirred for 2 days at 180 °C. Thereafter, water (5 mL) and CH_2Cl_2 (5 mL) were added and the organic layer was washed with 10% HCl (5 mL) and brine (5 mL). The combined organic layers were dried over Na_2SO_4 and evaporated to dryness prior to purification by column chromatography.

N-(4-Fluorophenyl)-2-nitroaniline (**29**). Yield: 68%, dark orange crystals, mp. 82–83 °C, purification: silica gel 60, petroleum ether/ethyl acetate 9:1 and RP-18 silica gel, methanol/water 7:3, analytical data are in complete accordance with literature values [29].

2-Fluoro-N-(2-nitrophenyl)aniline (**30**). Yield: 52%, orange crystals, mp. 79–80 °C, purification: silica gel 60, petroleum ether/ethyl acetate 9:1, analytical data are in complete accordance with literature values [40].

3.2.4. Alternative Procedure for the Synthesis of **29**

4-Fluoroaniline (7.89 g, 6.73 mL, 70.87 mmol), anhydrous KF (4.13 g, 70.87 mmol), and K_2CO_3 (9.81 g, 70.87 mmol) were well powdered with a mortar and a pestle, then 1-fluoro-2-nitrobenzene (10.00 g, 7.47 mL, 70.87 mmol) was added and the mixture was irradiated in the microwave (900 W, 10 min). Thereafter, water (8 mL) and CH_2Cl_2 (10 mL) were added and the organic layer was washed with 10% HCl (5 mL) and brine (5 mL). The combined organic layers were dried over Na_2SO_4 and evaporated to dryness prior to purification by column chromatography (silica gel 60, petroleum ether/ethyl acetate 9.5:0.5). Yield: 9.51 g (58%), dark orange crystals, mp. 82 °C–83 °C.

3.2.5. General Procedure for the Synthesis of **14** and **15**

To a solution of Zn^0 (13.8 mmol) in glacial acetic acid (1 mL) was added starting material **28** or **29** (1 mmol) at 0 °C under argon atmosphere. After the addition, the mixture was allowed to warm to room temperature and was stirred for 2 h. Zn^0 was filtered off and the pH of the solution was adjusted to pH 9 with 2N NaOH. Thereafter, the aqueous layer was extracted three times with CH_2Cl_2, the combined organic layers were dried over $MgSO_4$ and evaporated to dryness.

N-(4-Fluorophenyl)benzene-1,2-diamine (**14**). Yield: 93%, dark orange-reddish oil, purification: silica gel 60, petrol ether/ethyl acetate 9:1, analytical data are in complete accordance with literature values [30].

N-(2-Fluorophenyl)benzene-1,2-diamine (**15**). The crude reaction product was subjected to the next reaction step without further purification.

3.2.6. General Procedure for the Synthesis of **16–18**

To a solution of starting materials **13**, **14**, or **15** (1 mmol) in THF was added 1,1'-carbonyldiimidazole (1.4 mmol) under argon atmosphere and the mixture was stirred at room temperature overnight. Thereafter, the crude reaction product was purified by column chromatography.

1-Phenyl-1,3-dihydro-2H-benzimidazol-2-one (**16**). Yield: 0.85 g (74%), pink crystals, mp. 201 °C–202 °C, THF: 10 mL, purification: silica gel 60, petroleum ether/ethyl acetate 1:1, analytical data are in complete accordance with literature values [41].

1-(4-Fluorophenyl)-1,3-dihydro-2H-benzimidazol-2-one (**17**). Yield: 61%, brown resin, THF: 25 mL, purification: silica gel 60, petroleum ether/ethyl acetate 9:1, ^{1}H-NMR (200 MHz, CDCl$_3$): δ (ppm) 6.74–6.81 (m, 2H), 6.89–7.11 (m, 2H), 7.14–7.23 (m, 2H), 7.48–7.59 (m, 1H), 7.73–7.78 (m, 1H), 9.03 (br s, 1H), ^{13}C-NMR (50 MHz, CDCl$_3$): δ (ppm) 108.5, 110.0, 116.3, 116.8, 121.4, 121.7, 122.3, 128.0, 128.2, 130.4, 135.1, 155.1, 159.3, 164.2, MS: *m/z* (%) 228 (M$^+$, 100), 199 (31), 172 (8), 114 (9), 95 (10), 75 (17), 51 (10), HRMS: *m/z* calculated for C$_{13}$H$_{10}$FN$_2$O [M + H]$^+$: 229.0772. Found: 229.0769.

1-(2-Fluorophenyl)-1,3-dihydro-2H-benzimidazol-2-one (**18**). Yield: 69%, brown resin, THF: 20 mL, purification: silica gel 60, petroleum ether/ethyl acetate 9:1, ^{1}H-NMR (200 MHz, CDCl$_3$): δ (ppm) 6.82–6.85 (m, 1H), 7.00–7.19 (m, 3H), 7.26–7.88 (m, 2H), 7.43–7.60 (m, 2H), 10.54 (br s, 1H), ^{13}C-NMR (50 MHz, CDCl$_3$): δ (ppm) 108.9, 110.1, 117.0, 117.4, 121.5, 121.9, 122.4, 124.9, 125.0, 128.2, 129.6, 130.4, 154.8, 155.4, 160.5, MS: *m/z* (%) 228 (M$^+$, 33), 199 (9), 181 (15), 149 (17), 111 (22), 97 (20), 71 (41), 69 (100), 55 (53), 43 (56), HRMS: *m/z* calculated for C$_{13}$H$_9$FN$_2$NaO [M + Na]$^+$: 251.0597. Found: 251.0592.

3.2.7. Alternative Procedure for the Synthesis of **17** and **18**

A solution of 1,1'-carbonyldiimidazole (1.4 mmol) in DMF (4 mL) was slowly added to a mixture of **14** or **15** (1 mmol) in DMF (4 mL) under argon atmosphere. The resulting solution was stirred at 90 °C for 2 h. After completion of the reaction, the solvent was evaporated *in vacuo*, the slurry was taken up in water, filtered and dried.

1-(4-Fluorophenyl)-1,3-dihydro-2H-benzimidazol-2-one (**17**). Yield: 80%, brown resin.

1-(2-Fluorophenyl)-1,3-dihydro-2H-benzimidazol-2-one (**18**). Yield: 80%, brown resin.

3.2.8. General Procedure for the Synthesis of **19–21**

Starting materials **16–18** (1 mmol) and K$_2$CO$_3$ (2 mmol) were suspended in DMF (1.8 mL) and stirred at 25 °C for 30 min. **11** and **12**, respectively (1.5 mmol) were added after 30 min and the solution was stirred at room temperature overnight. To the mixture was added ethyl acetate (5 mL) and water (5 mL). The aqueous layer was extracted several times with ethyl acetate (10 mL) and the combined organic layers were washed with brine, dried over MgSO$_4$ and evaporated to dryness.

1-(3-Chloro-1-(4-fluorophenyl)propyl)-3-phenyl-1,3-dihydro-2H-benzimidazol-2-one (**19**). Yield: 32%, white oil, purification: silica gel 60, petroleum ether/ethyl acetate 8:2, ^{1}H-NMR (200 MHz, CDCl$_3$): δ (ppm) 2.70–2.87 (m, 1H), 3.12–3.30 (m, 1H), 3.60–3.66 (m, 2H), 5.73 (m, 1H), 7.00–7.11 (m, 6H), 7.38–7.57 (m, 7H), ^{13}C-NMR (50 MHz, CDCl$_3$): δ (ppm) 34.3, 42.0, 53.8, 108.7, 109.0, 115.5, 115.9, 121.7, 122.0, 126.0, 127.8, 129.2, 129.4, 129.5, MS: *m/z* (%) 380 (M$^+$, 21), 210 (100) (M$^+$-C$_9$H$_9$ClF), 181 (8), 167 (12), 135 (9), 115 (5), 109 (58), 77 (12), HRMS: *m/z* calculated for C$_{22}$H$_{18}$ClFN$_2$ONa [M + Na]$^+$: 403.0989. Found: 403.0989.

1-(3-Chloro-1-phenylpropyl)-3-(4-fluorophenyl)-1,3-dihydro-2H-benzimidazol-2-one (**20**). Yield: 63%, dark orange resin, purification: silica gel 60, petroleum ether/ethyl acetate 9:1 and RP-18 silica gel, methanol, ^{1}H-NMR (400 MHz, CDCl$_3$): δ (ppm) 2.80–2.88 (m, 1H, 2'-CH$_2$), 3.17–3.26 (m, 1H, 2'-CH$_2$), 3.62–3.70 (m, 2H, 3'-CH$_2$), 5.79 (dd, *J* = 10.0 Hz and 5.6 Hz, 1H, 1'-CH), 7.04–7.09 (m, 4H, benzim 4-CH, benzim 5-CH, benzim 6-CH, benzim 7-CH), 7.21–7.26 (m, 2H, f-phen 3-CH, f-phen 5-CH), 7.31–7.33 (m, 1H, phen 4-CH), 7.36–7.40 (m, 2H, phen 3-CH, phen 5-CH), 7.52–7.56 (m, 4H, f-phen 2-CH, f-phen 6-CH, phen 2-CH, phen 6-CH), ^{13}C-NMR (100 MHz, CDCl$_3$): δ (ppm) 34.1 (2'-CH$_2$), 42.0 (3'-CH$_2$), 54.4 (1'-CH), 108.6 (benzim 4-CH), 109.0 (benzim 7-CH), 116.4 (d, *J* = 22.9 Hz, f-phen 3-CH, 116.4 (d, *J* = 22.9 Hz, f-phen 5-CH), 121.6 (benzim 5-CH), 122.1 (benzim 6-CH), 127.4 (phen 2-CH), 127.4 (phen 6-CH), 127.9 (d, *J* = 8.6 Hz, f-phen 2-CH), 127.9 (d, *J* = 8.6 Hz, f-phen 6-CH), 128.1 (phen 4-CH), 128.7 (benzim 7a-C), 128.8 (phen 3-CH), 128.8 (phen 5-CH), 129.4 (benzim 3a-C), 130.3 (d, *J* = 3.1 Hz, f-phen 1-C), 138.4 (phen 1-C), 153.2 (benzim 2-CO), 161.6 (d, *J* = 247.7 Hz, f-phen 4-CF), ^{19}F-NMR (471 MHz, CDCl$_3$): δ (ppm) -113.31 (m, 5-CF), MS: *m/z* (%) 380 (M$^+$, 2), 228 (100) (M$^+$-C$_9$H$_{10}$Cl), 199 (11), 185 (16), 153 (6), 117 (14), 91 (73), 75 (8), HRMS: *m/z* calculated for C$_{22}$H$_{19}$ClFN$_2$O [M + H]$^+$: 381.1170. Found: 381.1176.

1-(3-Chloro-1-phenylpropyl)-3-(2-fluorophenyl-1,3-dihydro-2H-benzimidazol-2-one (**21**). Yield: 55%, orange resin, purification: silica gel 60, petroleum ether/ethyl acetate 9:1, ^{1}H-NMR (400 MHz, CDCl$_3$): δ (ppm) 2.78–2.86 (m, 1H, 2'-CH$_2$), 3.23 (br s, 1H, 2'-CH$_2$), 3.64–3.68 (m, 2H, 3'-CH$_2$), 5.79 (br s, 1H, 1'-CH), 6.85–6.87 (m, 1H, benzim 4-CH) 7.05–7.06 (m, 3H, benzim 5-CH, benzim 6-CH, benzim 7-CH), 7.28–7.40 (m, 5H, f-phen 3-CH, f-phen 6-CH, phen 3-CH, phen 4-CH, phen 5-CH), 7.43–7.48 (m, 1H, f-phen 4-CH), 7.52–7.56 (m, 3H, f-phen 5-CH, phen 2-CH, phen 6-CH), ^{13}C-NMR (100 MHz, CDCl$_3$): δ (ppm) 34.2 (2'-CH$_2$), 41.9 (3'-CH$_2$), 54.5 (1'-CH), 108.8 (d, *J* = 1.7 Hz, benzim 4-CH), 109.0 (benzim 7-CH), 117.1 (d, *J* = 19.5 Hz, f-phen 3-CH), 121.6 (benzim 5-CH), 121.9 (f-phen 1-C), 122.0 (benzim 6-CH), 124.8 (d, *J* = 3.9 Hz, f-phen 6-CH), 127.3 (phen 2-CH), 127.3 (phen 6-CH), 128.1 (phen 4-CH), 128.8 (phen 3-CH), 128.8 (phen 5-CH), 129.4 (benzim 3a-C), 129.5 (f-phen 5-CH), 130.2 (d, *J* = 7.8 Hz, f-phen 4-CH), 138.4 (phen 1-C), 153.0 (benzim 2-CO), 157.9 (d, *J* = 253.2 Hz, f-phen 2-CF), due to limited resolution of the measuring apparatus, quaternary carbon benzim 7a-C could not be detected, ^{19}F-NMR (471 MHz, CDCl$_3$): δ (ppm) -118.39 (m, f-phen 2-CF), MS: *m/z* (%) 380 (M$^+$, 12), 228 (100) (M$^+$-C$_9$H$_{10}$Cl), 199 (5), 153 (3), 117 (7), 91 (49), 75 (5), HRMS: *m/z* calculated for C$_{22}$H$_{19}$ClFN$_2$O [M + H]$^+$: 381.1170. Found: 381.1164.

3.2.9. General Procedure for the Synthesis of 22–24

A solution of starting materials **19**, **20**, or **21** (1 mmol) and NaI (1.03 g, 6.89 mmol) in acetone (7 mL) was refluxed for 24 h. The precipitate formed was filtered and the solvent was removed *in vacuo*.

1-(1-(4-Fluorophenyl)-3-iodopropyl)-3-phenyl-1,3-dihydro-2H-benzimidazol-2-one (**22**). Yield: 82%, yellow resin, ^{1}H-NMR (200 MHz, CDCl$_3$): δ (ppm) 2.77–2.92 (m, 1H), 3.14–3.31 (m, 3H), 5.63–5.70 (m, 1H), 7.01–7.10 (m, 6H), 7.36–7.56 (m, 7H), ^{13}C-NMR (50 MHz, CDCl$_3$): δ (ppm) 2.3, 35.3, 57.0, 108.8, 108.9, 115.5, 115.9, 121.6, 122.0, 126.0, 127.7, 128.5, 129.1, 129.3, 129.4, 134.0, 134.1, 134.3, 153.0, 159.9, 164.8, MS: *m/z* (%) 472 (M$^+$, 32), 317 (8), 210 (100) (M$^+$-C$_9$H$_9$F), 181 (11), 167 (23), 140 (3), 135 (43), 115 (8), 109 (34), 77 (15), 51 (7), HRMS: *m/z* calculated for C$_{22}$H$_{18}$FIN$_2$ONa [M + Na]$^+$: 495.0346. Found: 495.0353.

1-(4-Fluorophenyl)-3-(3-iodo-1-phenylpropyl)-1,3-dihydro-2H-benzimidazol-2-one (**23**). Yield: 76%, yellow crystals, mp. 39 °C–41 °C, purification: silica gel 60, petrol ether/ethyl acetate 9:1, ^{1}H-NMR (200 MHz, CDCl$_3$): δ (ppm) 2.81–2.99 (m, 1H, 2'-CH$_2$), 3.11–3.35 (m, 3H, 2'-CH$_2$, 3'-CH$_2$), 5.70 (dd, *J* = 6 Hz and 2 Hz, 1H, 1'-CH), 7.04–7.09 (m, 4H, benzim 4-CH, benzim 5-CH, benzim 6-CH, benzim 7-CH), 7.17–7.42 (m, 5H, f-phen 3-CH, f-phen 5-CH, phen 4-CH, phen 2-CH, phen 5-CH), 7.51–7.57 (m, 4H, f-phen 2-CH, f-phen 6-CH, phen 2-CH, phen 6-CH), ^{13}C-NMR (50 MHz, CDCl$_3$): δ (ppm) 2.4 (2'-CH$_2$), 35.2 (3'-CH$_2$), 57.6 (1'-CH), 108.6 (benzim 4-CH), 109.2 (benzim 7-CH), 116.4 (d, *J* = 23 Hz, f-phen 3-CH), 116.4 (d, *J* = 23 Hz, f-phen 5-CH), 121.6 (benzim 5-CH), 122.1 (benzim 6-CH), 127.3 (phen 2-CH), 127.3 (phen 6-CH), 127.8 (phen 4-CH), 128.1 (d, *J* = 5 Hz, f-phen 2-CH), 128.1 (d, *J* = 5 Hz, f-phen 6-CH), 128.5 (benzim 7a-C), 128.8 (phen 3-CH), 128.8 (phen 5-CH), 129.4 (benzim 3a-C), 130.3 (d, *J* = 3Hz, f-phen 1-C), 138.1 (phen 1-C), 153.2 (benzim 2-CO), 161.6 (d, *J* = 247 Hz, f-phen 4-CF), MS: *m/z* (%) 472 (M$^+$, 13), 317 (5), 228 (100) (M$^+$-C$_9$H$_{10}$I), 199 (8), 185 (15), 117 (47), 103 (2), 91 (40), 75 (8), 55 (5), HRMS: *m/z* calculated for C$_{22}$H$_{19}$FIN$_2$O [M + H]$^+$: 473.0526. Found: 473.0506.

1-(2-Fluorophenyl)-3-(3-iodo-1-phenylpropyl)-1,3-dihydro-2H-benzimidazol-2-one (**24**). Yield: 53%, yellow crystals, mp. 38 °C–39 °C, purification: silica gel 60, petrol ether/ethyl acetate 9:1, ^{1}H-NMR (200 MHz, CDCl$_3$): δ (ppm) 2.80–2.98 (m, 1H, 2'-CH$_2$), 3.13–3.36 (m, 3H, 2'-CH$_2$, 3'-CH$_2$), 5.67–5.74 (m, 1H, 1'-CH), 6.85–6.88 (m, 1H, benzim 4-CH) 6.99–7.07 (m, 3H, benzim 5-CH, benzim 6-CH, benzim 7-CH), 7.26–7.45 (m, 5H, f-phen 3-CH, f-phen 6-CH, phen 3-CH, phen 4-CH, phen 5-CH), 7.47–7.58 (m, 4 H, f-phen 4-CH, f-phen 5-CH, phen 2-CH, phen 6-CH), ^{13}C-NMR (50 MHz, CDCl$_3$): δ (ppm) 2.3 (2'-CH$_2$), 35.4 (3'-CH$_2$), 57.6 (1'-CH), 108.9 (d, *J* = 1.5 Hz, benzim 4-CH), 109.1 (benzim 7-CH), 117.1 (d, *J* = 19 Hz, f-phen 3-CH), 121.6 (benzim 5-CH), 121.9 (f-phen 1-C), 122.0 (benzim 6-CH), 124.8 (d, *J* = 3.5 Hz, f-phen 6-CH), 127.3 (phen 2-CH), 127.3 (phen 6-CH), 128.1 (phen 4-CH), 128.8 (phen 3-CH), 128.8 (phen 5-CH), 129.5 (f-phen 5-CH), 130.2 (d, *J* = 8 Hz, f-phen 4-CH), 138.2 (phen 1-C), 153.0 (benzim 2-CO), 157.8 (d, *J* = 251.5 Hz, f-phen 2-CF), MS: *m/z* (%) 472 (M$^+$, 9), 317 (5), 241 (4), 228 (100) (M$^+$-C$_9$H$_{10}$I), 199 (5), 185 (10), 117 (37), 91 (29), 75 (7), HRMS: *m/z* calculated for C$_{22}$H$_{19}$FIN$_2$O [M + H]$^+$: 473.0526. Found: 473.0532.

3.2.10. General procedure for the synthesis of *1-(3-amino-1-(4-fluorophenyl)propyl)-3-phenyl-1,3-dihydro-2H-benzimidazol-2-one* (**25**)

1-(1-(4-fluorophenyl)-3-iodopropyl)-3-phenyl-1,3-dihydro-2*H*-benzimidazol-2-one (0.26 g, 0.55 mmol) and a solution of NH$_3$ in isopropanol (2 M, 22 mL) were heated in a sealed tube for 3 h at 80 °C. After evaporation of the solvent, the crude product was purified by column chromatography (silica gel 60, CH$_2$Cl$_2$/MeOH 9:1). Yield: 0.10 g (50%), light brown crystals, mp. 87–88 °C. ^{1}H-NMR (400 MHz, CDCl$_3$): δ (ppm) 2.74–2.84 (m, 3H, 2'-CH$_2$, 3'-CH$_2$), 2.98–3.04 (m, 1H, 3'-CH$_2$), 5.74–5.78 (m, 1H, 1'-CH), 6.79–6.81 (m, 1H, benzim 7-CH), 6.96–7.01 (m, 5H, benzim 4-CH, benzim 5-CH, benzim 6-CH, f-phen 3-CH, f-phen 5-CH), 7.30–7.33 (m, 1H, phen 4-CH), 7.40–7.48 (m, 4H, f-phen 2-CH, f-phen 6-CH, phen 3-CH, phen 5-CH), 7.52–7.54 (m, 2H, phen 2-CH, phen 6-CH), due to limited resolution of the instrumentation, the NH$_2$ protons could not be detected, ^{13}C-NMR (100 MHz, CDCl$_3$): δ (ppm) 30.3 (2'-CH$_2$), 37.8 (3'-CH$_2$), 52.7 (1'-CH), 109.2 (benzim 4-CH), 110.0 (benzim 7-CH), 115.7 (d, *J* = 21.5 Hz, f-phen 3-CH), 115.7 (d, *J* = 8.2 Hz, f-phen 5-CH), 122.0 (benzim 5-CH), 122.3 (benzim 6-CH), 126.4 (phen 2-CH), 126.4 (phen 6-CH), 127.5 (benzim 7a-C), 128.1 (phen 4-CH), 129.2 (d, *J* = 8.2 Hz, f-phen 2-CH), 129.2 (d, *J* = 8.2 Hz, f-phen 6-CH), 129.5 (benzim 3a-C), 129.7 (phen 3-CH), 129.7 (phen 5-CH), 133.3 (d, *J* = 3.2 Hz, f-phen 1-C), 134.0 (phen 1-CH), 153.8 (benzim 2-CO), 162.3 (d, *J* = 247.4 Hz, f-phen 4-CF), ^{19}F-NMR (471 MHz, CDCl$_3$): δ (ppm) -113.68 (m, f-phen CF), MS: *m/z* (%) 361 (M$^+$, 17), 331 (10), 210 (100) (M$^+$-C$_9$H$_{11}$FN), 181 (15), 167 (16), 149 (29), 128 (17), 77 (19), 57 (20), 43 (12), HRMS: *m/z* calculated for C$_{22}$H$_{21}$FN$_3$O [M + H]$^+$: 362.1669. Found: 362.1674.

3.2.11. General Procedure for the Synthesis of **4–6**

Starting materials **18**, **19** or **20** (1 mmol) and a solution of methylamine in EtOH (12.5 mL, 8 M) were heated in a sealed tube for 3 h at 80 °C. After evaporation of the solvent, the crude reaction product was purified by column chromatography.

1-(1-(4-Fluorophenyl)-3-(methylamino)propyl)-3-phenyl-1,3-dihydro-2H-benzimidazol-2-one (**4**). Yield: 48%, light orange resin, purification: silica gel 60, dichloromethane/methanol 9:1 and dichloromethane/ethyl acetate/methanol 7:2:1, ^{1}H-NMR (400 MHz, CDCl$_3$): δ (ppm) 2.42 (s, 3H, NHCH$_3$), 2.57–2.74 (m, 4H, 2'-CH$_2$, 3'-CH$_2$), 3.15 (br s, 1H, NHCH$_3$), 5.76–5.79 (m, 1H, 1'-CH), 6.88–6.90 (m, 1H, benzim 7-CH), 6.97–7.05 (m, 4H, benzim 5-CH, benzim 6-CH, f-phen 3-CH, f-phen 5-CH), 7.07–7.10 (m, 1H, benzim 4-CH), 7.39–7.43 (m, 1H, phen 4-CH), 7.46–7.57 (m, 6H, f-phen 2-CH, f-phen 6-CH, phen 2-CH, phen 3-CH, phen 5-CH, phen 6-CH), ^{13}C-NMR (100 MHz, CDCl$_3$): δ (ppm) 30.6 (2'-CH$_2$), 35.9 (NHCH$_3$), 48.3 (3'-CH$_2$), 53.1 (1'-CH), 108.9 (benzim 4-CH), 109.5 (benzim 7-CH), 115.5 (d, *J* = 21.5 Hz, f-phen 3-CH), 115.5 (d, *J* = 21.5 Hz, f-phen 5-CH) 121.4 (benzim 5-CH), 121.8 (benzim 6-CH), 126.0 (phen 2-CH), 126.0 (phen 6-CH), 127.7 (phen 4-CH), 128.0 (benzim 7a-C), 129.0 (d, *J* = 8.1 Hz, f-phen 2-CH), 129.0 (d, *J* = 8.1 Hz, f-phen 6-CH), 129.4 (benzim 3a-C), 129.5 (phen 3-CH), 129.5 (phen 5-CH), 134.4 (phen 1-CH), 134.6 (d, *J* = 3.4 Hz, f-phen 1-CH), 153.5 (benzim 2-CO), 162.1 (d, *J* = 246.7 Hz, f-phen 4-CF), ^{19}F-NMR (471 MHz, CDCl$_3$): δ (ppm) -114.36 (m, f-phen 4-CF), MS: *m/z* (%) 375 (M$^+$, 16), 210 (57) (M$^+$-C$_{10}$H$_{13}$FN), 181 (10), 167 (12), 150 (10), 109 (22), 97 (16), 71 (27), 57 (78), 44 (100), HRMS: *m/z* calculated for C$_{23}$H$_{23}$FN$_3$O [M + H]$^+$: 376.1825. Found: 376.1821.

1-(4-Fluorophenyl)-3-(3-(methylamino)-1-phenylpropyl)-1,3-dihydro-2H-benzimidazol-2-one (**5**). Yield: 29%, light yellow crystals, mp. 100 °C–102 °C, purification: silica gel 60, dichloromethane/methanol 9:1 and RP-18 silica gel methanol/water 9:1 and 7:3, ^{1}H-NMR (200 MHz, CDCl$_3$): δ (ppm) 2.53 (s, 3H, NHC$\underline{H}_3$), 3.13 (br s, 4H, 2'-C$\underline{H}_2$, 3'-C$\underline{H}_2$), 5.77–5.81 (m, 1H, 1'-C$\underline{H}$), 6.96–7.03 (m, 4H, benzim 4-C$\underline{H}$, benzim 5-C$\underline{H}$, benzim 6-C$\underline{H}$, benzim 7-C$\underline{H}$), 7.16–7.34 (5H, f-phen 3-C$\underline{H}$, f-phen 5-C$\underline{H}$, phen 4-C$\underline{H}$, phen 2-C$\underline{H}$, phen 5-C$\underline{H}$), 7.49–7.56 (m, 4H, f-phen 2-C$\underline{H}$, f-phen 6-C$\underline{H}$, phen 2-C$\underline{H}$, phen 6-C$\underline{H}$), due to limited resolution of the instrumentation, the N$\underline{H}$ proton could not be detected, ^{13}C-NMR (50 MHz, CDCl$_3$): δ (ppm) 27.6 (2'-$\underline{C}$H$_2$), 33.1 (NH$\underline{C}$H$_3$), 47.0 (3'-$\underline{C}$H$_2$), 54.2 (1'-$\underline{C}$H), 108.8 (benzim 4-$\underline{C}$H), 109.9 (benzim 7-$\underline{C}$H), 116.6 (d, *J* = 23 Hz, f-phen 3-$\underline{C}$H), 116.6 (d, *J* = 23 Hz, f-phen 5-$\underline{C}$H), 122.1 (benzim 5-$\underline{C}$H), 122.6 (benzim 6-$\underline{C}$H), 127.2 (phen 2-$\underline{C}$H), 127.2 (phen 6-$\underline{C}$H), 127.7 (phen 4-$\underline{C}$H), 128.3 (d, *J* = 6 Hz, f-phen 2-$\underline{C}$H), 128.3 (d, *J* = 6 Hz, f-phen 6-$\underline{C}$H), 128.4 (benzim 7a-C), 128.9 (phen 3-$\underline{C}$H), 128.9 (phen 5-$\underline{C}$H), 129.2 (benzim 3a-C), 129.7 (d, *J* = 3 Hz, f-phen 1-C), 136.9 (phen 1-C), 153.4 (benzim 2-$\underline{C}$O), 161.7 (d, *J* = 248 Hz, f-phen 4-$\underline{C}$F), MS: *m/z* (%) 375 (M$^+$, 11), 330 (7), 228 (50) (M$^+$-C$_{10}$H$_{14}$N), 199 (7), 185 (9), 147 (17), 128 (26), 117 (8), 91 (13), 58 (34), 44 (100), HRMS: *m/z* calculated for C$_{23}$H$_{23}$FN$_3$O [M + H]$^+$: 376.1825. Found: 376.1828.

1-(2-Fluorophenyl)-3-(3-(methylamino)-1-phenylpropyl)-1,3-dihydro-2H-benzimidazol-2-one (**6**). Yield: 30%, yellow crystals, mp. 92 °C–93 °C, purification: silica gel 60, dichloromethane/methanol 9:1 and RP-18 silica gel methanol/water 9:1 and 7:3, ^{1}H-NMR (200 MHz, CDCl$_3$): δ (ppm) 2.57 (s, 3H, NHC$\underline{H}_3$), 3.01–3.15 (m, 4H, 2'-C$\underline{H}_2$, 3'-C$\underline{H}_2$), 5.75–5.86 (m, 1H, 1'-C$\underline{H}$), 6.83–6.87 (m, 1H, benzim 4-C$\underline{H}$) 6.98–7.05 (m, 3H, benzim 5-C$\underline{H}$, benzim 6-C$\underline{H}$, benzim 7-C$\underline{H}$), 7.25–7.42 (m, 5H, f-phen 3-C$\underline{H}$, f-phen 6-C$\underline{H}$, phen 3-C$\underline{H}$, phen 4-C$\underline{H}$, phen 5-C$\underline{H}$), 7.48–7.61 (m, 4 H, f-phen 4-C$\underline{H}$, f-phen 5-C$\underline{H}$, phen 2-C$\underline{H}$, phen 6-C$\underline{H}$), due to limited resolution of the instrumentation, the N$\underline{H}$CH$_3$ proton could not be detected, ^{13}C-NMR (50 MHz, CDCl$_3$): δ (ppm) 27.7 (2'-$\underline{C}$H$_2$), 33.2 (NH$\underline{C}$H3), 46.9 (3'-$\underline{C}$H$_2$), 53.5 (1'-$\underline{C}$H), 109.0 (benzim 4-$\underline{C}$H), 110.1 (benzim 7-$\underline{C}$H), 117.1 (d, *J* = 19.5 Hz, f-phen 3-$\underline{C}$H), 122.2 (benzim 5-$\underline{C}$H), 122.7 (benzim 6-$\underline{C}$H), 125.1 (d, *J* = 3.0 Hz, f-phen 6-$\underline{C}$H), 127.3 (phen 2-$\underline{C}$H), 127.1 (phen 6-$\underline{C}$H), 128.3 (phen 4-$\underline{C}$H), 128.9 (phen 3-$\underline{C}$H), 128.9 (phen 5-$\underline{C}$H), 129.5 (f-phen 5-$\underline{C}$H), 130.7 (d, *J* = 4.5 Hz, f-phen 4-$\underline{C}$H), 136.5 (phen 1-C), 153.6 (benzim 2-$\underline{C}$O), 157.6 (d, *J* = 250.5 Hz, f-phen 2-$\underline{C}$F), due to limited resolution of the measuring apparatus, quaternary carbon f-phen 1-C could not be detected, MS: *m/z* (%) 375 (M$^+$, 17), 318 (10), 228 (82) (M$^+$-C$_{10}$H$_{14}$N), 199 (9), 185 (9), 147 (16), 128 (35), 117 (9), 91 (20), 58 (43), 44 (100), HRMS: *m/z* calculated for C$_{23}$H$_{23}$FN$_3$O [M + H]$^+$: 376.1825. Found: 376.1822.

3.3. Computational Methods

The ligand structures were built in the protonated form using Molecular Operating Environment (MOE) 2013 [42]. Homology models of human NET, SERT and DAT were obtained from the *Drosophila* dopamine transporter template (dDAT$_{cryst}$, PDB id 4M48 [32]), by selecting the model with the most favorable Discrete Optimized Protein Energy (DOPE) of 250 generated by Modeller 9.11 [43]. The co-crystallized inhibitor nortriptyline was retained during model generation and the compounds were docked in the same site using Genetic Optimization for Ligand Docking (GOLD) 5.2 [44]. One hundred poses per ligand (*i.e.*, five hundred poses per protein target) were generated based on the GoldScore scoring function, while keeping the ligand flexible and the protein rigid.

The common chemical scaffold, *i.e.*, the reference compound, was extracted from the resulting poses, analogous to the methods of our previous study [35]. Cluster analysis was performed based on Euclidian distance and complete linkage of the root-mean square deviation of the ligand's heavy atoms matrix using XLStat [45]. The dendrogram was cut at eight clusters and the ones containing all five ligands were selected (Figure 5).

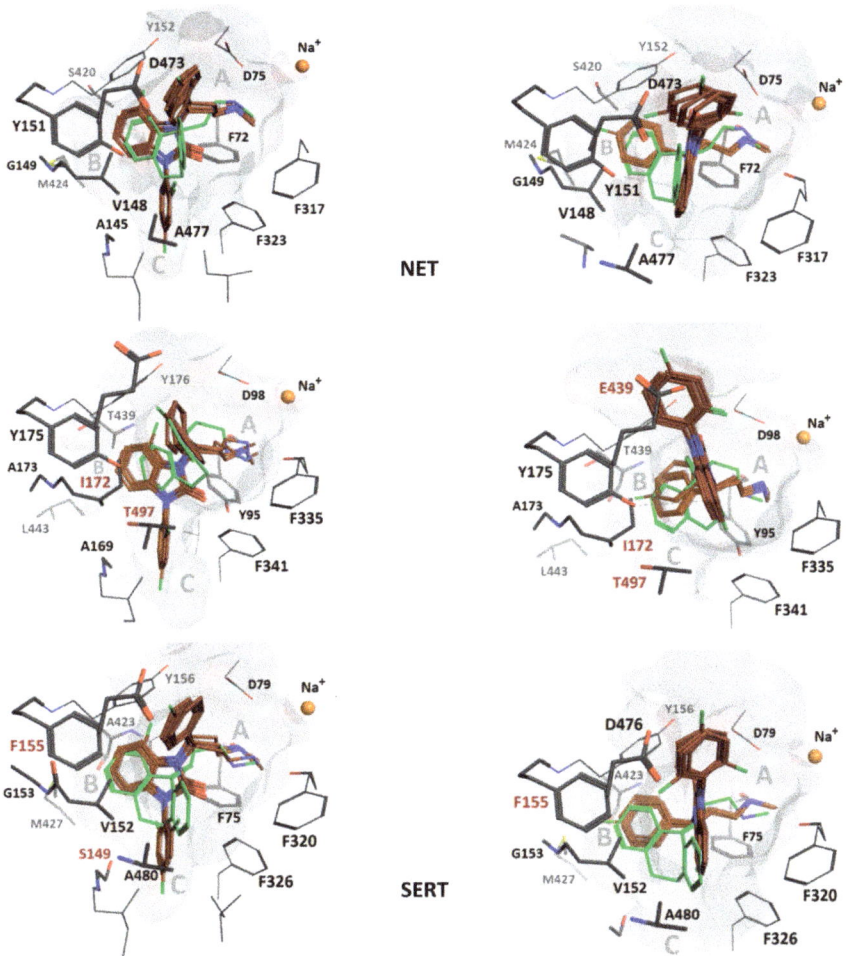

Figure 5. *Left column*: Overlay of compounds **1**, **2**, and **4**–**6** (maroon) in binding hypothesis 1 and comparison with the co-crystal pose of nortriptyline (**31**) (green). In NET, V148 allows more space than I172 in SERT. The angle between ligand ring 2 and 3 is *ca.* 60° in all poses. *Right column*: Overlay of compounds **1**, **2**, and **4**–**6** (maroon) in binding hypothesis 2. In NET, V148 still allows more space than in hSERT, whereas E439 might disrupt ligand ring 3. DAT lacks a more potent stacking interaction due to F155 as compared to NET[Y151] and SERT[Y175]. Binding mode 2 poses are more in agreement with the co-crystal pose. The angle between ligand ring 2 and 3 is almost 90° in all poses. The extracellular space is located above in all figures.

4. Conclusions

In conclusion, ten new compounds have been synthesized within the scope of this work, which aimed at the development of new, selective, high affinity references for the imaging of the NET system via PET. Four of these new compounds (**4–6** and **25**) will be employed in future studies. Whilst methylamines **4–6** (FAPPI:1-3) represent reference compounds for their later prepared radioactive analogs, additionally prepared free amine **25** (APPI:1) will serve as precursor for radiolabeling. Since docking studies indicate that fluorinated methyl amines **4–6** (FAPPI:1-3) bind in an analogous way to the NET as reference compound **1**, these compounds **4–6** are promising candidates for biological evaluation.

Acknowledgments

AS, AJ, and GFE acknowledge financial support provided by the Austrian Science Fund, grants AW/0123221 and F3502. This article was supported by the Open Access Publishing Fund of the University of Vienna.

Author Contributions

Catharina Neudorfer: Responsible for the performance of the syntheses and writing; Amir Seddik: Structure activity relationships calculations; Karem Shanab: Contributions to syntheses and experimental procedures; Andreas Jurik: Structure activity relationships calculations; Christina Rami-Mark: Participated in design and performance of the experiments; Wolfgang Holzer: Performance of the NMR analyses; Gerhard Ecker: Conceived and supervised the SAR experiments; Markus Mitterhauser: Designed parts of the research and proofread the manuscript; Wolfgang Wadsak: Designed parts of the research and proofread the manuscript; Helmut Spreitzer: Conceived and supervised the syntheses.

Conflicts of Interest

The authors declare no conflict of interest.

References

1. Sung, U.; Apparsundram, S.; Galli, A.; Kahlig, K.M.; Savchenko, V.; Schroeter, S.; Quick, M.W.; Blakely, R.D. A regulated interaction of syntaxin 1A with the antidepressant-sensitive norepinephrine transporter establishes catecholamine clearance capacity. *J. Neurosci.* **2003**, *23*, 1697–1709.
2. Kim, C.H.; Hahn, M.K.; Joung, Y.; Anderson, S.L.; Steele, A.H.; Mazei-Robinson, M.S.; Gizer, I.; Teicher, M.H.; Cohen, B.M.; Robertson, D.; *et al*. A polymorphism in the norepinephrine transporter gene alters promoter activity and is associated with attention-deficit hyperactivity disorder. *Proc. Natl. Acad. Sci. USA* **2006**, *103*, 19164–19169.
3. Hahn, M.K.; Robertson, D.; Blakely, R.D. A mutation in the human norepinephrine transporter gene (SLC6A2) associated with orthostatic intolerance disrupts surface expression of mutant and wild-type transporters. *J. Neurosci.* **2003**, *23*, 4470–4478.

4. Mirbolooki, M.R.; Upadhyay, S.K.; Constantinescu, C.C.; Pan, M.L.; Mukherjee, J. Adrenergic pathway activation enhances brown adipose tissue metabolism: A [^{18}F]FDG PET/CT study in mice. *Nucl. Med. Biol.* **2014**, *41*, 10–16.
5. Lin, S.L.; Fan, X.; Yeckel, C.W.; Weinzimmer, D.; Mulnix, T.; Gallezot, J.D.; Carson, R.E.; Sherwin, R.S.; Ding, Y.S. Ex vivo and in vivo Evaluation of the Norepinephrine Transporter Ligand [^{11}C]MRB for Brown Adipose Tissue Imaging. *Nucl. Med. Biol.* **2012**, *39*, 1081–1086.
6. Stöber, G.; Nöthen, M.M.; Pörzgen, P.; Brüss, M.; Bönisch, H.; Knapp, M.; Beckmann, H.; Propping, P. Systematic search for variation in the human norepinephrine transporter gene: Identification of five naturally occurring missense mutations and study of association with major psychiatric disorders. *Am. J. Med. Genet. (Neuropsychiatr. Genet.)* **1996**, *67*, 523–532.
7. Young, J.B.; Landsberg, L. Catecholamines and the adrenal medulla. In *Williams Textbook of Endocrinology*, 9th ed.; Wilson, J.D., Foster, D.W., Eds.; W.B. Saunders Co.: Philadelphia, PA, USA, 1998; p. 680.
8. Tellioglu, T.; Robertson, D. Genetic or acquired deficits in the norepinephrine transporter: Current understanding of clinical implications. *Exp. Rev. Mol. Med.* **2001**, 1–10.
9. Blakely, R.D.; Bauman, A.L. Biogenic amine transporters: regulation in flux. *Curr. Opin. Neurobiol.* **2000**, *10*, 328–336.
10. Zhu, M.Y.; Shamburger, S.; Li, J.; Ordway, G.A. Regulation of the Human Norepinephrine Transporter by Cocaine and Amphetamine. *J. Pharmacol. Exp. Ther.* **2000**, *295*, 951–959.
11. Moron, J.A.; Brockington, A.; Wise, R.A.; Rocha, B.A.; Hope, B.T. Dopamine Uptake through the Norepinephrine Transporter in Brain Regions with Low Levels of the Dopamine Transporter: Evidence from Knock-Out Mouse Lines. *J. Neurosci.* **2002**, *22*, 386–395.
12. Schroeter, S.; Apparsundaram, S.; Wiley, R.G.; Miner, L.H.; Sesack, S.R.; Blakely, R.D. Immunolocalization of the cocaine- and antidepressant-sensitive l-norepinephrine transporter. *J. Comp. Neurol.* **2000**, *420*, 211–232.
13. Torres, G.E.; Gainetdinov, R.R.; Caron, M.G. Plasma membrane monoamine transporters: Structure, regulation and function. *Nat. Rev. Neurosci.* **2003**, *4*, 13–25.
14. Ordway, G.A.; Stockmeier, C.A.; Cason, G.W.; Klimek, V. Pharmacology and Distribution of Norepinephrine Transporters in the Human Locus Coeruleus and Raphe Nuclei. *J. Neurosci.* **1997**, *17*, 1710–1719.
15. Smith, H.R.; Beveridge, T.J. R.; Porrino, L.J. Distribution of norepinephrine transporters in the non-human primate brain. *Neuroscience* **2006**, *138*, 703–714.
16. Zhou, J. Norepinephrine transporter inhibitors and their therapeutic potential. *Drugs Future* **2004**, *29*, 1235–1244.
17. Curatolo, P.; D'Agati, E.; Moavero, R. The neurobiological basis of ADHD. *Ital. J. Pediatr.* **2010**, *36*, 79.
18. Mash, D.C.; Ouyang, Q.; Qin, Y.; Pablo, J. Norepinephrine transporter immunoblotting and radioligand binding in cocaine abusers. *Neurosci. Methods* **2005**, *143*, 79–85.
19. Barr, C.L.; Kroft, J.; Feng, Y.; Wigg, K.; Roberts, W.; Malone, M.; Ickowicz, A.; Schachar, R.; Tannock, R.; Kennedy, J.L. The Norepinephrine Transporter Gene and Attention-Deficit Hyperactivity Disorder. *Am. J. Med. Genet. (Neuropsychiatr. Genet.)* **2002**, *114*, 255–229.

20. Klimek, V.; Stockmeier, C.; Overholser, J.; Meltzer, H.Y.; Kalka, S.; Dilley, G.; Ordway, G.A. Reduced Levels of Norepinephrine Transporters in the Locus Coeruleus in Major Depression. *J. Neurosci.* **1997**, *17*, 8451–8458.
21. Nedergaard, J.; Cannon, B. The changed metabolic world with human brown adipose tissue: Therapeutic visions. *Cell Metab.* **2010**, *11*, 268–272.
22. Gulyas, B.; Brockschnieder, D.; Nag, S.; Pavlova, E.; Kasa, P.; Beliczai, Z.; Legradi, A.; Gulya, K.; Thiele, A.; Dyrks, T.; *et al.* The norepinephrine transporter radioligand [18F]FD2MeNER shows significant decreases in NET density in the locus coeruleus and the thalamus in Alzheimer's disease: A post-mortem autoradiographic study in human brains. *Neurochem. Int.* **2010**, *56*, 789–798.
23. Wilson, A.A.; Johnson, D.P.; Mozley, D.; Hussey, D.; Ginovart, N.; Nobrega, J.; Garcia, A.; Meyer, J.; Houle, S. Synthesis and *in vivo* evaluation of novel radiotracers for the *in vivo* imaging of the norepinephrine transporter. *Nucl. Med. Biol.* **2003**, *30*, 85–92.
24. Takano, A.; Gulyas, B.; Varrone, A.; Halldin, C. Saturated norepinephrine transporter occupancy by atomoxetine relevant to clinical doses: a rhesus monkey study with (S,S)-[(18)F]FMeNER-D (2). *Eur. J. Nucl. Med. Mol. Imaging* **2009**, *36*, 1308–1314.
25. Schou, M.; Zoghbi, S.S.; Shetty, H.U.; Shchukin, E.; Liow, J.S.; Hong, J.; Andrée, B.A.; Gulyás, B.; Farde, L.; Innis, R.B.; *et al.* Investigation of the metabolites of [^{11}C](S,S)-MeNER in humans, monkeys and rats. *Mol. Imaging Biol.* **2009**, *11*, 23–30.
26. Zhang, P.; Terefenko, E.A.; McComas, C.C.; Mahaney, P.E.; Vu, A.; Trybulski, E.; Koury, E.; Johnston, G.; Bray, J.; Deecher, D. Synthesis and activity of novel 1- or 3-(3-amino-1-phenyl propyl)-1,3-dihydro-2H-benzimidazol-2-ones as selective norepinephrine reuptake inhibitors. *Bioorg. Med. Chem. Lett.* **2008**, *18*, 6067–6070.
27. Mark, C.; Bornatowicz, B.; Mitterhauser, M.; Hendl, M.; Nics, L.; Haeusler, D.; Lanzenberger, R.; Berger, M.L.; Spreitzer, H.; Wadsak, W. Development and automation of a novel NET-PET tracer: [C-11]Me@APPI. *Nucl. Med. Biol.* **2013**, *40*, 295–303.
28. Varney, M.D.; Romines, W.H.; Boritzki, T.; Margosiak, S.A.; Barlett, C.; Howland, E.J. Synthesis and biological evaluation of -n[4-(2-trans-[([2,6-diamino-4(3H)-oxopyrimidin-5-yl]methyl)thio]cyclobutyl)benzoyl] -l-glutamic acid a novel 5-thiapyrimidinone inhibitor of dihydrofolate reductase. *J. Heterocycl. Chem.* **1995**, *32*, 1493–1498.
29. Xu, Z.B.; Lu, Y.; Guo, Z.R. An Efficient and Fast Procedure for the Preparation of 2-Nitrophenylamines under Microwave Conditions. *Synlett* **2003**, *4*, 564–566.
30. Wang, X.J.; Xi, M.Y.; Fu, J.H.; Zhang, F.R.; Cheng, G.F.; Yin, D.L.; You, Q.D. Synthesis, biological evaluation and SAR studies of benzimidazole derivatives as H_1-antihistamine agents. *Chin. Chem. Lett.* **2012**, *23*, 707–710.
31. Jona, H.; Shibata, J.; Asai, M.; Goto, Y.; Arai, S.; Nakajima, S.; Okamoto, O.; Kawamoto, H.; Iwasawa, Y. Efficient and practical asymmetric synthesis of 1-tert-butyl 3-methyl (3*R*,4*R*)-4-(2-oxo-2,3-dihydro-1*H*-benzimidazol-1-yl)piperidine-1,3-dicarboxylate, a useful intermediate for the synthesis of nociceptin antagonists. *Tetrahedron: Asymmetry* **2009**, *20*, 2439–2446.
32. Penmatsa, A.; Wang, K.H.; Gouaux, E. X-ray structure of dopamine transporter elucidates antidepressant mechanism. *Nature* **2013**, *503*, 85–90.
33. Wang, H.; Goehring, A.; Wang, K.H.; Penmatsa, A.; Ressler, R.; Gouaux, E. Structural basis for action by diverse antidepressants on biogenic amine transporters. *Nature* **2013**, *503*, 141–145.

34. Tatsumi, M.; Groshan, K.; Blakely, R.D.; Richelson, E. Pharmacological profile of antidepressants and related compounds at human monoamine transporters. *Eur. J. Pharmacol.* **1997**, *340*, 249–258.
35. Richter, L.; de Graaf, C.; Sieghart, W.; Varagic, Z.; Morzinger, M.; de Esch, I.J.; Ecker, G.F.; Ernst, M. Diazepam-bound GABAA receptor models identify new benzodiazepine binding-site ligands. *Nat. Chem. Biol.* **2012**, *8*, 455–464.
36. Andersen, J.; Olsen, L.; Hansen, K.B.; Taboureau, O.; Jorgensen, F.S.; Jorgensen, A.M.; Bang-Andersen, B.; Egebjerg, J.; Stromgaard, K.; Kristensen, A.S. Mutational Mapping and Modeling of the Binding Site for (*S*)-Citalopram in the Human Serotonin Transporter. *J. Biol. Chem.* **2010**, *285*, 2051–2063.
37. Chelli, R.; Gervasio, F.L.; Procacci, P.; Schettino, V. Stacking and T-shape Competition in Aromatic–Aromatic Amino Acid Interactions. *J. Am. Chem. Soc.* **2002**, *124*, 6133–6143.
38. Yu, F.; Zhou, J.N.; Zhang, X.C.; Sui, Y.Z.; Wu, F.F.; Xie, L.J.; Chan, A.S.C.; Wu, J. Copper(II)-Catalyzed Hydrosilylation of Ketones Using Chiral Dipyridylphosphane Ligands: Highly Enantioselective Synthesis of Valuable Alcohols. *Chemistry* **2011**, *17*, 14234–14240.
39. La Regina, G.; Diodata D'Auria, F.; Tafi, A.; Piscitelli, F.; Olla, S.; Caporuscio, F.; Nencioni, L.; Cirilli, R.; La Torre, F.; Rodrigues De Melo, N.; *et al.* 1-[(3-Aryloxy-3-aryl)propyl]-1H-imidazoles, new imidazoles with potent activity against Candida albicans and dermatophytes. Synthesis, structure-activity relationship, and molecular modeling studies. *J. Med. Chem.* **2008**, *51*, 3841–3855.
40. Panagopoulos, A.M.; Steinman, D.; Goncharenko, A.; Geary, K.; Schleisman, C.; Spaargaren, E.; Zeller, M.; Becker, D.P. Apparent Alkyl Transfer and Phenazine Formation via an Aryne Intermediate. *J. Org. Chem.* **2013**, *78*, 3532–3540.
41. Liu, P.; Wang, Z.; Hu, X. Highly Efficient Synthesis of Ureas and Carbamates from Amides by Iodosylbenzene-Induced Hofmann Rearrangement. *Eur. J. Org. Chem.* **2012**, *10*, 1994–2000.
42. *Molecular Operating Environment (MOE)*; Chemical Computing Group Inc.: Montreal, QC, Canada, 2013.
43. Sali, A.; Potterton, L.; Yuan, F.; van Vlijmen, H.; Karplus, M. Evaluation of comparative protein modeling by MODELLER. *Proteins* **1995**, *23*, 318–326.
44. Jones, G.; Willett, P.; Glen, R.C.; Leach, A.R.; Taylor, R. Development and validation of a genetic algorithm for flexible docking. *J. Mol. Biol.* **1997**, *267*, 727–748.
45. *XLStat*; Addinsoft Inc.: New York, NY, USA, 2009.

Sample Availability: Samples of the compounds **4–6**, **9–25**, and **29** and **30** are available from the authors.

© 2015 by the authors; licensee MDPI, Basel, Switzerland. This article is an open access article distributed under the terms and conditions of the Creative Commons Attribution license (http://creativecommons.org/licenses/by/4.0/).

Review

Methods to Increase the Metabolic Stability of ^{18}F-Radiotracers

Manuela Kuchar and Constantin Mamat *

Helmholtz-Zentrum Dresden-Rossendorf, Institut für Radiopharmazeutische Krebsforschung, Bautzner Landstraße 400, Dresden D-01328, Germany; E-Mail: m.kuchar@hzdr.de

* Author to whom correspondence should be addressed; E-Mail: c.mamat@hzdr.de; Tel.: +49-351-260-2805; Fax: +49-351-260-3232.

Academic Editor: Svend Borup Jensen

Received: 16 June 2015 / Accepted: 26 August 2015 / Published: 3 September 2015

Abstract: The majority of pharmaceuticals and other organic compounds incorporating radiotracers that are considered foreign to the body undergo metabolic changes *in vivo*. Metabolic degradation of these drugs is commonly caused by a system of enzymes of low substrate specificity requirement, which is present mainly in the liver, but drug metabolism may also take place in the kidneys or other organs. Thus, radiotracers and all other pharmaceuticals are faced with enormous challenges to maintain their stability *in vivo* highlighting the importance of their structure. Often in practice, such biologically active molecules exhibit these properties *in vitro*, but fail during *in vivo* studies due to obtaining an increased metabolism within minutes. Many pharmacologically and biologically interesting compounds never see application due to their lack of stability. One of the most important issues of radiotracers development based on fluorine-18 is the stability *in vitro* and *in vivo*. Sometimes, the metabolism of ^{18}F-radiotracers goes along with the cleavage of the C-F bond and with the rejection of [^{18}F]fluoride mostly combined with high background and accumulation in the skeleton. This review deals with the impact of radiodefluorination and with approaches to stabilize the C-F bond to avoid the cleavage between fluorine and carbon.

Keywords: fluorine-18; metabolism; stability; deuterium

1. Introduction

Positron emission tomography (PET) and the combined techniques PET/MRT and PET/CT are outstanding imaging instruments and allow for the quantification and localization of physiological as well as pathophysiological processes *in vivo*, which were analyzed by tracing the appropriate biochemical fundamentals [1]. The basics of PET originate in the coincidental detection of annihilation photons emitted 180° apart, which originate from the radiotracer emitting positron, which again collides with electrons in the surrounding tissue. Measurement and quantification of the tracer distribution were obtained noninvasively in living organisms [2]. Fluorine-18 is an ideal radionuclide due to its favorable nuclear decay properties. It has a half-life of 109.8 min, which provides sufficient time to radiolabel the molecule of interest and localize it *in vivo*. Additionally, it emits a positron of low kinetic energy, which only travels a short range in tissue leading to high image resolution. However, tracers for PET imaging are always restricted by the kind of molecules that researchers can prepare and label. A summary of commonly used PET radionuclides is found in Table 1.

Table 1. Most commonly used PET radionuclides with their radiochemical details [3].

Nuclide	$t_{\frac{1}{2}}$ (min)	Production Route	Average Range in H$_2$O (mm)	E$_{av.}$ (β$^+$) (keV)
^{11}C	20.4	^{14}N(p,α)^{11}C	1	385
^{13}N	10	^{16}O(p,α)^{13}N	1.5	491
^{15}O	2	^{15}N(d,n)^{15}O	2.7	735
^{18}F	109.8	^{20}Ne(d,α)^{18}F ^{18}O(p,n)^{18}F	0.3	242
^{68}Ga	67.6	^{68}Ge-^{68}Ga generator	3.7	740
^{124}I	250.6	^{124}Te(p,n)^{124}I	3	188

Fluorine-18 is a unique radionuclide for PET imaging. In contrast to other β$^+$ emitting organic radionuclides like ^{11}C, ^{13}N, and ^{15}O, which are inclined to isotopic labeling, fluorine-18 is most commonly incorporated leading to an alteration of the original compound [4,5]. Due to the absence of fluorine in nearly all naturally occurring biomolecules [6], radiolabeling is often accomplished by a formal replacement of a proton or an OH group with ^{18}F (isosteric and isopolar) which is known as bioisosteric labeling (Table 2) [7]. However, in medicinal chemistry, the role of fluorine in drug design and development is expanding rapidly and a wide variety of small compounds/drugs were developed in the past with pharmacological relevance still having one or more fluorine atoms inside [8]. These molecules can serve as brilliant lead structures for ^{18}F-radiotracers. The other variant deals with the connection of small ^{18}F-building blocks or ^{18}F-prosthetic groups like [^{18}F]SFB or [^{18}F]FBAM [9], but this is mostly used with biomacromolecules like peptides, proteins, or antibodies. Both methods come along with changes of biological and/or pharmacological properties of the tracer molecule compared to the original compound. In general, smaller molecules exhibit a larger change in their properties by the introduction of a radionuclide to the considered molecule.

1.1. Nature of the C-F Bond

The similarity in size of fluorine (147 pm), hydrogen (120 pm) and oxygen (152 pm) makes fluorine-18 an appropriate candidate for the preparation of radiotracers, due to its longer half-life time compared to ^{11}C, ^{13}N or ^{15}O (Table 1) [10]. The substitution of single hydrogen or a hydroxyl group by fluorine induces only a slight steric perturbation [11]. The similarity of the C-F to the C-O bond length (Table 2) and the similar electronic properties like the induced dipole due to the inductive effect allows the isoelectronic replacement of an OH group by fluorine [12]. However, fluorine is only a (poor) hydrogen bond acceptor, while an OH group is both a hydrogen donor and an acceptor.

Table 2. Van der Waals radii [13], electronegativity and aliphatic C-X bond lengths of selected atoms.

Element X	Van der Waals Radius (pm)	Electronegativity (Pauling Scale)	Bond Length of C-X (pm)
H	120	2.1	109
C	170	2.5	154
O	152	3.5	143
F	147	4.0	135

Of all the atoms, the fluorine atom possesses the highest electronegativity; therefore, biological aspects have to be considered and can be of advantage in pharmaceutical as well as in radiotracer design. A favorable feature of fluorine is the strong but highly polarized σ bond to carbon [11]; this should make the fluorine a perfect leaving group in case of nucleophilic displacement reactions. However, the fluorine unexpectedly does not show good donor ability despite the high polarization of the C-F bond. This fact can be explained by the strong interaction of the partially positively charged carbon (residue) and the partially negatively charged fluorine which results in the strongest known σ bond in organic chemistry. Additionally, the highest bond dissociation energy (BDE) of approx. 441.3 kJ/mol is found for an aliphatic C-F bond compared to other carbon single bonds [14].

Interestingly, the average BDE differs with the number of covalently bound fluorine. More fluorine atoms bound to carbon increases the BDE and diminishes the C-F bond length [15]. The series of fluoromethane compounds in Scheme 1 demonstrate this trend, which can be explained by each of the C-F bonds pulling p-orbital electron density from the sp^3 carbon to the low lying sp^2 orbitals of fluorine (Bent's rule [16]), making the carbon more sp^2 in character [17].

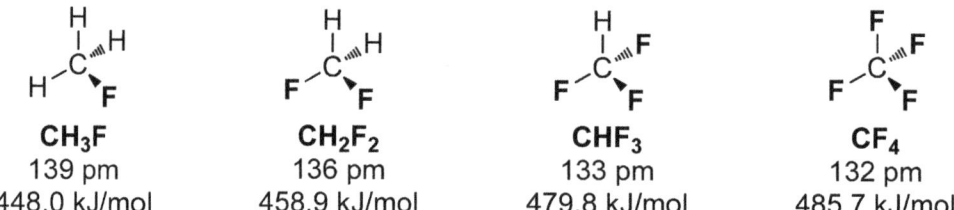

Scheme 1. Comparison of fluoromethanes with increasing number of bound fluorine and their associated bond length and bond dissociation energy.

Particular attention has to be made for fluorine in the benzylic position. It was found that the bond enthalpies (DH$_{298}$) are lower for benzyl fluorides (413.2 kJ/mol) compared to aliphatic compounds (439.2 kJ/mol for methyl fluoride) or aromatic fluorine derivatives (472.7 kJ/mol for fluorobenzene) [11]. Furthermore, the C-F bond length is also extended (138 pm compared to 135 pm average) [18]. The remaining substituents on the aromatic ring also have crucial influence on the stability of the benzylic fluorine especially concerning steric and inductive effects [19]. The introduction of [^{18}F]fluoride into the benzylic position of precursors occurs as previously described for aliphatic compounds, under milder reaction conditions in most of the cases due to the comparatively higher reactivity of the benzylic position. In this regard, the metabolic stability of benzyl [^{18}F]fluorides is also decreased [20]. Additionally, similar observations were made for allyl fluorides [21].

Aromatic $C_{Ar}(sp^2)$-F bonds are stronger than aliphatic $C(sp^3)$-F bonds [22] resulting in their bond lengths being shorter: 140 pm (aliphatic) *vs.* 136 pm (aromatic) [23]. This finding can be explained by the high polarization of the σ-C-F bond, the possible delocalization of the (partial) positive charge of the carbon in the aromatic moiety, and that fluorine possibly acts as a π-electron donor [24,25], which strengthens the C-F bond additionally. Thus, the ^{18}F radiolabeling of aromatic systems should be favored over aliphatic systems.

Bonding of fluorine to a sp-hybridized carbon is also possible, but this results in a highly reactive species due to the repulsion of the free electron pairs of the fluorine and the π-electron system of the triple bond [17]. This trend strongly follows Bent's rule, which states that the s-character of an atom concentrates in orbitals directed toward electropositive substituents [16]. No ^{18}F-radiotracer with a direct connection of ^{18}F to a triple bond has been developed to date.

To sum up, to produce a stabilized C-F bond, the most important criterion is the hybridization of the carbon. Moreover, inductive and steric effects of further substituents and organic residues influence the (metabolic) stability of the C-F bond as well.

1.2. Possibilities to Introduce Fluorine-18—Short Overview

The radionuclide ^{18}F is produced by a cyclotron using the nuclear reactions shown in Table 1. Once the radionuclide is produced, it must quickly be incorporated in the molecule of interest. Normally, the introduction of ^{18}F into aliphatic molecules (sp^3-hybridized carbon) is accomplished using no-carrier-added (n.c.a.) [^{18}F]fluoride and a precursor with a good leaving group (Br, I, OMs, OTs, ONs, OTf, NR$_3^+$) in a S$_N$2 reaction. This method has the advantage of preparing radiotracers with high specific activity (As). A challenging aspect of this labeling procedure is to eliminate traces of water to remove the hydration shell around the fluoride. Polar organic solvents (ACN, DMF, DMSO) were used with a cryptand (Krypofix K2.2.2.) to function as a phase transfer catalyst and to further separate the charge of the cation and fluoride (producing what is called naked fluoride) [26].

Introduction of fluorine-18 into aromatic systems can be performed by several reaction pathways. The classical Balz-Schiemann reaction is only rarely used for this purpose [27]. Commonly, two ways are applied: the nucleophilic aromatic substitution (S$_N$Ar) and the electrophilic aromatic substitution (S$_E$Ar). The major drawback when using the first variant is the necessity to activate the respective aromatic precursor with electron withdrawing groups (CN, halogens, NO$_2$, C=O) as well as good leaving groups. An isotopic exchange of ^{19}F by ^{18}F is also possible, but this results in a low As. This is reasoned by the disability to separate the ^{19}F-precursor from the ^{18}F-radiotracer. This is because the As is always influenced by the applied amount of the ^{19}F-compound. Other appropriate leaving groups for the nucleophilic aromatic displacement are halogens, NO$_2$ or Me$_3$N$^+$. Newer developments are based on iodonium [28] or sulfonium salts [29,30] as precursors and can be used for non-activated aromatic systems as well [31,32].

Using S$_E$Ar, [^{18}F]F$_2$ was applied consisting of both ^{18}F and ^{19}F (carrier added, c.a.), thus, the labeling will proceed in an "electrophilic" manner. As a consequence, a minimum of 50% of the elemental fluorine is ^{19}F and therefore not β^+-decaying. This pathway leads to a reduced As of the radiotracers due to the incorporation of ^{19}F. Usually, stannylated precursors, in which the carbon has the partial negative charge, are required for the labeling with [^{18}F]F$_2$ [33].

2. Radiodefluorination

Today's arsenal of radiotracers comprises more and more complex molecules ranging from small organic and pharmacologically active derivatives such as carbohydrates, amino acids or steroids to high molecular weight compounds like peptides, proteins or oligonucleotides. The development of new radiotracers for molecular imaging has to address important questions on target selection and radiobiological validation. These special requirements are encountered in radiotracer synthesis such as choice of the appropriate radionuclide and suitable labeling position. In this regard, a radiotracer has to meet different criteria to be delivered to the target area of interest such as an adequate lipophilicity, high selectivity to the biological target and a high metabolic stability *in vivo* [34]. Hence, special attention should be paid to implement fast and highly selective labeling reactions for radiotracers which tolerate other functional groups. One of the most important aspects in the design of new radiopharmaceuticals is the development of metabolically stable tracers to meet the desired requirements and characteristics as mentioned above [35]. The radiolytic decomposition of ^{18}F-radiotracers is also an important issue, especially during isolation and formulation of the tracer. This drawback can be avoided using additives like anti-oxidant stabilizers [36].

Drug metabolism, also known as xenobiotic metabolism, involves the biochemical modification of substances (pharmaceuticals, drugs, poisons, radiotracers). Drugs often are foreign compounds to the organism's normal biochemistry. This metabolism usually occurs through specialized enzymatic systems by living organisms. Because of this mechanism, lipophilic substances are often converted into more readily hydrophilic derivatives, which are then excreted. The rate of metabolism determines the duration and efficacy of a drug, also known as the biological half-life [37]. In the case of radiopharmaceuticals, the physical half-life of the appendant radionuclide influences this mechanism supplementary.

The reactions in these biochemical pathways are of particular interest in medicine as part of drug metabolism and as a factor contributing to multidrug resistance in infectious diseases, cancer chemotherapy or radiopharmacy. The speed of the homing process of a radioactive drug has to be relatively fast compared to the biological and physical half live of the drug to be able to obtain good signal to background ratio.

Drug metabolism in general is divided into three phases. In phase I, enzymes such as cytochrome P450 oxidase (oxidative metabolism: CYP, FMO, MAO, Mo-CO, aldehyde oxidase, peroxidases, xanthine oxidase; hydrolytic metabolism: esterase, amidases, epoxide hydrolases) introduce reactive or polar groups into the xenobiotics. Afterwards, these modified compounds are conjugated to yield more polar compounds in phase II reactions. These reactions are catalyzed by transferase enzymes (UGT, ST, NAT, GST, MT) [38]. Finally, in phase III, the conjugated xenobiotics may be further processed, before being recognized by efflux transporters and eliminated from the cells. Radiotracers follow this method of degradation as well with one large difference; they are administered in concentrations, which are significantly lower than "normal" pharmaceuticals.

PET radiotracers are typically injected intravenously in contrast to the orally administered "normal" pharmaceuticals. While circulating in the blood and tissues prior to localizing at the target site, a portion of the drug may be metabolized. The major organs involved in this metabolism process are the kidneys and liver. The biotransformation may happen within minutes of administration and the resulting radiometabolites are generally less lipophilic than the original radiotracer. Possible metabolic degradation pathways are illustrated in Scheme 2. Radiodefluorination is known to be a phase I reaction occurring primarily through the action of cytochrome P450 2E1 (CYP2E1) isozyme in liver microsomes [39–41].

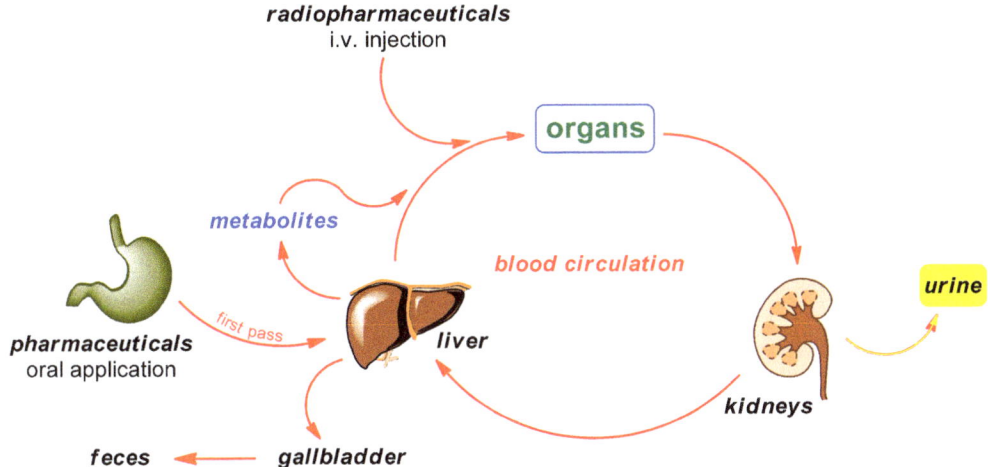

Scheme 2. Possible pathways of radiopharmaceuticals in contrast to pharmaceuticals in the body.

Mechanisms of Radiodefluorination

Radiotracers, independent on their corresponding radionuclide, that resist extensive metabolism *in vivo* over the period of time of a PET scanning session are seldom [42]. Metabolic paths that derivatize rather than disintegrate the respective tracer into small, more polar fragments, can produce unpleasant radiometabolites. Alongside the formation of more polar radiometabolites, the cleavage of [^{18}F]fluoride from the tracer, also known as radiodefluorination, is a major way for several ^{18}F-radiotracers to degrade despite the high strength of the C-F bond [43]. Afterwards, [^{18}F]fluoride as the main metabolite binds primarily to bone and skull. Especially while imaging the central nervous system, [^{18}F]fluoride binding to the skull is problematic [42].

There are several metabolic pathways discussed and proposed for the degradation of [^{18}F]fluoroalkyl chains. Two major factors affect the method of degradation: the location of the fluoroalkyl chain in the molecules and their length. In 1988, Welch and co-workers showed the difference in the metabolism between *N*-[^{18}F]fluoroethylated and *N*-[^{18}F]fluoropropylated spiperones [^{18}F]**1** and [^{18}F]**2** [44]. They proposed a metabolization by *N*-dealkylation followed by oxidation to the respective [^{18}F]fluoroaldehydes [^{18}F]**3b** and [^{18}F]**4b**. As a result, 2-[^{18}F]fluoroacetaldehyde ([^{18}F]**3b**) is a stable lipophilic metabolite whereas 3-[^{18}F]fluoropropanal ([^{18}F]**4b**) is unstable towards elimination of [^{18}F]fluoride (retro Michael addition). Further oxidation led to 3-[^{18}F]fluoropropionate ([^{18}F]**4c**), which eliminated subsequently to

[^{18}F]fluoride, too. In general, this kind of metabolism occurs when [$^{18/19}$F]fluoroalkyl chains are bound to heteroatoms like oxygen, nitrogen or sulfur. Both pathways are shown in Scheme 3.

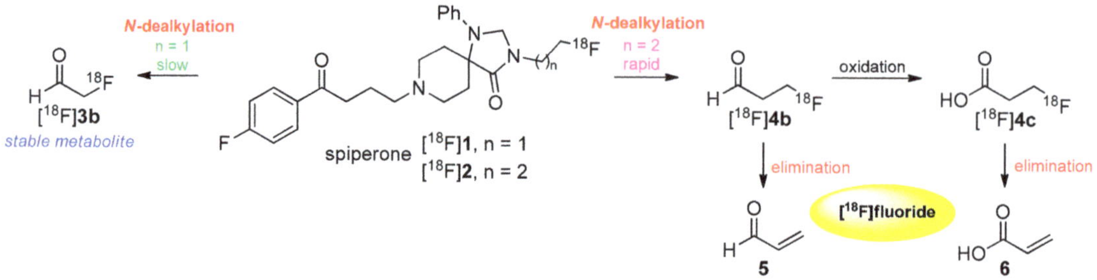

Scheme 3. Mechanism of degradation of [^{18}F]**1** and [^{18}F]**2** leading to *N*-dealkylation of the radiotracer.

In many cases, 2-[^{18}F]fluoroethanol [^{18}F]**3a** or its metabolites 2-[^{18}F]fluoroacetaldehyde [^{18}F]**3b** and 2-[^{18}F]fluoroacetate [^{18}F]**3c** have been observed during metabolic degradation of several PET tracers containing a 2-[^{18}F]fluoroethyl group such as [^{18}F]FECNT [^{18}F]**7** [45], [^{18}F]FETO [^{18}F]**8** [46], [^{18}F]FDDNP [^{18}F]**9** [47], [^{18}F]FFMZ [^{18}F]**10** [48], and [^{18}F]FERhB [^{18}F]**11** [49]. The metabolites (presumably 2-[^{18}F]fluoroacetaldehyde [^{18}F]**3b** and 2-[^{18}F]fluoroacetate [^{18}F]**3c**) from *N*-defluoroethylation of [^{18}F]FECNT [^{18}F]**7** and [^{18}F]FDDNP [^{18}F]**9** have been shown to distribute evenly in the brain, confounding data analysis [46,47]. The metabolic behavior of [^{18}F]FECNT [^{18}F]**7** is shown in Scheme 4.

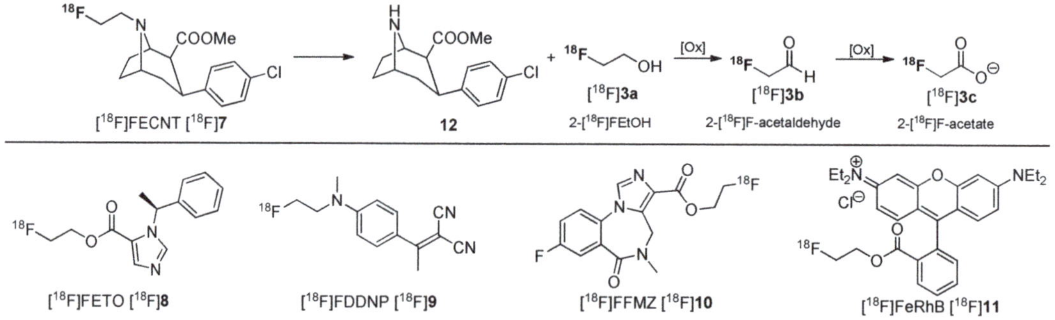

Scheme 4. Selected [^{18}F]fluoroethylated tracers and assumed metabolic pathway for degradation of [^{18}F]FECNT [^{18}F]**7**.

The behavior of 2-[^{18}F]fluoroethanol ([^{18}F]**3a**) and 3-[^{18}F]fluoropropanol ([^{18}F]**4a**) as possible radiometabolites was further investigated during the process of radiodefluorination [50]. The basis of these experiments was the assumption that fluoroalkyl ethers and ester were cleaved to give both aforementioned radiometabolites. Further, it was suggested that 2-[^{18}F]fluoroethanol ([^{18}F]**3a**) is converted to 2-[^{18}F]fluoroacetaldehyde ([^{18}F]**3b**), which is then metabolized to 2-[^{18}F]fluoroacetate ([^{18}F]**3c**). After formation of 2-[^{18}F]fluoroacetyl-CoA, it remains trapped inside the cell [51]. No activity was found in the bone, but it was stated that 2-[^{18}F]fluoroethanol ([^{18}F]**3a**) behaves like H$_2$[^{15}O]O.

In contrast, when investigating 3-[^{18}F]fluoropropanol ([^{18}F]**4a**) *in vivo* as a potential radiometabolite of 3-[^{18}F]fluoropropylated PET tracers a rapid accumulation in the skeleton was observed. This can be explained by free [^{18}F]fluoride being generated from 3-[^{18}F]fluoropropionaldehyde ([^{18}F]**4b**) or 2-[^{18}F]fluoropropionate ([^{18}F]**4c**). Both metabolites leading to further β-elimination under release of [^{18}F]fluoride (vide supra).

Furthermore, Lee and co-workers demonstrated the radiodefluorination of [^{18}F]fluoroalkyl groups bound to an aromatic system [52]. Several biphenyl derivatives ([^{18}F]**13**, [^{18}F]**15**, [^{18}F]**17**) and their degradation behavior were investigated. In all evaluated reactions, the first step consists of the oxidation of the carbon next to the aromatic ring. In the case of the fluoromethyl group of [^{18}F]**13**, a fast elimination step followed to give the respective aldehyde and [^{18}F]fluoride. In the case of the fluoroethyl residue of [^{18}F]**15**, a slow α-elimination occurred to give [^{18}F]fluoride and the remaining enol (ketone). In the third case, β-elimination took place after oxidation of [^{18}F]**17** to an α,β-unsaturated system **20** and elimination of [^{18}F]fluoride. The different metabolic degradation of these compounds is investigated (Scheme 5).

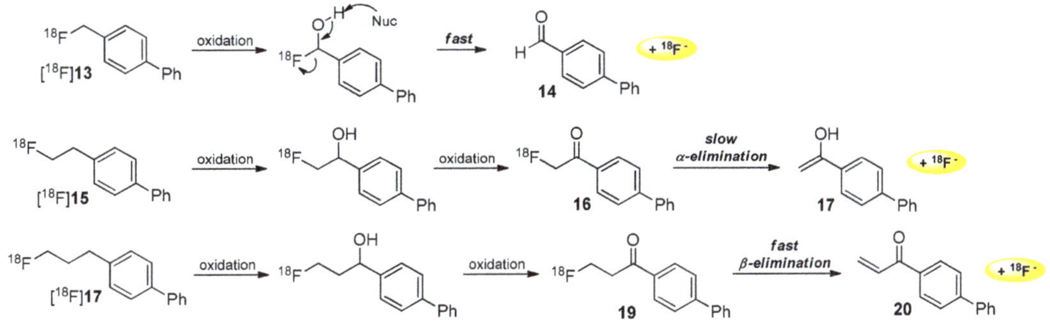

Scheme 5. Different metabolic behavior of [^{18}F]fluoroalkylated aromatic compounds.

Schibli and co-workers gave an alternative explanation for the mechanism of radiodefluorination. They assumed an oxidation of the carbon next to the [^{18}F]fluorine of tracer [^{18}F]PSS223 [^{18}F]**21** involved by cytochrome P450 enzyme (CYP) leading to a separation of [^{18}F]fluoride (Scheme 6). Experiments performed *in vivo* showed an accumulation of [^{18}F]fluoride in the bone [53] as reported above.

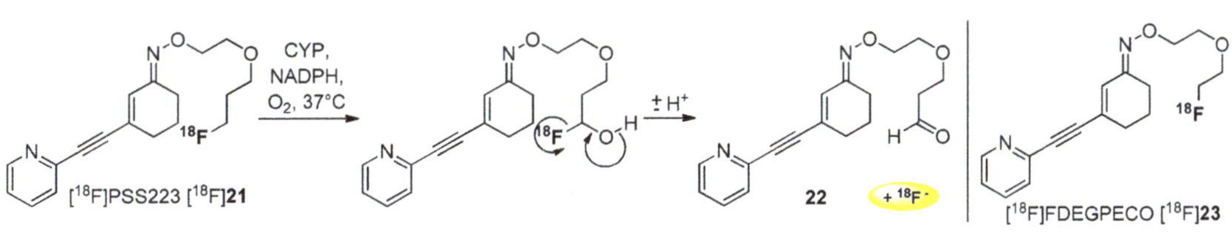

Scheme 6. Release of [^{18}F]fluoride from [^{18}F]PSS223 [^{18}F]**21** during degradation with cytochrome P450 (CYP).

Stability determinations with [^{18}F]PSS223 [^{18}F]**21** using rat and human liver microsomal enzymes were executed and pointed out two more polar radiometabolites as demonstrated by radio-UPLC measurements. The degradation process is shown to be NADPH-dependent, which implied the involvement of oxidoreductases. Amongst others, the fluorine-containing carbon atom was oxygenated leading to the release of [^{18}F]fluoride.

The difference in the *in vivo* behavior between [^{18}F]PSS223 [^{18}F]**21** and [^{18}F]FDEGPECO [^{18}F]**23** could be explained by the β-heteroatom effect [54,55], by which primary aliphatic bound [^{18}F]fluorine in β-position to heteroatoms (e.g., ROCH$_2$CH$_2$[^{18}F]F) is found to be metabolized at a slower rate. This rationale supports the absence of defluorination for [^{18}F]FDEGPECO [^{18}F]**23** containing only a [^{18}F]fluoroethyl group.

3. Methods to Avoid Radiodefluorination

The probably best alternative to avoid radiodefluorination consists of the direct connection of fluorine-18 to a phenyl moiety instead of aliphatic residues wherever applicable [56]. This is consistent with the higher stability of a C$_{Ar}$-F bond compared to a C(sp^3)-F bond as previously described. The fluoroaryl groups are stable to metabolism and do not lead to a considerable radiodefluorination. Otherwise, the [^{18}F]fluoroalkyl moiety has to be modified to reduce or avoid rapid metabolic degradation by the following methods.

Deuteration in Direct Neighborhood of Fluorine-18

A fundamental approach in medical chemistry is the application of deuterium to increase the stability of active pharmaceutical ingredients [57], which is useable to raise the metabolic stability of ^{18}F-radiotracers by means of the deuterium-proton exchange at carbon atoms close to the ^{18}F-atom. This procedure can sometimes suppress but not completely prevent the process of radiodefluorination.

The method of action of this effect is explained by the kinetic isotope effect that reduces the rate of metabolic degradation. The deuterium is not only twice as heavy as hydrogen, but also the zero-point energy is significantly lower than the energy of hydrogen. Due to these differences, the activation energy of the C-D-bond in chemical or biochemical reactions is significantly higher than for C-H-bond. Therefore, reactions on the C-D-bond will proceed considerably slower than the same reactions with a C-H-bond at the same position. In general, the cleavage rate of a C-H bond is 6.7 times faster compared to a C-D bond at 25 °C and it is postulated that the break of the C-H bond is the rate-determining step in this kind of defluorination [58,59]. A successful example of the stabilization by means of deuteration consists of the preparation of [^{18}F]FE-DTBZ-D$_4$ [^{18}F]**26**, which is pointed out in Scheme 7.

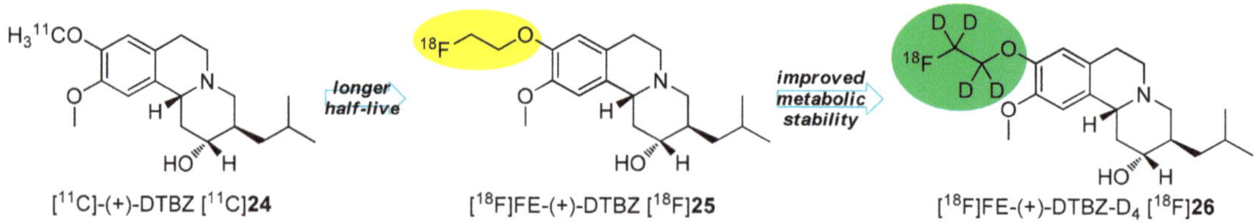

Scheme 7. Enhancement of the half-life as well as of the metabolic stability of DTBZ **24**.

[^{11}C]-(+)-DTBZ [^{11}C]**24** was initially used to studied dementia and Parkinson in the clinic [60–62]. Further improvements were necessary including the change of the radionuclide to elongate the half-live, which led to the development of [^{18}F]FE-(+)-DTBZ [^{18}F]**25** [63,64]. Successful *in vitro* studies with this ^{18}F-tracer were accomplished followed by *in vivo* studies showing a high accumulation of radioactivity in joints and bones. To improve the metabolic stability, [^{18}F]FE-(+)-DTBZ-D$_4$ [^{18}F]**26** was developed and showed enhanced properties. The main improvement resulted in the considerably reduced bone uptake when comparing both tracers. The defluorination rate ($k_{defluorination}$) was determined for both tracers to be 0.012 for [^{18}F]**25** and 0.0016 for [^{18}F]**26** resulting in an elongated plasma-t$_{1/2}$ from 46.2 min to 438.7 min [65].

The example in Scheme 7 exhibits the introduction of a deuterated [^{18}F]fluoroalkyl residue via a building block strategy using a fluorine-18 containing deuterated building block. The second general method consists of the gradual introduction of the deuterium followed by ^{18}F-labeling as the last step. In this case, the precursor already possesses the deuterium.

One of the first reports regarding the introduction of deuterium into precursors to prepare ^{18}F-radiotracers was presented by Ding, Fowler and Wolf in 1993 [66]. They introduced deuterium in different positions of the alkyl chain of 6-[^{18}F]fluorodopamine (6-[^{18}F]FDA) regioselectively to execute mechanistic studies regarding the degradation of these derivatives ([^{18}F]**30**, [^{18}F]**32**, [^{18}F]**33**) by monoamine oxidase B (MAO B) and dopamine β-hydrolase (DBH) via PET. The reaction path to precursors and resulting radiotracers is shown in Scheme 8.

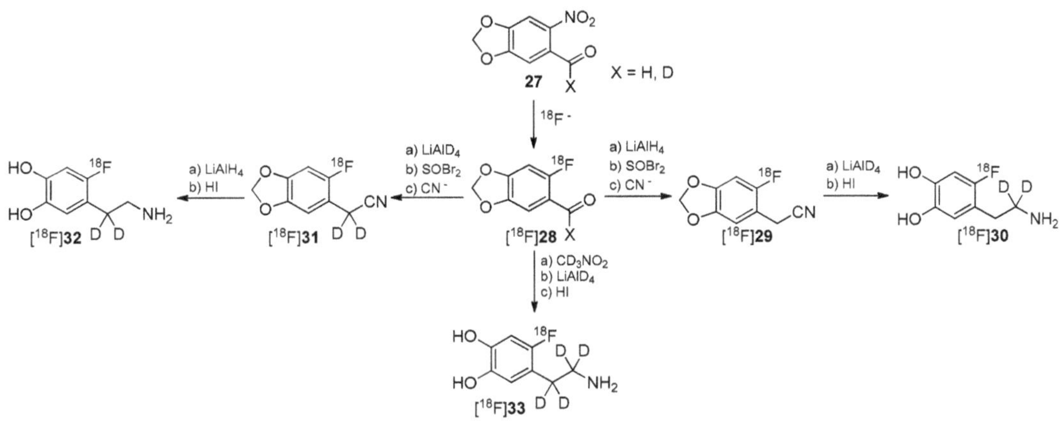

Scheme 8. Synthesis of different regioselectively deuterated [^{18}F]fluorodopamine derivatives.

It has recently been shown that MAO B and DBH stereoselectively remove only the pro-R hydrogen of the non-deuterated compounds [67–70]. Both dopamine compounds [^{18}F]**30** and [^{18}F]**32** with two deuterium atoms on one carbon were prepared to further verify this finding. Such specifically deuterated derivatives are therefore the most appropriate candidates for unambiguously assessing the contribution of metabolism by MAO and DBH on the kinetics of 6-[^{18}F]fluorodopamine.

In a following paper, it was shown that [^{18}F]**30** has a reduced rate of clearance, consistent with MAO-catalyzed cleavage of the α-C-D bond, whereas [^{18}F]**32** showed no change, indicating that cleavage of the β-C-D bond (DBH) is not rate limiting [71]. Both pathways of degradation are shown in Scheme 9. Furthermore, the rate of metabolism was also significantly reduced by pretreatment with pargyline (MAO inhibitor).

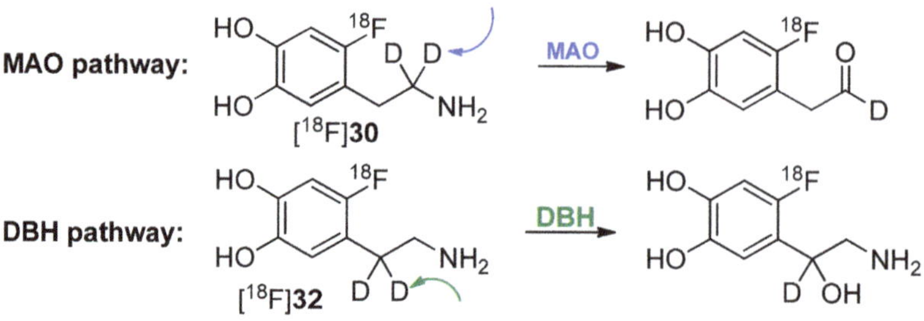

Scheme 9. Metabolic conversion of [^{18}F]**30** and [^{18}F]**32** by DBH and MAO, respectively.

The most often applied approach of the stabilization with deuterium consists of the use of deuterated [^{18}F]fluoroalkyl building blocks. For preparation, dihalogens, disulfonates or derivatives with mixed functions were used as starting material with deuterated methylene (-CD$_2$-) or ethylene (-CD$_2$CD$_2$-) groups. The introduction of ^{18}F follows standard labeling conditions (K222, anhydrous acetonitrile, 80–100 °C, 15–30 min). Examples for building blocks and most common labeling conditions to prepare these building blocks are shown in Scheme 10 [55,65,72–80].

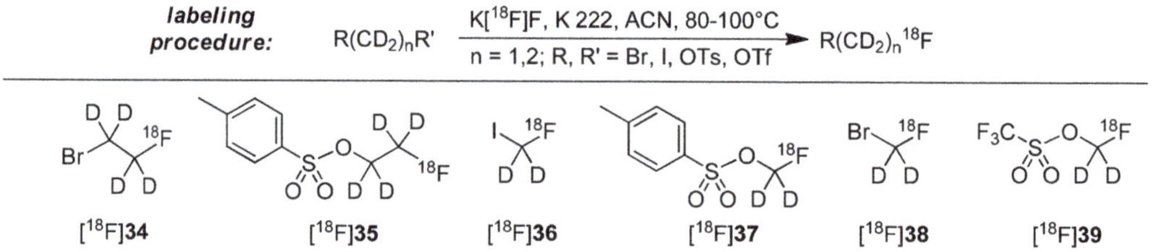

Scheme 10. General labeling procedure to create the deuterated building blocks and known building blocks [^{18}F]**34**–[^{18}F]**39**.

The subsequent labeling procedure with the above mentioned building blocks [^{18}F]**34**–[^{18}F]**39** represents a nucleophilic displacement at the carbon of the building block. Normally, sulfonate leaving groups are superior to halogens, but Schou and co-workers demonstrated that the type of the leaving group has only a small influence on the radiochemical yield (RCY) of the resulting radiotracer [72]. Mostly, the final alkylation reaction of building block with precursor proceeds rapidly (approx. 5 min reaction time) [65,72].

Amongst others, this building block strategy was extensively investigated for MeNER **41** (Scheme 11), which was identified as high-affinity ligand (IC$_{50}$: 2.5 nM *in vitro*) for the norepinephrine transporter (NET). Imaging of the NET moved into focus of research to investigate several neuropsychiatric and neurodegenerative disorders. The first successful PET images were obtained with carbon-11 labeled derivative [^{11}C]MeNER [^{11}C]**41**, which was synthesized by the use of NER **40** as precursor and radiolabeled with [^{11}C]MeOTf. Unfortunately, the binding of this MeNER derivate [^{11}C]**41** to the NET proceeded within a range of about 90 min *in vivo*, which was too long for a carbon-11 labeled tracer ($t_½$ = 20.4 min) [81]. This, led to the development of [^{18}F]FMeNER [^{18}F]**42**, an improved tracer with fluorine-18 ($t_½$ = 109.77 min) on the methyl group. [^{18}F]**42** still bound to the receptor with high affinity, while providing a sufficient half-life for imaging. This tracer was synthesized from the same precursor using bromo[^{18}F]fluoromethane and [^{18}F]fluoromethyl triflate with similar results. In contrast, the signal-to-background ratio and the bone uptake was increased compared to PET images from [^{11}C]MeNER [^{11}C]**41**. Fortunately, this effect was nearly completely suppressed by the use of the deuterated derivative [^{18}F]FMeNER-D$_2$ [^{18}F]**43**, which shows the impact of the isotope effect for the development of radiotracers [72].

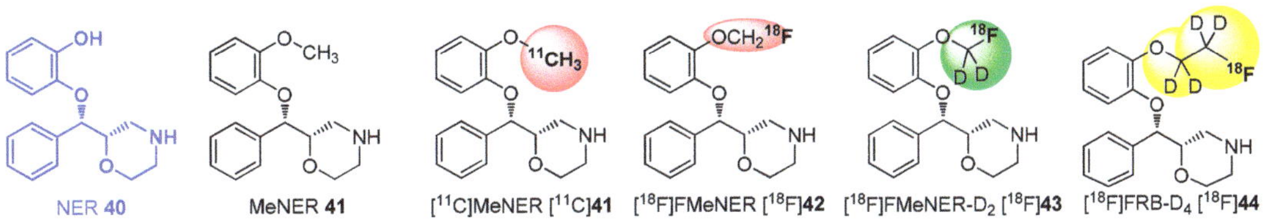

Scheme 11. Carbon-11, fluorine-18 and deuterated derivatives of NER **40** to increase metabolic stability.

This successful procedure was also applied for the preparation of [^{18}F]FRB, the ethoxy derivative of MeNER **41** based on Reboxetine (IC$_{50(NET)}$: 8.23 nM). For this purpose, the precursor NER **40** was successful labeled with [^{18}F]fluoroethyl bromide and [^{18}F]fluoroethyl bromide-D$_4$ ([^{18}F]**34**), to give [^{18}F]FRB and [^{18}F]FRB-D$_4$ [^{18}F]**44**, respectively. Due to the better pharmacological properties of [^{18}F]FMeNER-D$_2$ [^{18}F]**43** compared to [^{18}F]FRB-D$_4$ [^{18}F]**44**, a fully automated synthesis was developed for [^{18}F]FMeNER-D$_2$ [^{18}F]**43** in 2013 [80].

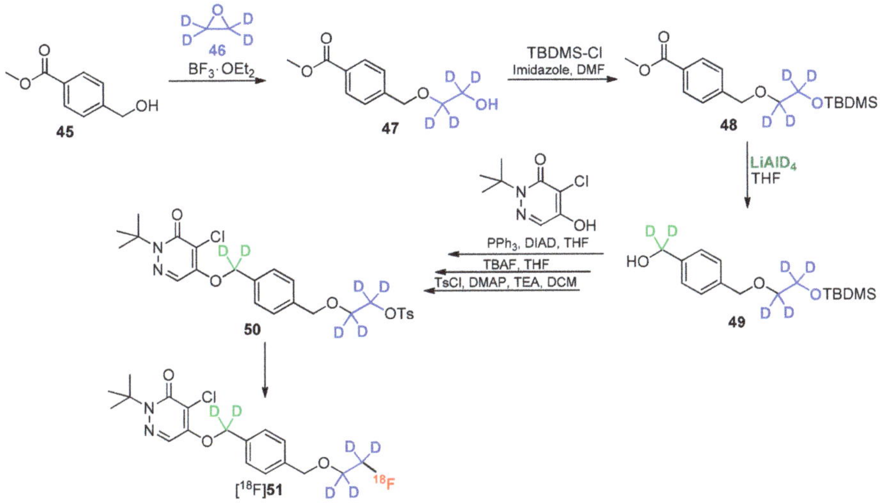

Scheme 12. Direct approach to introduce deuterium and fluorine-18 into radiotracer [^{18}F]**51**.

Another promising approach to use deuterated building blocks was shown by Casebier and colleagues [82]. In contrast to the previously discussed approaches, the deuterium containing residue was directly connected to the precursor molecule prior to radiolabeling to avoid a two-step-synthesis of radiotracer. The interesting task of this work was the use of fully deuterated ethylene oxide as building block (Scheme 12), which was introduced via ring-opening reaction. The next steps required the protection of the OH group with TBDMS-Cl followed by reduction of the methyl ester with LiAlD$_4$. The obtained deuterated methylene group is mandatory for a further stabilization of the tracer in terms of metabolic degradation. Upon completion of the basic structure of the molecule, the hydroxyl group was selectively deprotected using TBAF and functionalized with *p*-tosylchloride for labeling with fluorine-18.

Several other ^{18}F-tracers are known which are stabilized with deuterium. Selected examples are given in the following overview in Scheme 13 [83–93]. As stated before, the radiodefluorination process cannot fully be avoided, but it can be delayed considerably.

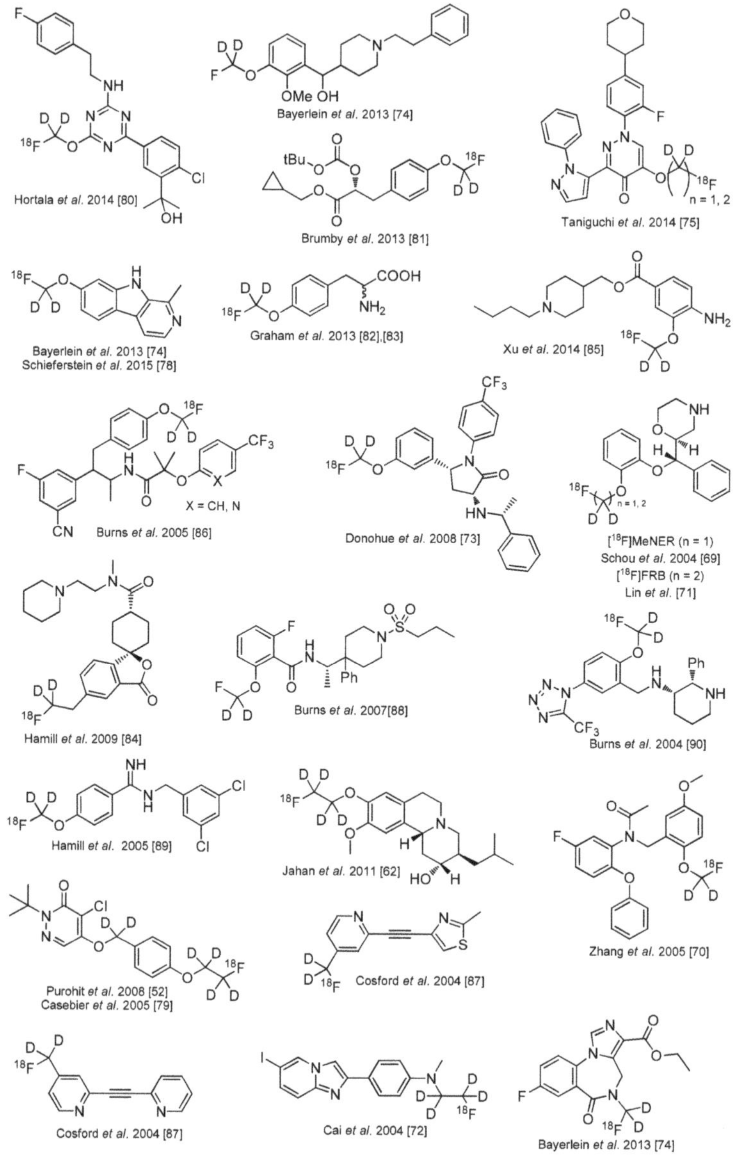

Scheme 13. Overview over ^{18}F-radiotracers stabilized with deuterium in direct neighborhood to ^{18}F.

4. Deuteration on other Parts of the Molecule to Avoid Degradation

As already mentioned, hydrogen–deuterium exchange will not only be adopted in direct proximity to fluorine-18 to avoid radiodefluorination or other metabolic degradation. A similar effect could be achieved by the use of deuterium connected to endangered areas in the tracer molecule, which are prone to metabolic degradation. Such an additional stabilization was already shown for compound [^{18}F]**51** by Casebier and colleagues in Scheme 12 [82]. Furthermore, the application of deuterium leads occasionally to another metabolic pathways as it was figured out by Leyton, Smith and co-workers [94,95]. Some examples for ^{18}F-radiotracers deuterated on other parts of the molecule are shown in Scheme 14.

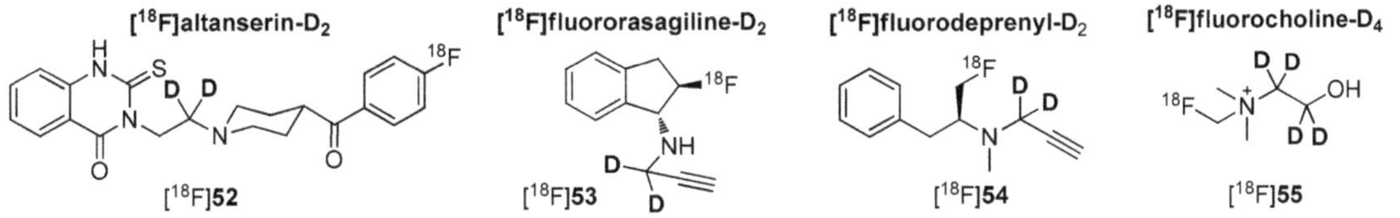

Scheme 14. Examples for radiotracers deuterated on other parts of the molecule.

[^{18}F]Fluororasagiline-D$_2$ ([^{18}F]**53**) and [^{18}F]fluorodeprenyl-D$_2$ ([^{18}F]**54**) (Scheme 14) are two examples of a successful enhancement of the metabolic stability using deuterium. Both tracers are known to be inhibitors of monoamnooxidase (MAO) and were applied for detection of psychiatric and neurological disorders such as depression, Alzheimer, and Parkinson diseases [96]. Although both lead structures (rasagiline and L-deprenyl) contain a propargyl group, there are significant differences in their metabolic products [97]. Inhibition experiments *in vitro* pointed out a high selectivity of MAO-B compared to MAO-A for both above mentioned radiotracers as well as for their non-deuterated derivatives. Moreover, it was found that the alkynyl chain of these molecules was cleaved under *in vivo* conditions. Thus, deuterium was selectively introduced into this residue to stabilize these molecules.

Comparing the deuterated and non-deuterated tracers in terms of their radiopharmacological behavior *in vivo*, [^{18}F]fluorodeprenyl showed a fast and irreversible binding to the enzyme limited by blood flow rather than by the MAO-B enzyme concentration, whereas [^{18}F]fluororasagiline expressed continuous increase of the radioactivity in the brain indicating a blood–brain barrier penetrating radiometabolite. In contrast, [^{18}F]fluororasagiline-D$_2$ ([^{18}F]**53**) and [^{18}F]fluorodeprenyl-D$_2$ ([^{18}F]**54**) exhibited fast clearance from the brain and less accumulation in cortical and sub-cortical regions. Furthermore, both deuterated analogues were more stable in monkey plasma compared to their non-deuterated analogues [98]; metabolic degradation was almost completely reduced. Thus, the deuterated tracers seemed to be more suitable for an application over the non-deuterated derivatives.

Another interesting example is the metabolic behavior of radiolabeled cholines. Known radiolabeled derivatives are exemplified in Scheme 15. In general, two main metabolic pathways are known for choline derivatives. The first pathway is based on the phosphorylation of choline via choline kinase (E.C. 2.7.1.32) to phosphocholine which is further transformed to phosphatidylcholine, a key component of the plasma membrane. This way is also known as The Kennedy pathway [99]. Once phosphorylated, phosphocholine is trapped within the cell, which is crucial for PET imaging with ^{11}C and ^{18}F radiotracers based on choline. The second main pathway of choline metabolism is based on the oxidation of choline to betaine. It was first described by Ikuta and co-workers in 1977 [100] and involves the conversion of choline to betain by choline oxidase (E.C. 1.1.3.17) via a four-electron oxidation using two sequential FAD-dependent reactions [101]. However, the second pathway is not preferred for PET imaging applications using radiolabeled choline derivatives. To overcome this obstacle, ^{11}C and ^{18}F labeled choline derivatives, which are deuterated at the ethylene moiety and not in the immediate neighborhood of the desired radionuclide were applied due to their altered pharmacological behavior.

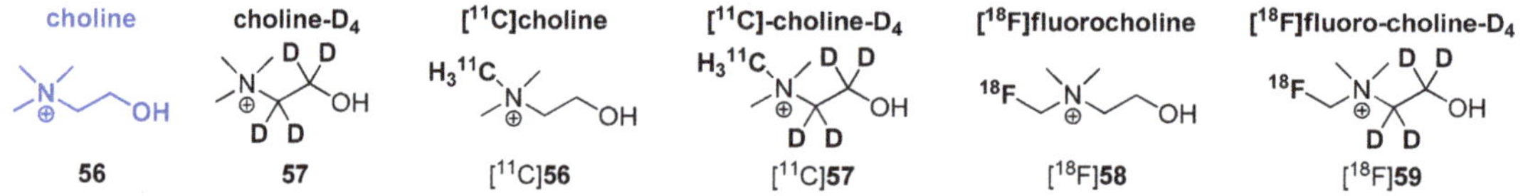

Scheme 15. Several choline derivatives with and without deuterium labeled with carbon-11 or fluorine-18.

In 2003, Gadda investigated enzyme kinetics for choline oxidase with choline (**56**) and choline-D$_4$ (**57**) as substrates to evaluate the impact of the kinetic isotope effect. It was shown that the oxidation of deuterated choline **57** was reduced to a minimum [101], which led to the successful development of choline-based radiotracers.

In 2009, Aboagye and colleagues compared the relative oxidation rates of the two isotopically radiolabeled choline species, [^{18}F]fluorocholine ([^{18}F]**58**) and [^{18}F]fluorocholine-D$_4$ ([^{18}F]**59**) with respect to their metabolites [94]. Both betaine metabolites from [^{18}F]**58** and [^{18}F]**59** were obtained from mouse plasma after intravenous injection of both radiotracers. As a result, it was pointed out that [^{18}F]**59** was remarkably more stable to oxidation than [^{18}F]**58** with ~40% conversion of [^{18}F]**59** to the betaine at 15 min after intravenous injection into mice compared to ~80% conversion of [^{18}F]**58** to the respective betaine-metabolite.

In 2012, both ^{11}C-labeled choline derivatives [^{11}C]**56** and [^{11}C]**57** as well as ^{18}F-choline-D$_4$ ([^{18}F]**59**) were synthesized to compare their biodistribution and metabolic behavior. Additionally, the same group performed small-animal PET studies and kinetic analyses to evaluate the tracer uptake in human colon HCT116 xenograft-bearing mice [102]. It was found that the simple substitution of deuterium for hydrogen and the presence of ^{18}F improves the stability and reduces degradation of the parent tracers. Furthermore, the availability is increased for phosphorylation and trapping within cells, which leads to a better signal-to-background contrast, thus improving tumor detection sensitivity of PET. In addition, deuterated ^{11}C choline was demonstrated to have a higher stability compared to non-deuterated ^{11}C-choline, but an increased rate of oxidation of betaine compared to ^{18}F-D$_4$-choline was observed. In 2014, the first promising human studies with healthy volunteers were accomplished [103].

4.1. Cycloalkyl Derivatives and Fluorine Connected to a Secondary Carbon Atom

Several literature sources reported that the replacement of an alkyl chain by a cycloalkyl ring resulted in more metabolically stable compounds [104–107]. Examples are given in Scheme 16. Despite this increased stability, only a few reports exist on PET radiotracers containing cycloalkyl rings. One example describes a potential radiotracer for assessing myocardial fatty acid metabolism, [^{18}F]FCPHA [^{18}F]**60**, containing a cyclopropyl moiety which allows the tracer to be trapped in the cells [108]. Another example describes non-natural ^{18}F-amino acids with fluorine-18 located at the cycloalkyl residue [109,110]. [^{18}F]**61** and [^{18}F]**62** show an increased metabolic stability compared to their non-cyclic counterparts. The placement of ^{18}F is especially important for compound [^{18}F]**63**, because the methoxy group itself and also the introduction of a fluoroalkoxy moiety instead of the methoxy group at the phenol part of the molecule lead to a fast cleavage [111].

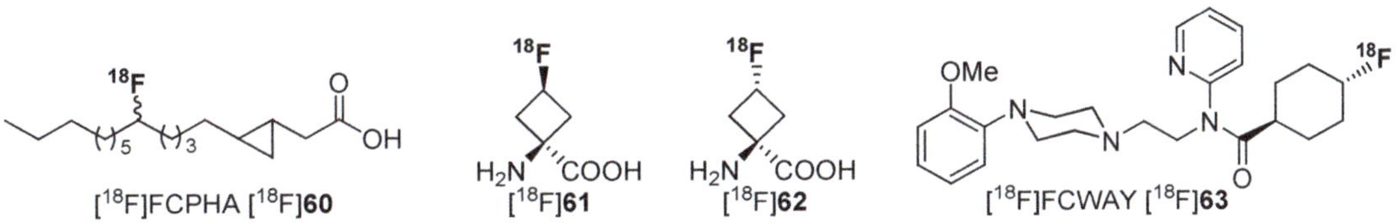

Scheme 16. Selected examples of fluorine-18 bound to secondary carbon for stabilization.

Both non-natural amino acids were used as brain tumor imaging agent and W. Yu *et al.* [112] found that the newly developed amino acid [^{18}F]**61** is comparable to [^{18}F]**62**. However, the cyclic unnatural amino acids are not metabolized [113]. The major drawback of this approach is the stereoselective construction of the amino acid skeleton. Thus, Franck and colleagues reported a diverse approach using cyclic building blocks bearing the ^{18}F-label. The research was focused on the metabolism of ^{18}F-tracers with [^{18}F]fluoroalkyl chains attached to hetereoatoms such as O, N, and S. Biotransformation (radiodefluorination) of these radiotracers was avoided by the utilization of cyclobutyl groups containing fluorine-18. Hence, cyclobutyl 1,3-ditosylate (**64**) was used as starting material. Radiofluorination was performed under standard conditions using [^{18}F]F$^-$, K 222 in anhydrous acetonitrile. After successful synthesis of the building block [^{18}F]**65**, L-tyrosine was used and labeled. The concept and the full reaction path including radiolabeling are pointed out in Scheme 17.

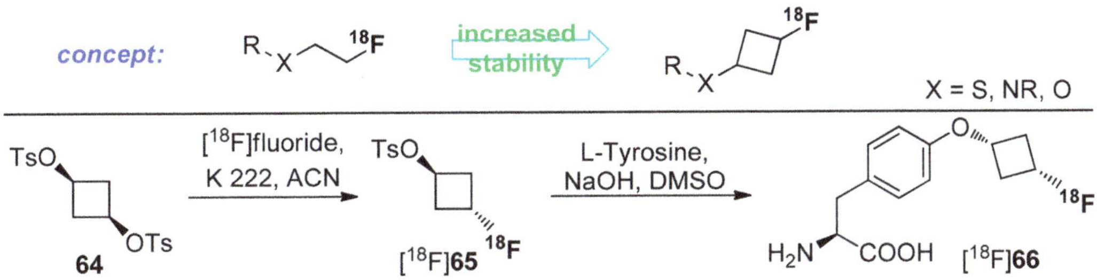

Scheme 17. Labeling concept to avoid radiodefluorination and radiolabeling of L-tyrosine with [^{18}F]fluorocyclobutyl tosylate ([^{18}F]**65**).

The obtained [^{18}F]fluorocyclobutyl derivative [^{18}F]**66** is comparable with the well-known amino acid O-(2-[^{18}F]fluoroethyl)-L-tyrosine ([^{18}F]FET) in the case of cell uptake and blocking and showed an excellent metabolic stability in phosphate buffer and in human and rat plasma for 120 min [114,115].

Further, the connection of fluorine to a secondary carbon could also help to prevent radiodefluorination in some cases. However, when using ^{18}F-FCWAY [^{18}F]**63** (Scheme 16), the defluorination process is a major issue. To prevent degradation, the responsible enzyme (cytochrome P450 2E1 (CYP2E1) isozyme) is suppressed with miconazole nitrate prior to the injection of the radiotracer [116]. With this method it was possible to substantially avoid radiodefluorination and the combined uptake of [^{18}F]fluoride in the skull.

4.2. SiFA-Techniology

The Si-F bond represents one of the strongest single bonds with a corresponding bond energy of 565 kJ/mol, which is 80 kJ/mol higher than the Si-C bond and suggest a high thermodynamically stability [117]. This fact led to the development of fluorine-18-radiotracers based on organosilanes, which should be unaffected against radiodefluoroination commonly associated with alkylfluorides. In 1985, Rosenthal and colleagues were the first who successfully radiolabeled [^{18}F]fluorotrimethylsilane [118]. The reaction was performed using chlorotrimethylsilane as precursor with a yield of 65% and high radiochemical purity. However, subsequent *in vivo* investigation of [^{18}F]fluorotrimethylsilane indicated a rapid hydrolysis followed by an enrichment of radioactivity in bones. For this reason, this concept was ineffective for the preparation of ^{18}F-radiotracers [119,120].

In 2000, Walsh and co-workers tried to induce the stabilization of Si-F-bond with bulky substituents such as phenyl or *tert*-butyl groups and confirmed the assumption of Rosenthal, who predicted the use of bulky substituents on silicon diminishes the hydrolysis of Si-F bond [121]. Furthermore, Choudhry and Blower investigated the behavior of different sized alkyl groups (Me, Ph, *tert*-Bu) and their combinations connected to fluorosilanes. The results showed that *tert*-butyldiphenyl[^{18}F]fluorosilane ([^{18}F]**68**) contained the highest stabilized Si-F-bond [122]. Contemporaneously, Schirrmacher and Jurkschat carried out comparable experiments and found di-*tert*-butylphenylfluorosilane ([^{18}F]**69**) with the highest stability against hydrolysis of Si-F-bond and called this compound class SiFA (silicon-based fluoride-acceptor). The hydrolytic stability in dependence of the alkyl group is expressed in Scheme 18 [123].

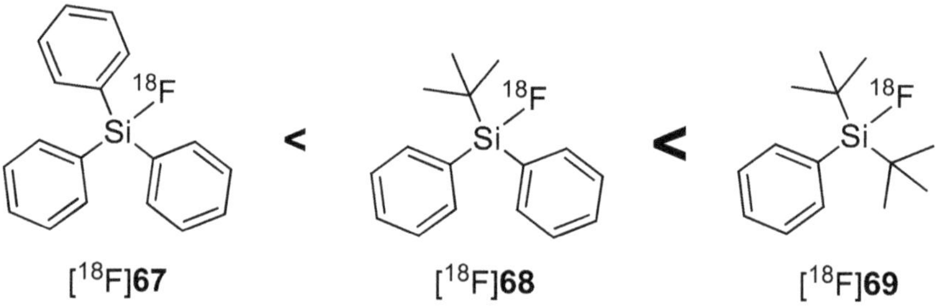

Scheme 18. *In vitro* hydrolytic stability of [^{18}F]fluorosilanes in dependence of their organic groups in human serum.

The high polarization of Si-F bond results in the kinetic instability of Si-F-bond [124] and allows an exchange under mild conditions. Due to the low energy of vacant d-orbitals tetravalent silicon as weak Lewis acid reacts with Lewis bases [125], which allows a nucleophilic attack by hydroxyl-groups in the case of aqueous conditions. Generally, nucleophilic displacement reactions on silicon proceed by the predicted S$_N$2 mechanism in Scheme 19 [126,127].

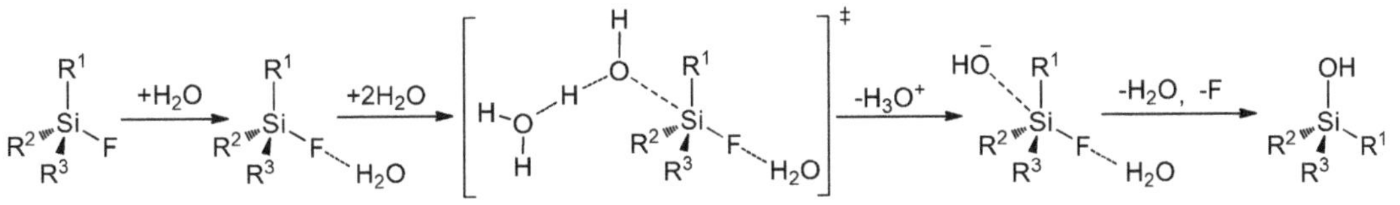

Scheme 19. Suggested S_N2 mechanism of the hydrolysis reaction of organofluorosilanes.

Contrary to carbon, a real pentagonal transition state including hypervalent silicon is formed and assists this substitution. The larger covalent radius of silicon compared to carbon contributes to this nucleophilic substitution [128], which led to the poor kinetic stability of Si-F-bond despite the high thermodynamic stability. Thus, a stabilization of Si-F bond to prevent a nucleophilic attack is only possible by raising the sterical bulkiness of the substituents. This fact explains the weak impact (plain structure) of phenyl moieties and they are also responsible for the augmented Lewis acid properties of silanes.

Only the use of *tert*-butyl groups located in direct neighborhood of Si-F-bond prevents hydrolysis due to their bulky three-dimensional structure. The third substituent on the silicon is utilized for further derivatization. Hence, the phenyl group seems to be the perfect choice for a functionalization with groups such as aldehydes, NCS-, or -SH in mainly para-position to the silyl residue. These resulting building blocks were often used for labeling of peptides and proteins [123,129–131]. An overview is given in Scheme 20. Furthermore, the use of alkyl-groups as third substituent with supplemental functionalization was proven, but exhibited a reduced hydrolytic stability compared to the phenyl tracers [126,132].

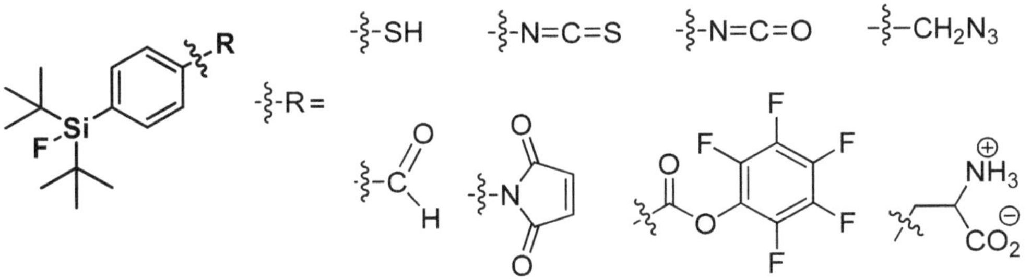

Scheme 20. A summary of applied SiFA building blocks taken from the review by Bernard-Gauthier *et al.*, 2014 [128].

Next, the introduction of fluorine-18 was evaluated by the use of different leaving groups such as alkoxycarbonyl-groups [122,133], by halogen and isotope exchange [118,123] and by applying hydrosilanes [134]. Manifold examples for the use of the SiFA concept were octreotide, bombesin, RGD, PSMA, antibodies, simple molecules, carbohydrates, and biotin. An excellent overview is provided by an outstanding review, see [128]. In most of the cases, the respective building blocks were applied especially for the biomacromolecules, but also a direct introduction of fluorine-18 was accomplished. However, the direct introduction exhibited a rather low yield compared to the building block approaches. Examples of hypoxia tracers [^{18}F]**70**–[^{18}F]**72** with rising metabolic stability and [^{18}F]**73** as SiFA-labeled peptide is found in Scheme 21.

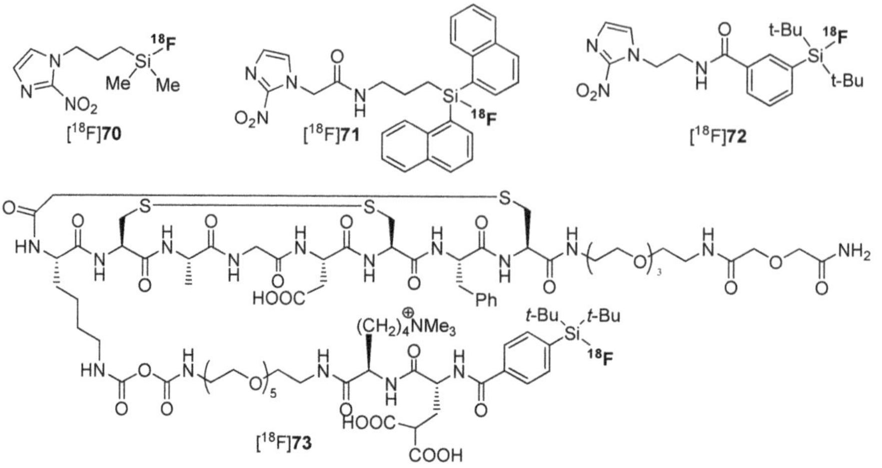

Scheme 21. Different hypoxia tracers [^{18}F]70–[^{18}F]72 with diverging metabolic stability [135] and an example for biomacromalecule [^{18}F]73 labeled with SiFa [131].

5. Miscellaneous

5.1. Fluorosulfonamides

Metabolically stable building blocks also referred to as prosthetic groups were required especially for the radiolabeling of peptides or other biomacromolecules. Conventional building blocks such as [^{18}F]SFB [^{18}F]74 were used to radiolabel particularly with primary amine residues of peptides (N-terminus or lysine) under formation of amide (peptide) bonds. A selection of these ^{18}F-building blocks is pointed out in Scheme 22.

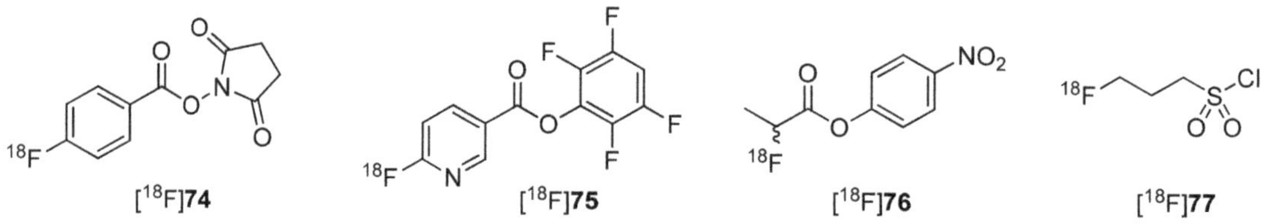

Scheme 22. Selected examples for ^{18}F building blocks for radiolabeling of peptides.

However, this kind of radiofluorinated aromatic fluoroacetamides turned out to be unstable *in vivo* and undergoes *N*-defluoroacylation [136]. It was reported that this degradation may be caused by the involvement of carboxylesterase (E.C. 3.1.1.1) or other hydrolases [137,138]. As an alternative to these acyl-based prosthetic groups, the 3-[^{18}F]fluoropropanesulfonyl chloride ([^{18}F]77) was introduced by Li *et al.* [139] and by Löser and co-workers [140]. They substantiate the metabolic integrity of fluorinated sulfonamide: *N*-(4-fluorophenyl)-3-fluoropropane-1-sulfonamide (**80**) compared to the aromatic acyl derivative *N*-(4-fluorophenyl)-fluoroacetamide (**79**) in a spectrophotometric enzyme assay using pig liver esterase. Both compounds are shown in Scheme 23.

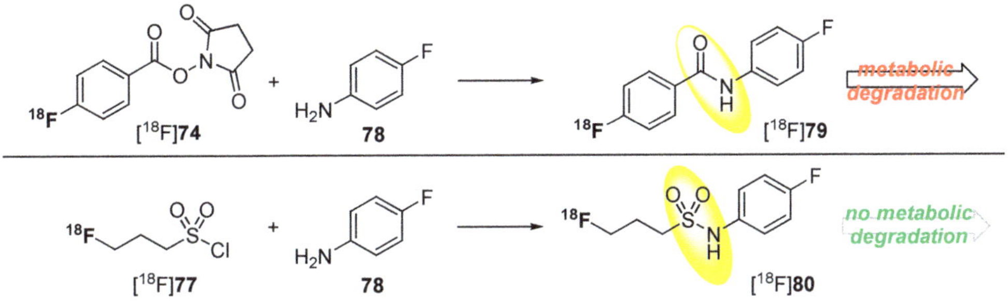

Scheme 23

Scheme 23. Comparison in radiofluorination and metabolic stability for [^{18}F]SFB [^{18}F]**74** and [^{18}F]fluoropropylsulfonyl chloride [^{18}F]**77**.

After 120 min (approx. one half-life of ^{18}F), only 20% of the starting ^{19}F-compound **79** was intact whereas, at the same time point, over 95% of the sulfonamide **80** was still detectable. Furthermore, pseudo-first order kinetics for the degradation of the acylamide could be determined.

5.2. Click-Chemistry

A further approach to avoid degradation was using triazoles [141], which were obtained by click chemistry [142,143]. Two different research groups investigated [^{18}F]fluoroalkyl groups bound at position N-3 of the triazole moiety of thymidine derivatives like [^{18}F]**82** with conflicting results and uncertainty over the metabolic stability of the radiotracers *in vivo*. The prepared ^{18}F-tracers are shown in Scheme 24. Smith and colleagues postulated that 1,4-disubstituted triazoles have a higher metabolic stability *in vivo* due to the greater steric bulk of the triazole. The metabolic stability is increased relative to simple fluoroalkyl substituents to thymidine-phosphorylase-mediated cleavage [144,145].

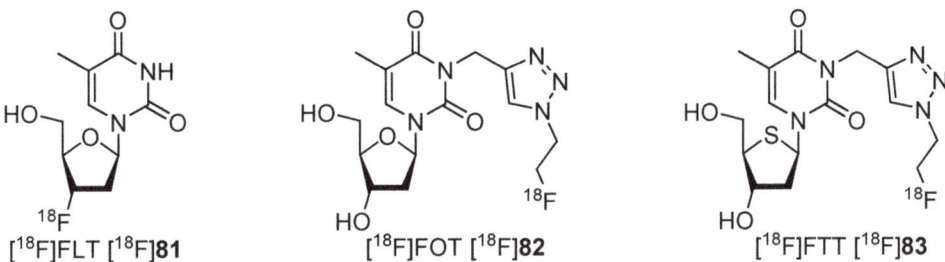

Scheme 24. Presentation of ^{18}F-labeled thymidine derivatives in the case of their labeling position.

Initial studies of the ability of these nucleosides to undergo phosphorylation demonstrated that [^{18}F]FLT [^{18}F]**81** was phosphorylated to approx. 7%–8% after 60 min incubation, whereas no phosphorylation was observed with [^{18}F]FOT [^{18}F]**82** over the same time period. Comparison with [^{18}F]FLT [^{18}F]**81** showed that [^{18}F]FOT [^{18}F]**82** was poorly phosphorylated at the 5-position of the deoxyribose residue. The poor thymidine kinase 1 (TK1) substrate tolerance due to substitution at nitrogen N-3 was given as a possible reason for this finding.

The working group of Choe developed ^{18}F-Labeled styryltriazole and resveratrol derivatives such as [^{18}F]**85** and [^{18}F]**87** for β-amyloid plaque imaging [146]. Compounds **84** and **86** were labeled under standard labeling conditions (*n*-Bu$_4$N[^{18}F]F, acetonitrile, 90 °C or 110 °C, 10 min) and yielded both tracers in 20%–30% RCY for [^{18}F]**85** and 56% RCY for [^{18}F]**87** with a As ≈ 38 GBq/μmol and a RCP > 99% (Scheme 25).

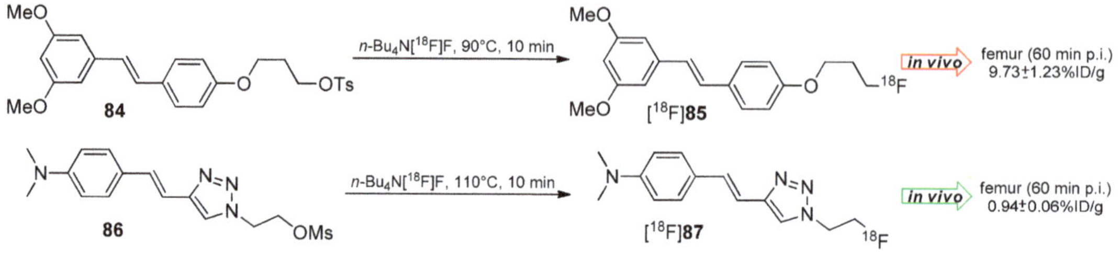

Scheme 25. Radiolabeling and selected *in vivo* results of reservatrol [^{18}F]**85** and styryltriazole [^{18}F]**87**.

In vivo studies of both tracers showed a remarkable metabolic degradation of reservatrol derivative [^{18}F]**85** under elimination of [^{18}F]fluoride which was accumulated in the femur (16.15% ± 3.10% ID/g after 120 min). Conversely, the styryltriazole compound [^{18}F]**87** showed almost no cleavage of [^{18}F]fluoride (1.54% ± 0.02% ID/g after 120 min).

5.3. CF$_3$-Derivatives

As stated in the introduction (Section 1.1), the use of CF$_3$ groups could increase the metabolic stability of pharmacologically relevant compounds and radiotracers [147] due to the increased bond strength of the C-F bond in this group compared to single fluorine connected to carbon and due to the higher steric shielding of the carbon center. Furthermore, the trifluoromethyl group is present in a large number of agrochemicals, biologically active drugs and anesthetics, which led to attempts to introduce fluorine-18 to yield [^{18}F]CF$_3$ group containing radiotracers; see an excellent review by Lien and Riss [148].

Normally, the introduction of [^{18}F]fluoride was accomplished via ^{18}F/^{19}F isotopic exchange [149–151], Lewis acid mediated reactions [152,153], halogen for ^{18}F exchange [154–156] or H^{18}F addition [157] and electrophilic reactions with [$^{18/19}$F]F$_2$ [158,159], but most of these reactions suffer from low specific activities due to the carrier added reactions and/or rough conditions.

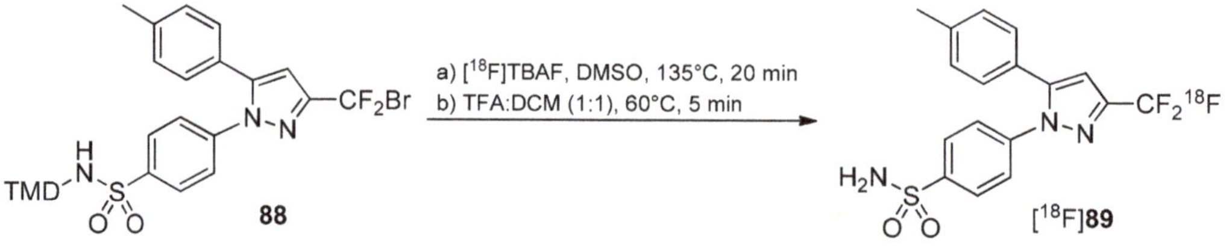

Scheme 26. Radiolabeling of [^{18}F]CF$_3$ containing Celecoxib derivative [^{18}F]**89**.

An example is presented regarding the synthesis of 4-[5-(4-methylphenyl)-3-([^{18}F]trifluoromethyl)-1*H*-pyrazol-1-yl]benzenesulfonamide ([^{18}F]Celecoxib) ([^{18}F]**89**) which is known to be a selective COX-2 inhibitor [160]. The labeling procedure was accomplished exchanging bromide with [^{18}F]F$^-$ using [^{18}F]TBAF in DMSO at 135 °C (Scheme 26). [^{18}F]Celecoxib was achieved in 10% ± 2% RCY (end of synthesis) with >99% chemical and radiochemical purity and a specific activity, which was 4.40 ± 1.48 GBq/µmol (end of bombardment). *In vitro* stability experiments showed only a small amount of [^{18}F]fluoride coming from radiodefluorination in 10% ethanol-saline (6.5% after 4 h). However, *in vivo* experiments of [^{18}F]**89** with Wistar rats showed a higher skeleton uptake compared to brain or heart; regions where COX-2 is known to be present due to the radiodefluorination process. In contrast, no uptake in skull and skeleton was observed in baboon indicating only a low degree of defluorination of [^{18}F]**89** *in vivo*. In addition, metabolite analyses show that [^{18}F]**89** undergoes fast metabolism. Polar metabolites were found in baboon plasma and 17.0% of unmetabolized tracer was determined at 60 min after injection; no evidence was obtained for free [^{18}F]fluoride.

5.4. ^{18}F-Fluoroborates

An impressive stability was found for the B-F bond (645 kJ/mol) in BF$_3$ [117]. Thus, the introduction of fluorine-18 directly connected to boron represents a further promising alternative to avoid radiodefluorination. The non-binding electrons of fluorine atoms in BF$_3$ form π-bonds with boron, which represent partial double bonds with an average bond length of 130 pm. Based on this fact, the still electron demanding boron center is less hydrolytic unstable. The previously sp^2-hybridized boron center is changed to sp^3 by accepting an electron pair of an additional fluoride in the former p$_z$-orbital to form a tetrafluoroborate anion (BF$_4^-$). Thereupon, the bonds in BF$_4^-$ are single bonds with also a high hydrolytic stability [117,161].

This basic principle is used for the creation of fluourine-18-containing boron derivatives. Exchange of F$^-$ in these species is rare due to the aforementioned high bond strength of the B-F bond. The fluorine atoms in BF$_4^-$ are substitutable but the exchange should be advisedly chosen. In general, the exchange of fluorine by other halogens leads to weaker bonds [162]. Calculations for triarylfluoroborates predicted a weakening of remaining B-F bond [163]. Therefore, comparable functionalizations will be necessary to apply this concept for the development of radiotracers.

The application of bodipy derivatives represents a promising approach. Several methylated compounds such as **91** and **92** were described for the first time by Treibs and Kreuzer in 1968 in addition with the excellent fluorescence properties of these dyes [164]. However, the synthesis of the core structure **90** succeeded first in 2009 [165–167], see Scheme 27.

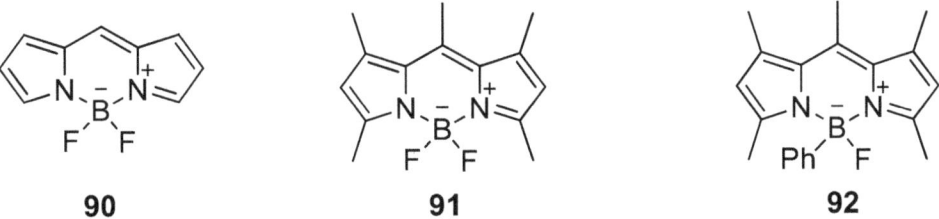

Scheme 27. Core structure **90** of selected bodipy derivatives **91** and **92**.

In 2011, bodipy dyes moved into focus for radiolabeling with fluorine-18. For that purpose, a radiolabeling building block based on a modified bodipy was created by Li and co-workers [168]. The radiosynthesis of the BPh[^{18}F]F core containing derivative [^{18}F]96 was simply realized by using KHF$_2$/[^{18}F]F$^-$ (directly from target water without drying) in water/methanol starting from BPhOH-precursor 95 with a As = 0.9 GBq/µmol. The solubility of the precursor 95 and the desired bodipy derivative [^{18}F]96 was achieved due to the ammonium triflate moiety ArN$^+$. The results are summarized in Scheme 28.

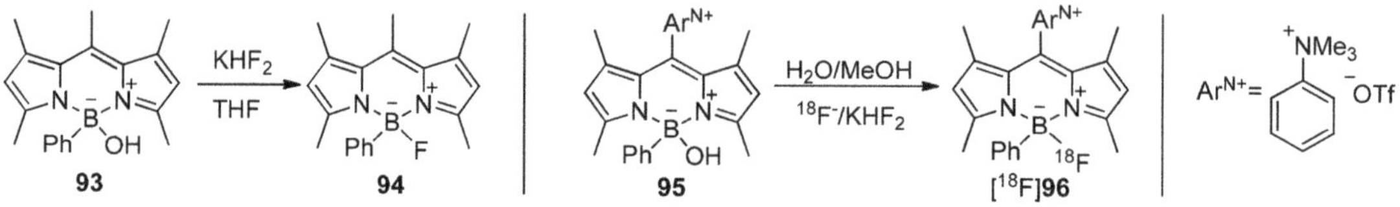

Scheme 28. Conversion of OH compound **93** to **94** and radiolabeling of **95** to yield [^{18}F]**96**.

In 2012, an alternative radiolabeling building block based on the B[^{18}F]F$_2$ core containing bodipy derivative was prepared [169]. Thus, the respective precursors were either synthesized via an exchange of one of the fluorides of **97** by a DMAP leaving group using TMS triflate/DMAP to yield **98** or directly by changing one fluoride to triflate with TMS triflate to yield **99**. The desired DMAP/^{18}F exchange of **98** to [^{18}F]**100** succeed but the triflate precursor **99** was proven to be more effective due to the higher radiochemical yields of [^{18}F]**100**, milder reaction conditions and a short reaction time during the labeling reaction (Scheme 29). A labeling building block based on [^{18}F]**100** was further used for the successful labeling of Trastuzumab with fluorine-18.

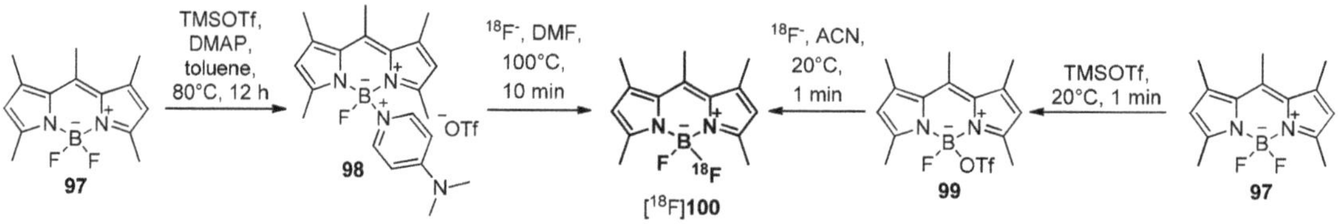

Scheme 29. Preparation of the building block [^{18}F]**100** from compound **97** via either precursor **98** or **99** under mild labeling conditions.

Both groups demonstrated the high metabolic stability of the desired bodipy derivatives *in vitro* as well as *in vivo*. No radiodefluorination in terms of an accumulation of activity in the skeleton due to free [^{18}F]fluoride was observed. Based on these results, this concept has great potential to create stable radiotracers having a B-[^{18}F]F bond.

6. Conclusions

Radiodefluorination is one of the most important metabolic degradation processes for ^{18}F-radiotracers due to the release of [^{18}F]fluoride *in vivo*, which is then accumulated in the skull and bones. This undesired accumulation leads to PET images that are false-positive in terms of skeleton imaging or comprise a bad signal to background ratio.

Several efforts have been made in the past to avoid this defluorination or to considerably reduce it. The insertion of deuterium to stabilize the C-F bond seems to be the most successful approach. Thus, a building block strategy was developed using small deuterated molecules with ^{18}F-label. In many cases, radiodefluorination could be reduced in an appreciable manner.

Other approaches can show reduced radiodefluorination in a remarkable manner as well. The introduction of deuterium in other positions relative to ^{18}F (or ^{11}C) is also promising. In this case, the stabilization is used to reduce cleavage of other parts of the molecule. Furthermore, the introduction of C[^{18}F]F$_3$ groups leads to a reduced degradation since the bond strength of the C-F bond is increased in the CF$_3$ group. Finally, the insertion of a [^{18}F]fluorocyclobutyl moiety is favored over open fluoroalkyl chains due to the increased steric demand and, therefore, reduced metabolism. The same effect can be reached by the utilization of special functional groups to avoid cleavage on this position.

Binding [^{18}F]fluorine to heteroatoms like silicon or boron offers also the possibility to obtain radiotracers, which show reduced radiodefluorination. Though, in the case of silicon based ^{18}F-radiotracers, the additional protection of the Si-[^{18}F]F center with bulky substituents is mandatory. The use of [^{18}F]bodipy derivatives offers the chance to use the same molecule for PET as for optical imaging.

Acknowledgments

The authors are grateful to Matthew D. Gott for proofreading the manuscript.

Author Contributions

Both authors (Manuela Kuchar and Constantin Mamat) contributed equally to this work in the case of writing this manuscript.

Conflicts of Interest

The authors declare no conflict of interest.

References and Notes

1. Valk, P.E.; Baily, D.L.; Townsend, D.W.; Maisey, M.N. *Positron Emission Tomography: Basic Science and Clinical Practice*; Springer: London, UK, 2003.
2. Fowler, J.S.; Ding, Y.-S. Radiotracer Chemistry. In *Principles and Practice of PET and PET/CT*, 2nd ed.; Wahl, R.L., Ed.; Lippincott Williams & Wilkins: Philadelphia, PA, USA, 2009.
3. Ross, T.L.; Wester, H.J. ^{18}F: Labeling Chemistry and Labeled Compounds. In *Handbook of Nuclear Chemistry*, 2nd ed.; Vértes, A., Nagy, S., Klencsár, Z., Lovas, R.G., Rösch, F., Eds.; Springer: Dordrecht, The Netherlands, 2011; Volume 4, pp. 2023–2025.
4. Miller, P.W.; Long, N.J.; Vilar, R.; Gee, A.D. Synthese von ^{11}C-, ^{18}F-, ^{15}O- und ^{13}N-Radiotracern für die Positronenemissionstomographie. *Angew. Chem.* **2008**, *120*, 9136–9172.
5. Pretze, M.; Große-Gehling, P.; Mamat, C. Cross-Coupling Reactions as Valuable Tool for the Preparation of PET Radiotracers. *Molecules* **2011**, *16*, 1129–1165.
6. O'Hagan, D.; Harper, D.B. Fluorine-containing natural products. *J. Fluor. Chem.* **1999**, *100*, 127–133.

7. Reddy, V.P. *Organofluorine Compounds in Biology and Medicine*, 1st ed.; Elsevier: Amsterdam, The Netherlands, 2015; pp. 1–23.
8. Gillis, E.P.; Eastman, K.J.; Hill, M.D.; Donnelly, D.J.; Meanwell, N.A. Applications of Fluorine in Medicinal Chemistry. *J. Med. Chem.* **2015**, *58*, doi:10.1021/acs.jmedchem.5b00258.
9. Richter, S.; Wuest, F. ^{18}F-Labeled Peptides: The Future Is Bright. *Molecules* **2014**, *19*, 20536–20556.
10. Pimlott, S.L.; Sutherland, A. Molecular tracers for the PET and SPECT imaging of disease. *Chem. Soc. Rev.* **2011**, *40*, 149–162.
11. O'Hagan, D. Understanding organofluorine chemistry. An introduction to the C-F bond. *Chem. Soc. Rev.* **2008**, *37*, 308–319.
12. Müller, K.; Faeh, C.; Diederich, F. Fluorine in Pharmaceuticals: Looking Beyond Intuition. *Science* **2007**, *317*, 1881–1886.
13. Bondi, A. van der Waals Volumes and Radii. *J. Phys. Chem.* **1964**, *48*, 441–451.
14. Sharpe, A.G. The physical properties of the carbon-fluorine bond. In *Ciba Foundation Symposium 2—Carbon-Fluorine Compounds: Chemistry, Biochemistry and Biological Activities*; Elliot, K., Birch, J., Eds.; Associated Scientific Publishers: Amsterdam, The Netherlands, 1972; pp. 33–54.
15. Peters, D. Problem of the Lengths and Strengths of Carbon-Fluorine Bonds. *J. Chem. Phys.* **1963**, *38*, 561–563.
16. Bent, H.A. An Appraisal of Valence-bond Structures and Hybridization in Compounds of the First-row elements. *Chem. Rev.* **1961**, *61*, 275–311.
17. Lemal, D.M. Perspective on Fluorocarbon Chemistry. *J. Org. Chem.* **2004**, *69*, 1–11.
18. Tozer, D.J. The conformation and internal rotational barrier of benzyl fluoride. *Chem. Phys. Lett.* **1999**, *308*, 160–164.
19. Kochi, J.K.; Hammond, G.S. Benzyl Tosylates. II. The Application of the Hammett Equation to the Rates of their Solvolysis. *J. Am. Chem. Soc.* **1953**, *75*, 3445–3451.
20. Wüst, F.; Müller M.; Bergmann, R. Synthesis of 4-([^{18}F]fluoromethyl)-2-chlorophenylisothiocyanate: A novel bifunctional ^{18}F-labelling agent. *Radiochim. Acta* **2004**, *92*, 349–353.
21. Zavitsas, A.A.; Rogers, D.W.; Matsunag, N. Remote Substituent Effects on Allylic and Benzylic Bond Dissociation Energies. Effects on Stabilization of Parent Molecules and Radicals. *J. Org. Chem.* **2007**, *72*, 7091–7101.
22. Hiyama, T. *Organofluorine Compounds: Chemistry and Applications*; Springer: Berlin, Germany, 2000; p. 126.
23. Frank, H. Allen, F.H.; Kennard, O.; Watson, D.G. Brammer, L.; Orpen, A.G., Taylor, R. Tables of Bond Lengths determined by X-ray and Neutron Diffraction. Part 1. Bond Lengths in Organic Compounds. *J. Chem. Soc. Perkin Trans.* **1987**, *2*, S1–S19.
24. Wiberg, K.B.; Rablen, P.R. Substituent Effects. 7. Phenyl Derivatives. When Is a Fluorine a π-Donor? *J. Org. Chem.* **1998**, *63*, 3722–3730.
25. Carroll, T.X.; Thomas, T.D.; Bergersen, H.; Børve, K.J.; Sæthre, L.J. Fluorine as a π Donor. Carbon 1s Photoelectron Spectroscopy and Proton Affinities of Fluorobenzenes. *J. Org. Chem.* **2006**, *71*, 1961–1968.
26. Liotta, C.L.; Harris, H.P. Chemistry of naked anions. I. Reactions of the 18-crown-6 complex of potassium fluoride with organic substrates in aprotic organic solvents. *J. Am. Chem. Soc.* **1974**, *96*, 2250–2252.

27. Tressaud, A.; Haufe, G. *Fluorine and Health: Molecular Imaging, Biomedical Materials and Pharmaceuticals*; Elsevier: Amsterdam, The Netherlands, 2008; pp. 35–42.
28. Ermert, J. ^{18}F-Labelled Intermediates for Radiosynthesis by Modular Build-Up Reactions: Newer Developments. *BioMed. Res. Int.* **2014**, *15*, doi:10.1155/2014/812973.
29. Mu, L.; Fischer, C.R.; Holland, J.P.; Becaud, J.; Schubiger, P.A.; Schibli, R.; Ametamey, S.M.; Graham, K.; Stellfeld, T.; Dinkelborg, L.M.; et al. ^{18}F-Radiolabeling of Aromatic Compounds Using Triarylsulfonium Salts. *Eur. J. Org. Chem.* **2012**, *2012*, 889–892.
30. Sander, K.; Gendron, T.; Yiannaki, E.; Cybulska, K.; Kalber, T.L.; Lythgoe, M.F.; Årstad, E. Sulfonium Salts as Leaving Groups for Aromatic Labelling of Drug-like Small Molecules with Fluorine-18. *Sci. Rep.* **2015**, *5*, 9941, doi:10.1038/srep09941.
31. Brandt, J.R.; Lee, E.; Boursalian, G.B.; Ritter, T. Mechanism of electrophilic fluorination with Pd(IV): Fluoride capture and subsequent oxidative fluoride transfer. *Chem. Sci.* **2014**, *5*, 169–179.
32. Brooks, A.F.; Topczewski, J.J.; Ichiishi, N.; Sanford, M.S.; Scott, P.J.H. Late-stage [^{18}F]fluorination: New solutions to old problems. *Chem. Sci.* **2014**, *5*, 4545–4553.
33. Ross, T.L.; Wester, H.J. ^{18}F: Labeling Chemistry and Labeled Compounds. In *Handbook of Nuclear Chemistry*, 2nd ed.; Vértes, A., Nagy, S., Klencsár, Z., Lovas, R.G., Rösch, F., Eds.; Springer: Dordrecht, The Netherlands, 2011; Volume 4, pp. 2026–2032.
34. Vallabhajosula, S. *Molecular Imaging: Radiopharmaceuticals for PET and SPECT*; Springer: Dordrecht, The Netherlands, 2009; pp. 142–144.
35. Tressaud, A.; Haufe, G. *Fluorine and Health: Molecular Imaging, Biomedical Materials and Pharmaceuticals*; Elsevier: Amsterdam, The Netherlands, 2008; p. 7.
36. Scott, P.J.H.; Hockley, B.G.; Kung, H.F.; Manchanda, R.; Zhang, W.; Kilbourn, M.R. Studies into radiolytic decomposition of fluorine-18 labeled radiopharmaceuticals for positron emission tomography. *Appl. Radiat. Isot.* **2009**, *67*, 88–94.
37. Langguth, P.; Seydel, J. Überarbeitetes Glossar zu Begriffen der Pharmazeutik. *Angew. Chem.* **2011**, *123*, 3635–3651.
38. Middleton, R.K. Drug Interactions. In *Textbook of Therapeutics: Drug and Disease Management*, 8th ed.; Helms, R.A., Herfindal, E.T., Quan, D.J., Eds.; Lipincott Williams & Wilkins: Philadelphia, PA, USA, 2006; p. 50.
39. Kharasch, E.D.; Thummel, K.E. Identification of cytochrome P450 2E1 as the predominant enzyme catalyzing human liver microsomal defluorination of sevoflurane, isoflurane, and methoxyflurane. *Anesthesiology* **1993**, *79*, 795–807.
40. Yin, H.; Anders, M.W.; Jones, J.P. Metabolism of 1,2-dichloro-1-fluoroethane and 1-fluoro-1,2,2-trichloroethane: Electronic factors govern the regioselectivity of cytochrome P450-dependent oxidation. *Chem. Res. Toxicol.* **1996**, *9*, 50–57.
41. Bier, D.; Holschbach, M.H.; Wutz, W.; Olsson, R.A.; Coenen, H.H. Metabolism of the a1 adenosine receptor positron emission tomography ligand [^{18}f]8-cyclopentyl-3-(3-fluoropropyl)-1-propylxanthine ([^{18}f]cpfpx) in rodents and humans. *Drug Metabol. Dispos.* **2006**, *34*, 570–576.
42. Pike, V.W. PET radiotracers: Crossing the blood–brain barrier and surviving metabolism. *Trends Pharm. Sci.* **2009**, *30*, 431–440.
43. Park, B.K.; Kitteringham, N.R. Effects of fluorine substitution on drug metabolism: Pharmacological and toxicological implications. *Drug Metab. Rev.* **1994**, *26*, 605–643.

44. Welch, M.J.; Katzenellenbogen, J.A.; Mathias, C.J.; Brodack, J.W.; Carlson, K.E.; Chi, D.Y.; Dence, C.S.; Kilbourn, M.R.; Perlmutter, J.S.; Raichle, M.E.; et al. N-(3-[^{18}F]fluoropropyl)-spiperone: The preferred ^{18}F labeled spiperone analog for positron emission tomographic studies of the dopamine receptor. Nucl. Med. Biol. **1988**, 15, 83–97.

45. Zoghbi, S.S.; Shetty, H.U.; Ichise, M.; Fujita, M.; Imaizumi, M.; Liow, J.S.; Shah, J.; Musachio, J.L.; Pike, V.W.; Innis, R.B. PET Imaging of the Dopamine Transporter with ^{18}F-FECNT: A Polar Radiometabolite Confounds Brain Radioligand Measurements. J. Nucl. Med. **2006**, 47, 520–527.

46. Ettlinger, D.E.; Wadsak, W.; Mien, L.-K.; Machek, M.; Wabnegger, L.; Rendl, G.; Karanikas, G.; Viernstein, H.; Kletter, K.; Dudczak, R.; et al. [^{18}F]FETO: Metabolic considerations. Eur. J. Nucl. Med. Mol. Imaging **2006**, 33, 928–931.

47. Agdeppa, E.D.; Kepe, V.; Liu, J.; Flores-Torres, S.; Satyamurthy, N.; Petric, A.; Cole, G.M.; Small, G.W.; Huang, S.-C.; Barrio, J.R. Binding Characteristics of Radiofluorinated 6-Dialkylamino-2-Naphthylethylidene Derivatives as Positron Emission Tomography Imaging Probes for β-Amyloid Plaques in Alzheimer's Disease. J. Neurosci. **2001**, 21, RC189.

48. Mitterhauser, M.; Wadsak, W.; Wabnegger, L.; Mien, L.K.; Tögel, S.; Langer, O.; Sieghart, W.; Viernstein, H.; Kletter, K.; Dudczak, R. Biological evaluation of 2'-[^{18}F]fluoroflumazenil ([^{18}F]FFMZ), a potential GABA receptor ligand for PET. Nucl. Med. Biol. **2004**, 31, 291–295.

49. Gottumukkala, V.; Heinrich, T.K.; Baker, A.; Dunning, P.; Fahey, F.H.; Treves, S.T.; Packard, A.B. Biodistribution and Stability Studies of [^{18}F]Fluoroethylrhodamine B, a Potential PET Myocardial Perfusion Agent. Nucl. Med. Biol. **2010**, 37, 365–370.

50. Pan, J.; Pourghiasian, M.; Hundal, N.; Lau, J.; Bénard, F.; Dedhar, S.; Lin, K.-S. 2-[^{18}F]Fluoroethanol and 3-[^{18}F]fluoropropanol: Facile preparation, biodistribution in mice, and their application as nucleophiles in the synthesis of [^{18}F]fluoroalkyl aryl ester and ether PET tracers. Nucl. Med. Biol. **2013**, 40, 850–857.

51. Tewson, T.J.; Welch, M.J. Preparation and preliminary biodistribution of "no carrier added" fluorine F-18 fluoroethanol. J. Nucl. Med. **1980**, 21, 559–564.

52. Lee, K.C.; Lee, S.-Y.; Choe, Y.S.; Chi, D.Y. Metabolic Stability of [^{18}F]Fluoroalkylbiphenyls. Bull. Korean Chem. Soc. **2004**, 25, 1225–1230.

53. Milicevic Sephton, S.; Dennler, P.; Leutwiler, D.S.; Mu, L.; Wanger-Baumann, C.A.; Schibli, R.; Krämer, S.D.; Ametamey, S.M. Synthesis, radiolabelling and in vitro and in vivo evaluation of a novel fluorinated ABP688 derivative for the PET imaging of metabotropic glutamate receptor subtype 5. Am. J. Med. Mol. Imaging **2012**, 2, 14–18.

54. French, A.N.; Napolitano, E.; van Brocklin, H.F.; Brodack, J.W.; Hanson, R.N.; Welch, M.J.; Katzenellenbogen, J.A. The β-heteroatom effect in metabolic defluorination: The interaction of resonance and inductive effects may be a fundamental determinant in the metabolic liability of fluorine-substituted compounds. J. Label. Compd. Radiopharm. **1991**, 30, 431–433.

55. Purohit, A.; Radeke, H.; Azure, M.; Hanson, K.; Benetti, R.; Su, F.; Yalamanshili, P.; Yu, M.; Hayes, M.; Guaraldi, M.; et al. Synthesis and biological evaluation of pyridazinone analogues as potential cardiac positron emission tomography tracers. J. Med. Chem. **2008**, 51, 2954–2970.

56. Dollé, F. Fluorine-18-labelled fluoropyridines: Advances in radiopharmaceutical design. Curr. Pharm. Des. **2005**, 11, 3221–3235.

57. Gant, T.G. Using Deuterium in Drug Discovery: Leaving the Label in the Drug. *J. Med. Chem.* **2014**, *57*, 3595–3611.
58. Kohen, A.; Limbach, H.-H. *Isotope Effects in Chemistry and Biology*; CRC Press: Boca Raton, FL, USA, 2006.
59. Roston, D.; Islam, Z.; Kohen, A. Isotope Effects as Probes for Enzyme Catalyzed Hydrogen-Transfer Reactions. *Molecules* **2013**, *18*, 5543–5567.
60. Fagerholm, V.; Mikkola, K.K.; Ishizu, T.; Arponen, E.; Kauhanen, S.; Nagren, K.; Solin, O.; Nuutila, P.; Haaparanta, M. Assessment of islet specificity of dihydrotetrabenazine radiotracer binding in rat pancreas and human pancreas. *J. Nucl. Med.* **2010**, *51*, 1439–1446.
61. Simpson, N.R.; Souza, F.; Witkowski, P.; Maffei, A.; Raffo, A.; Herron, A.; Kilbourn, M.; Jurewicz, A.; Herold, K.; Liu, E.; et al. Visualizing pancreatic-cell mass with [^{11}C]DTBZ. *Nucl. Med. Biol.* **2006**, *33*, 855–864.
62. Goland, R.; Freeby, M.; Parsey, R.; Saisho, Y.; Kumar, D.; Simpson, N.; Hirsch, J.; Prince, M.; Maffei, A.; Mann, J.J.; et al. ^{11}C-Dihydrotetrabenazine PET of the pancreas in subjects with long-standing type 1 diabetes and in healthy controls. *J. Nucl. Med.* **2009**, *50*, 382–389.
63. Eriksson, O.; Jahan, M.; Johnström, P.; Korsgren, O.; Sundin, A.; Halldin, C.; Johansson, L. In vivo and in vitro characterization of [^{18}F]-FE-(+)-DTBZ as a tracer for beta-cell mass. *Nucl. Med. Biol.* **2010**, *37*, 357–363.
64. Lin, K.J.; Weng, Y.H.; Wey, S.P.; Hsiao, I.T.; Lu, C.S.; Skovronsky, D.; Chang, H.P.; Kung, M.P.; Yen, T.C. Whole-body biodistribution and radiation dosimetry of ^{18}F-FP-(+)-DTBZ (^{18}F-AV-133): A novel vesicular monoamine transporter 2 imaging agent. *J. Nucl. Med.* **2010**, *51*, 1480–1485.
65. Jahan, M.; Eriksson, O.; Johnström, P.; Korsgren, O.; Sundin, A.; Johansson, L.; Halldin, C. Decreased defluorination using the novel betacell imaging agent [^{18}F]FE-DTBZ-d4 in pigs examined by PET. *EJNMMI Res.* **2011**, *1*, 33, doi:10.1186/2191-219X-1-33.
66. Ding, Y.-S.; Fowler, J.S.; Wolf, A.P. Rapid, regiospecific syntheses of deuterium substituted 6-[^{18}F]fluorodopamine (α,α-D$_2$; β,β-D$_2$ and α,α,β,β-D$_4$) for mechanistic studies with positron emission tomography. *J. Labelled Compd. Radiopharm.* **1993**, *33*, 645–654.
67. DeWolf, W.E.J., Jr.; Carr, S.A.; Varrichio, A.; Goodhart, P.J.; Mentzer M.A.; Roberts, G.D.; Southan, C.; Dolle, R.E.; Kurse, L.I. Inactivation of dopamine β-hydroxylase by *p*-cresol: Isolation and characterization of covalently modified active site peptides. *Biochemistry* **1988**, *27*, 9093–9101.
68. Coleman A.A.; Hindsgaul, O.; Palcic, M.M. Stereochemistry of copper amine oxidase reactions. *J. Biol. Chem.* **1989**, *264*, 19500–19505.
69. Yu, P.H. Three types of stereospecificity and the kinetic deuterium isotope effect in the oxidative deamination of dopamine as catalyzed by different amine oxidases. *Biochem. Cell Biol.* **1988**, *66*, 853–861.
70. Yu, P.H.; Bailey, B.A.; Durden, D.A.; Boulton, A.A. Stereospecific deuterium substitution at the alpha-carbon position of dopamine and its effect on oxidative deamination catalyzed by MAO-A and MAO-B from different tissues. *Biochem. Pharmacol.* **1986**, *35*, 1027–1036.
71. Ding, Y.-S.; Fowler, J.S.; Gatley, S.J.; Logan, J.; Volkow, N.D.; Shea, C. Mechanistic Positron Emission Tomography Studies of 6-[^{18}F]Fluorodopamine in Living Baboon Heart: Selective Imaging and Control of Radiotracer Metabolism Using the Deuterium Isotope Effect. *J. Neurochem.* **1995**, *65*, 682–690.

72. Schou, M.; Halldin, C.; Sóvágó, J.; Pike, V.W.; Hall, H.; Gulyás, B.; Mozley, P.D.; Dobson, D.; Shchukin, E.; Innis, R.B.; et al. PET evaluation of novel radiofluorinated reboxetine analogs as norepinephrine transporter probes in the monkey brain. *Synapse* **2004**, *53*, 57–67.
73. Zhang, M.-R.; Maeda, J.; Ito, T.; Okauchi, T.; Ogawa, M.; Noguchi, J.; Suhara, T.; Halldin, C.; Suzuki, K. Synthesis and evaluation of N-(5-fluoro-2-phenoxyphenyl)-N-(2-[^{18}F]fluoromethoxy-d2–5-methoxybenzyl)acetamide: A deuterium substituted radioligand for peripheral benzodiazepine receptor. *Bioorg. Med. Chem.* **2005**, *13*, 1811–1818.
74. Lin, K.-S.; Ding, Y.-S.; Kim, S.-W.; Kil, K.-E. Synthesis, enantiomeric resolution, F-18 labeling and biodistribution of reboxetine analogs: Promising radioligands for imaging the norepinephrine transporter with positron emission tomography. *Nucl. Med. Biol.* **2005**, *32*, 415–422.
75. Cai, L.; Chin, F.T.; Pike, V.W.; Toyama, H.; Liow, J.-S.; Zoghbi, S.S.; Modell, K.; Briard, E.; Shetty, H.U.; Sinclair, K.; et al. Synthesis and Evaluation of Two ^{18}F-Labeled 6-Iodo-2-(4-N,N-dimethylamino)phenylimidazo[1,2-a]pyridine Derivatives as Prospective Radioligands for β-Amyloid in Alzheimer's Disease. *J. Med. Chem.* **2004**, *47*, 2208–2218.
76. Donohue, S.R.; Krushinski, J.H.; Pike, V.W.; Chernet, E.; Phebus, L.; Chesterfield, A.K.; Felder, C.C.; Halldin, C.; Schaus, J.M. Synthesis, *Ex Vivo* Evaluation, and Radiolabeling of Potent 1,5-Diphenylpyrrolidin-2-one Cannabinoid Subtype-1 Receptor Ligands as Candidates for *in Vivo* Imaging. *J. Med. Chem.* **2008**, *51*, 5833–5842.
77. Beyerlein, F.; Piel, M.; Hoehnemann, S.; Roesch, F. Automated synthesis and purification of [^{18}F]fluoro-[di-deutero]methyl tosylate. *J. Label. Compd. Radiopharm.* **2013**, *56*, 360–363.
78. Taniguchi, T.; Miura, S.; Hasui, T.; Halldin, C.; Stepanov, V.; Takano, A. Radiolabeled Compounds and Their Use as Radiotracers for Quantitative Imaging of Phosphodiesterase (PDE10A) in Mammals. WO 2013027845 A1, 28 February 2013.
79. Schieferstein, H.; Piel, M.; Beyerlein, F.; Lueddens, H.; Bausbacher, N.; Buchholz, H.-G.; Ross, T.L.; Roesch, F. Selective binding to monoamine oxidase A: *In vitro* and *in vivo* evaluation of ^{18}F-labeled β-carboline derivatives. *Bioorg. Med. Chem.* **2015**, *23*, 612–623.
80. Rami-Mark, C.; Zhang, M.-R.; Mitterhauser, M.; Lanzenberger, R.; Hacker, M.; Wadsak, W. [^{18}F]FMeNER-D2: Reliable fully-automated synthesis for visualization of the norepinephrine transporter. *Nucl. Med. Biol.* **2013**, *40*, 1049–1054.
81. Schou, M.; Halldin, C.; Sóvágó, J.; Pike, V.W.; Gulyás, B.; Mozley, P.D.; Johnson, D.P.; Hall, H.; Innis, R.B.; Farde, L. Specific *in vivo* binding to the norepinephrine transporter demonstrated with the PET radioligand, (S,S)-[^{11}C]MeNER. *Nucl. Med. Biol.* **2003**, *30*, 707–714.
82. Casebier, D.S.; Robinson, S.P.; Purohit, A.; Radeke, H.S.; Azure, M.T.; Dischino, D.D. Contrast Agents for Myocardial Perfusion Imaging, WO 2005079391 A2, 1 September 2005.
83. Hortala, L.; Arnaud, J.; Roux, P.; Oustric, D.; Boulu, L.; Oury-Donat, F.; Avenet, P.; Rooney, T.; Alagille, D.; Barret, O.; et al. Synthesis and preliminary evaluation of a new fluorine-18 labelled triazine derivative for PET imaging of cannabinoid CB2 receptor. *Bioorg. Med. Chem. Lett.* **2014**, *24*, 283–287.
84. Brumby, T.; Graham, K.; Krueger, M. Direct Synthesis of [^{18}F]fluoromethoxy Compounds for PET Imaging and New Precursors for Direct Radiosynthesis of Protected Derivatives of O-([^{18}F]fluoromethyl)tyrosine. WO 2013001088 A1, 3 January 2013.

85. Graham, K.; Ede, S. Simplified Radiosynthesis of O-[^{18}F]fluoromethyl Tyrosine Derivatives. WO 2013026940 A1, 28 February 2013.
86. Graham, K.; Zitzmann-Kolbe, S.; Brumby, T. Preparation of Fluorodeuteriomethyl Tyrosine Derivatives. WO 2012025464 A1, 1 March 2012.
87. Hamill, T.G.; Sato, N.; Jitsuoka, M.; Tokita, S.; Sanabria, S.; Eng, W.; Ryan, C.; Krause, S.; Takenaga, N.; Patel, S.; et al. Inverse agonist histamine H3 receptor PET tracers labeled with carbon-11 or fluorine-18. *Synapse* **2009**, *63*, 1122–1132.
88. Xu, R.; Hong, J.; Morse, C.L.; Pike, V.W. Synthesis, Structure-Affinity Relationships, and Radiolabeling of Selective High-Affinity 5-HT4 Receptor Ligands as Prospective Imaging Probes for Positron Emission Tomography. *J. Med. Chem.* **2010**, *53*, 7035–7047.
89. Burns, D.H.; Chen, A.M.; Gibson, R.E.; Goulet, M.T.; Hagmann, W.K.; Hamill, T.G.; Jewell, J.P.; Lin, L.S.; Liu, P.; Peresypkin, A.V. Heterocyclic Radiolabeled Cannabinoid-1 Receptor Modulators. WO 2005009479 A1, 3 February 2005.
90. Cosford, N.D.P.; Govek, S.P.; Hamill, T.G.; Kamenecka, T.; Roppe, J.R.; Seiders, T.J. Alkyne Derivatives as Tracers for Metabotropic Glutamate Receptor Binding. WO 2004038374 A2, 6 May 2004.
91. Burns, H.D.; Hamill, T.G.; Lindsley, C.W. Radiolabeled Glycine Transporter Inhibitors. WO 2007041025 A2, 12 April 2007.
92. Hamill, T.G.; McCauley, J.A.; Burns, H.D. The Synthesis of A Benzamidine-containing NR2B-selective NMDA Receptor Ligand Labelled with Tritium or Fluorine-18. *J. Label. Compd. Radiopharm.* **2005**, *48*, 1–10.
93. Burns, H.D.; sEng, W.-S.; Gibson, R.E.; Hamill, T.G. Radiolabeled Neurokinin-1 Receptor Antagonists. WO 2004029024 A2, 8 April 2004.
94. Leyton, J.; Smith, G.; Zhao, Y.; Perumal, M.; Nguyen, Q.-D.; Robins, E.; Årstad, E.; Aboagye E.O. [^{18}F]Fluoromethyl-[1,2-^{2}H$_4$]-Choline: A Novel Radiotracer for Imaging Choline Metabolism in Tumors by Positron Emission Tomography. *Cancer Res.* **2009**, *69*, 7721–7728.
95. Smith, G.; Zhao, Y.; Leyton, J.; Shan, B.; de Nguyen, Q.; Perumal, M. Radiosynthesis and pre-clinical evaluation of [^{18}F]fluoro-[1,2-^{2}H$_4$]choline. *Nucl. Med. Biol.* **2011**, *38*, 39–51.
96. Nag, S.; Lehmann, L.; Kettschau, G.; Toth, M.; Heinrich, T.; Thiele, A.; Varrone, A.; Halldin C. Development of a novel fluorine-18 labeled deuterated fluororasagiline ([^{18}F]fluororasagiline-D$_2$) radioligand for PET studies of monoamino oxidase B (MAO-B). *Bioorg. Med. Chem.* **2013**, *21*, 6634–6641.
97. Abu-Raya, S.; Tabakman, R.; Blaugrund, E.; Trembovler, V.; Lazarovici, P. Neuroprotective and neurotoxic effects of monoamine oxidase-B inhibitors and derived metabolites under ischemia in PC12 cells. *Eur. J. Pharmacol.* **2002**, *434*, 109–116.
98. Nag, S. Development of Novel Fluorine-18 Labeled PET Radioligands for Monoamine Oxidase B (MAO-B). Ph.D. Thesis, Karolinska Institutet, Stockholm, Sweden, 2013.
99. Ramirez de Molina, A.; Gallego-Ortega, D.; Sarmentero-Estrada, J.; Lagares, D.; Bandrés, E.; Gomez del Pulgar, T.; García-Foncillas, J.; Lacal, J.C. Choline kinase as a link connecting phospholipid metabolism and cell cycle regulation: Implications in cancer therapy. *Int. J. Biochem. Cell Biol.* **2008**, *40*, 1753–1763.

100. Ikuta, S.; Imamura, S.; Misaki, H.; Horiuti, Y. Purification and Characterization of *Choline oxidase* from *Arthrobacter globiformis*. *J. Biochem.* **1977**, *82*, 1741–1749.
101. Gadda, G. pH and deuterium kinetic isotope effects studies on the oxidation of choline to betaine-aldehyde catalyzed by choline oxidase. *Biochim. Biophys. Acta* **2003**, *1650*, 4–9.
102. Witney, T.H.; Alam, I.S.; Turton, D.R.; Smith, G.; Carroll, L.; Brickute, D.; Twyman, F.J.; Nguyen, Q.-D.; Tomasi, G.; Awais, R.O.; et al. Evaluation of Deuterated ^{18}F- and ^{11}C-Labeled Choline Analogs for Cancer Detection by Positron Emission Tomography. *Clin. Cancer Res.* **2012**, *18*, 1063–1072.
103. Challapalli, A.; Sharma, R.; Hallett, W.A.; Kozlowski, K.; Carroll, L.; Brickute, D.; Twyman, F.; Al-Nahhas, A.; Aboagye, E.O. Biodistribution and Radiation Dosimetry of Deuterium-Substituted ^{18}F-Fluoromethyl-[1, 2-^{2}H$_{4}$]Choline in Healthy Volunteers. *J. Nucl. Med.* **2014**, *55*, 256–263.
104. Dounay, A.; Barta, N.; Bikker, J.; Borosky, S.; Campbell, B.; Crawford, T.; Denny, L.; Evans, L.; Gray, D.; Lee, P. Synthesis and pharmacological evaluation of aminopyrimidine series of 5-HT1A partial agonists. *Bioorg. Med. Chem. Lett.* **2009**, *19*, 1159–1163.
105. Manoury, P.M.; Binet, J.; Rousseau, J.; Lefevre-Borg, F.; Cavero, I. Synthesis of a series of compounds related to betaxolol, a new β$_1$-adrenoceptor antagonist with a pharmacological and pharmacokinetic profile optimized for the treatment of chronic cardiovascular diseases. *J. Med. Chem.* **1987**, *30*, 1003–1011.
106. Sorensen, B.; Rohde, J.; Wang, J.; Fung, S.; Monzon, K.; Chiou, W.; Pan, L.; Deng, X.; Stolarik, D.A.; Frevert, E.U. Adamantane 11-β-HSD-1 inhibitors: Application of an isocyanide multicomponent reaction. *Bioorg. Med. Chem. Lett.* **2006**, *16*, 5958–5962.
107. Wrobleski, M.; Reichard, G.; Paliwal, S.; Shah, S.; Tsui, H.; Duffy, R.; Lachowicz, J.; Morgan, C.; Varty, G.; Shih, N. Cyclobutane derivatives as potent NK$_1$ selective antagonists. *Bioorg. Med. Chem. Lett.* **2006**, *16*, 3859–3863.
108. Shoup, T.; Elmaleh, D.; Bonab, A.; Fischman, A. Evaluation of *trans*-9-1^{8}F-fluoro-3,4-methyleneheptadecanoic acid as a PET tracer for myocardial fatty acid imaging. *J. Nucl. Med.* **2005**, *46*, 297–304.
109. Martel, F.; Berlinguet, L. Impairment of tumor growth by unnatural amino acids. *Can. J. Biochem. Physiol.* **1959**, *37*, 433–439.
110. Connors, T.; Elson, L.; Haddow, A.; Ross, W. The pharmacology and tumour growth inhibitory activity of 1-aminocyclopentane-1-carboxylic acid and related compounds. *Biochem. Pharmacol.* **1960**, *5*, 108–129.
111. Wooten, D.W.; Moraino, J.D.; Hillmer, A.T.; Engle, J.W.; DeJesus, O.J.; Murali, D.; Barnhart, T.E.; Nickles, R.J.; Davidson, R.J.; Schneider, M.L.; et al. *In Vivo* Kinetics of [F-18]MEFWAY: A comparison with [C-11]WAY100635 and [F-18]MPPF in the nonhuman primate. *Synapse* **2011**, *65*, 592–600.
112. Yu, W.; Williams, L.; Camp, V.; Malveaux, E.; Olson, J.; Goodman, M. Stereoselective synthesis and biological evaluation of *syn*-1-amino-3-[^{18}F]fluorocyclobutyl-1-carboxylic acid as a potential positron emission tomography brain tumor imaging agent. *Bioorg. Med. Chem.* **2009**, *17*, 1982–1990.
113. Miyagawa, T.; Oku, T.; Uehara, H.; Desai, R.; Beattie, B.; Tjuvajev, J.; Blasberg, R. Facilitated amino acid transport is upregulated in brain tumors. *J. Celeb. Blood Flow. Metab.* **1998**, *18*, 500–509.

114. Franck, D.; Kniess, T.; Steinbach, J.; Zitzmann-Kolbe, S.; Friebe, M.; Dinkelborg, L.M.; Graham, K. Investigations into the synthesis, radiofluorination and conjugation of a new [^{18}F]fluorocyclobutyl prosthetic group and its *in vitro* stability using a tyrosine model system. *Bioorg. Med. Chem.* **2013**, *21*, 643–652.
115. Franck, D. Radiofluorinated Cyclobutyl Group for Increased Metabolic Stability Using Tyrosine Derivatives as Model System. Ph.D. Thesis, TU Dresden, Dresden, Germany, 2012,
116. Tipre, D.N.; Zoghbi, S.S.; Liow, J.S.; Green, M.V.; Seidel, J.; Ichise, M.; Innis, R.B.; Pike, V.W. PET Imaging of Brain 5-HT1A Receptors in Rat *in Vivo* with ^{18}F-FCWAY and Improvement by Successful Inhibition of Radioligand Defluorination with Miconazole. *J. Nucl. Med.* **2006**, *47*, 345–353.
117. Holleman, A.; Wiberg, N.; Wiberg, E. *Lehrbuch der Anorganischen Chemie*, 102nd ed.; De Gruyter Verlag: Berlin, Germany, 2008; pp. 1097–1103.
118. Rosenthal, M.S.; Bosch, A.L.; Nickles, R.J.; Gatley, S.J. Synthesis and some characteristics of no-carrier added [^{18}F]fluorotrimethylsilane. *Int. J. Appl. Radiat. Isot.* **1985**, *36*, 318–319.
119. Gatley, S.J. Rapid production and trapping of [F-18]fluorotrimethylsilane, and its use in nucleophilic F-18 labeling without an aqueous evaporation step. *Appl. Radiat. Isot.* **1989**, *40*, 541–544.
120. Mulholland, G.K. Recovery and purification of no-carrier-added [^{18}F]fluoride with bistrimethylsilylsulfate (BTMSS). *Int. J. Radiat. Appl. Instr.* **1991**, *42*, 1003–1008.
121. Walsh, J.C.; Fleming, L.M.; Satyamurthy, N.; Barrio, J.R.; Phelps, M.E.; Gambhir, S.S.; Toyokuni, T. Application of silicon-fluoride chemistry for the development of amine-reactive F-18-labeling agents for biomolecules. *J. Nucl. Med.* **2000**, *41*, 249.
122. Choudhry, U.; Martin, K.E.; Biagini, S.; Blower, P.J. Alkoxysilane groups for instant labeling of biomolecules with ^{18}F. *Nucl. Med. Commun.* **2006**, *27*, 293.
123. Schirrmacher, R.; Bradtmöller, G.; Schirrmacher, E.; Thews, O.; Tillmanns, J.; Siessmeier, T.; Buchholz, H.G.; Bartenstein, P.; Wängler, B.; Niemeyer, C.M.; *et al.* ^{18}F-labeling of peptides by means of an organosilicon-based fluoride acceptor. *Angew. Chem. Int. Ed.* **2006**, *45*, 6047–6050.
124. Schirrmacher, R.; Kostikov, A.; Wängler, C.; Jurkschat, K.; Bernard-Gauthier, V.; Schirrmacher, E.; Wängler, B. Silicon Fluoride Acceptors (SIFAs) for Peptide and Protein Labeling with ^{18}F. In *Radiochemical Syntheses*, 1st ed.; Scott, P.J.H., Ed.; J. Wiley & Sons: Hoboken, NJ, USA, 2015; Volume 2, pp. 149–162.
125. Fleischer, H. Molecular "Floppyness" and the Lewis Acidity of Silanes: A Density Functional Theory Study. *Eur. J. Inorg. Chem.* **2001**, 393–404.
126. Höhne, A.; Yu, L.; Mu, L.; Reiher, M.; Voigtmann, U.; Klar, U.; Graham, K.; Schubiger, P.A.; Ametamey, S.M. Organofluorosilanes as model compounds for ^{18}F-labeled silicon-based PET tracers and their hydrolytic stability: Experimental data and theoretical calculations (PET = Positron Emission Tomography). *Chemistry* **2009**, *15*, 3736–3743.
127. Bento, A.P.; Bickelhaupt, F.M. Nucleophilic Substitution at Silicon (S$_N$2@Si) via a Central Reaction Barrier. *J. Org. Chem.* **2007**, *72*, 2201–2207.
128. Bernard-Gauthier, V.; Wängler, C.; Schirrmacher, E.; Kostikov, A.; Jurkschat, K.; Wängler, B.; Schirrmacher, R. ^{18}F-Labeled Silicon-Based Fluoride Acceptors: Potential Opportunities for Novel Positron Emitting Radiopharmaceuticals. *BioMed. Res. Int.* **2014**, doi:10.1155/2014/45450.

129. Rosa-Neto, P.; Wängler, B.; Iovkova, L.; Boening, G.; Reader, A.; Jurkschat, K.; Schirrmacher, E. [^{18}F]SiFA-isothiocyanate: A new highly effective radioactive labeling agent for lysine-containing proteins. *ChemBioChem* **2009**, *10*, 1321–1324.

130. Wängler, C.; Niedermoser, S.; Chin, J.; Orchowski, K.; Schirrmacher, E.; Jurkschat, K.; Kostikov, A.P.; Iovkova-Berends, L.; Schirrmacher, R.; Wängler, B. One-step ^{18}F-labeling of peptides for positron emission tomography imaging using the SiFA methodology. *Nat. Prot.* **2012**, *7*, 1946–1955.

131. Lindner, S.; Michler, C.; Leidner, S.; Rensch, C.; Wängler, C.; Schirrmacher, R.; Bartenstein, P.; Wängler, B. Synthesis and *in Vitro* and *in Vivo* Evaluation of SiFA-Tagged Bombesin and RGD Peptides as Tumor Imaging Probes for Positron Emission Tomography. *Bioconjugate Chem.* **2014**, *25*, 738–749.

132. Balentova, E.; Collet, C.; Lamandé-Langle, S.; Chrétien, F.; Thonon, D.; Aerts, J.; Lemaire, C.; Luxen, A.; Chapleuret, Y. Synthesis and hydrolytic stability of novel 3-[^{18}F]fluoroethoxybis(1-methylethyl)silyl]propanamine-based prosthetic groups. *J. Fluor. Chem.* **2011**, *132*, 250–257.

133. Ting, R.; Adam, M.J.; Ruth, T.J.; Perrin, D.M. Arylfluoroborates and alkylfluorosilicates as potential PET imaging agents: High-yielding aqueous biomolecular^{18}F-labeling. *J. Am. Chem. Soc.* **2005**, *127*, 13094–13095.

134. Mu, L.; Höhne, A.; Schubiger, P.A.; Ametamey, S.M.; Graham, K.; Cyr, J.E.; Dinkelborg, L.; Stellfeld, T.; Srinivasan, A.; Voigtmann, U.; *et al.* Silicon-Based Building Blocks for One-Step ^{18}F-Radiolabeling of Peptides for PET Imaging. *Angew. Chem. Int. Ed.* **2008**, *47*, 4922–4925.

135. Bohn, P.; Deyine, A.; Azzouz R.; Bailly, L.; Fiol-Petit, C.; Bischoff, L.; Fruit, C.; Marsais, F.; Vera, P. Design of siliconbased misonidazole analogues and ^{18}F-radiolabelling. *Nucl. Med. Biol.* **2009**, *36*, 895–905.

136. Briard, E.; Zoghbi, S.S.; Siméon, F.G.; Imaizumi, M.; Gourley, J.P.; Shetty, H.U.; Lu, S.; Fujita, M.; Innis, R.B.; Pike, V.W. Single-Step High-Yield Radiosynthesis and Evaluation of a Sensitive ^{18}F-Labeled Ligand for Imaging Brain Peripheral Benzodiazepine Receptors with PET. *J. Med. Chem.* **2009**, *52*, 688–699.

137. Testa, B.; Mayer, J.M. *Hydrolysis in Drug and Prodrug Metabolism. Chemistry, Biochemistry and Enzymology*; Verlag Helvetica Chimica Acta: Zürich, Switzerland, 2003.

138. Fukami, T.; Yokoi, T. The Emerging Role of Human Esterases. *Drug Metabol. Pharmacokin.* **2012**, *27*, 466–477.

139. Li, Z.; Lang, L.; Ma, Y.; Kiesewetter, D.O. [^{18}F]Fluoropropylsulfonyl chloride: A new reagent for radiolabeling primary and secondary amines for PET imaging. *J. Label. Compd. Radiopharm.* **2008**, *51*, 23–27.

140. Löser, R.; Fischer, S.; Hiller, A.; Köckerling, M.; Funke, U.; Maisonial, A.; Brust, P.; Steinbach, J. Use of 3-[^{18}F]fluoropropanesulfonyl chloride as a prosthetic agent for the radiolabelling of amines: Investigation of precursor molecules, labelling conditions and enzymatic stability of the corresponding sulfonamides. *Beilstein J. Org. Chem.* **2013**, *9*, 1002–1011.

141. Smith, G.; Sala, R.; Carroll, L.; Behan, K.; Glaser, M.; Robins, E.; Nguyen, Q.-D.; Aboagye E.O. Synthesis and evaluation of nucleoside radiotracers for imaging proliferation. *Nucl. Med. Biol.* **2012**, *39*, 652–665.

142. Mamat, C.; Ramenda, T.; Wuest, F.R. Recent applications of click chemistry for the synthesis of radiotracers for molecular imaging. *Mini-Rev. Org. Chem.* **2009**, *6*, 21–34.
143. Pretze, M.; Pietzsch, D.; Mamat, C. Recent Trends in Bioorthogonal Click-Radiolabeling Reactions Using Fluorine-18. *Molecules* **2013**, *18*, 8618–8665.
144. Toyohara, J.; Hayashi, A.; Gogami, A.; Hamada, M.; Hamashima, Y.; Katoh, T.; Node, M.; Fujibayashi, Y. Alkyl- fluorinated thymidine derivatives for imaging cell proliferation I. The in vitro evaluation of some alkyl-fluorinated thymidine derivatives. *Nucl. Med. Biol.* **2006**, *33*, 751–764.
145. Mukhopadhyay, U.; Soghomonyan, S.; Yeh, H.H.; Flores, L.G.; Shavrin, A.; Volgin, A.Y.; Gelovani, J.G.; Alauddin, M.M. Synthesis and preliminary PET imaging of N^3-[^{18}F]fluoroethyl thymidine and N^3-[^{18}F]fluoropropyl thymidine. *Nucl. Med. Biol.* **2008**, *35*, 697–705.
146. *Fluorine in Medicinal Chemistry and Chemical Biology*; Ojima, I., Ed.; Wiley-Blackwell: Sussex, UK, 2009.
147. Lee, I.; Choe, Y.S.; Choi, J.Y.; Lee, K.-H.; Kim, B.-T. Synthesis and Evaluation of ^{18}F-Labeled Styryltriazole and Resveratrol Derivatives for β-Amyloid Plaque Imaging. *J. Med. Chem.* **2012**, *55*, 883–892.
148. Lien, V.T.; Riss, P.J. Radiosynthesis of [^{18}F]Trifluoroalkyl Groups: Scope and Limitations. *BioMed. Res. Int.* **2014**, *10*, doi:10.1155/2014/380124.
149. Ido, T.; Irie, T.; Kasida, Y. Isotope exchange with ^{18}F on superconjugate system. *J. Label. Compd. Radiopharm.* **1979**, *16*, 153–154.
150. Satter, M.R.; Martin, C.C.; Oakes, T.R.; Christian, B.; Nickles, R.J. Synthesis of the fluorine-18 labeled inhalation Anesthetics. *Appl. Radiat. Isot.* **1994**, *45*, 1093–1100.
151. Suehiro, M.; Yang, G.; Torchon G.; Ackerstaff, E.; Humm, J.; Koutcher, J.; Ouerfelliet, O. Radiosynthesis of the tumor hypoxia marker [^{18}F]TFMISO via O-[^{18}F]trifluoroethylation reveals a striking difference between trifluoroethyl tosylate and iodide in regiochemical reactivity toward oxygen nucleophiles. *Bioorg. Med. Chem.* **2011**, *19*, 2287–2297.
152. Angelini, G.; Speranza, M.; Shiue, C.-Y.; Wolf, A.P. H^{18}F + Sb$_2$O$_3$: A new selective radiofluorinating agent. *Chem. Commun.* **1986**, *12*, 924–925.
153. Angelini, G.; Speranza, M.; Wolf, A.P.; Shiue, C.-Y. Synthesis of N-(α,α,α-tri[^{18}F]fluoro-m-tolyl) piperazine. A potent serotonin agonist. *J. Label. Compd. Radiopharm.* **1990**, *28*, 1441–1448.
154. Kilbourn, M.R.; Pavia, M.R.; Gregor, V.E. Synthesis of fluorine-18 labeled GABA uptake inhibitors. *Appl. Radiat. Isot.* **1990**, *41*, 823–828.
155. Das, M.K.; Mukherjee, J. Radiosynthesis of [F-18]fluoxetine as a potential radiotracer for serotonin reuptake sites. *Appl. Radiat. Isot.* **1993**, *44*, 835–842.
156. Johnstrom, P.; Stone-Elander, S. The ^{18}F-labelled alkylating agent 2,2,2-trifluoroethyl triflate: Synthesis and specific activity. *J. Label. Compd. Radiopharm.* **1995**, *36*, 537–547.
157. Riss, P.J.; Aigbirhio, F.I. A simple, rapid procedure for nucleophilic radiosynthesis of aliphatic [^{18}F]trifluoromethyl groups. *Chem. Commun.* **2011**, *47*, 11873–11875.
158. Dolbier, W.R., Jr.; Li, A.-R.; Koch, C.J.; Shiue, C.-Y.; Kachur, A.V. [^{18}F]-EF5, a marker for PET detection of hypoxia: Synthesis of precursor and a new fluorination procedure. *Appl. Radiat. Isot.* **2001**, *54*, 73–80.
159. Prakash, G.K.S.; Alauddin, M.M.; Hu, J.; Conti, P.S.; Olah, G.A. Expedient synthesis of [^{18}F]-labeled α-trifluoromethyl ketones. *J. Label. Compd. Radiopharm.* **2003**, *46*, 1087–1092.

160. Prabhakaran, J.; Underwood, M.D.; Parsey, R.V.; Arango, V.; Majo, V.J.; Simpson, N.R.; van Heertum, R.; Mann, J.J.; Kumar, J.S.D. Synthesis and *in vivo* evaluation of [^{18}F]-4-[5-(4-methylphenyl)-3-(trifluoromethyl)-1*H*-pyrazol-1-yl]benzenesulfonamide as a PET imaging probe for COX-2 expression. *Bioorg. Med. Chem.* **2007**, *15*, 1802–1807.
161. Schilling, B.; Kaufmann, D.E. Organometallics: Boron Compounds. In *Science of Synthesis, Houben-Weyl Methods of Molecular Transformations*; Kaufmann, D.E., Matteson, D.S., Eds.; Georg Thieme Verlag: Stuttgart, Germany, 2004; Volume 6, pp. 247–256.
162. Hartman, J.S.; Miller, J.M. Adducts of Mixed Trihalides of Boron. In *Advances in Inorganic Chemistry and Radiochemistry*; Eméleus, H.J., Sharpe, A.G., Eds.; Academic Press: New York, NY, USA, 1978; Volume 21, pp. 147–177.
163. Wade, C.R.; Broomsgrove, E.J.; Aldridge, S.; Gabbaï, F.P. Fluoride Ion Complexation and Sensing Using Organoboron Compounds. *Chem. Rev.* **2010**, *110*, 3958–3984.
164. Treibs, A.; Kreuzer, F.-H. Difluorboryl-Komplexe von Di- und Tripyrrylmethenen. *Liebigs Ann. Chem.* **1968**, *718*, 208–223.
165. Schmitt, A.; Hinkeldey, B.; Wild, M.; Jung, G. Synthesis of the Core Compound of the BODIPY Dye Class: 4,4-Difluoro-4-bora-(3a,4a)-diaza-s-indacene. *J. Fluoresc.* **2009**, *19*, 755–758.
166. Tram, K.; Yan, H.; Jenkins, H.A.; Vassiliev, S.; Bruce, D. The synthesis and crystal structure of unsubstituted 4,4-difluoro-4-bora-3a,4a-diaza-s-indacene (BODIPY). *Dyes Pigments* **2009**, *82*, 392–395.
167. Arroyo, I.J.; Hu, R.; Merino, G.; Tang, B.Z.; Pena-Cabrera, E. The smallest and one of the brightest. Efficient preparation and optical description of the parent borondipyrromethene system. *J. Org. Chem.* **2009**, *74*, 5719–5722.
168. Li, Z.; Lin, T.-P.; Liu, S.; Huang, C.-W.; Hudnall, T.W.; Gabbaï, F.P.; Conti, P.S. Rapid aqueous [^{18}F]-labeling of a bodipy dye for positron emission tomography/fluorescence dual modality imaging. *Chem. Commun.* **2011**, *47*, 9324–9326.
169. Hendricks, J.A.; Keliher, E.J.; Wan, D.; Hilderbrand, S.A.; Weissleder, R.; Mazitschek, R. Synthesis of [^{18}F]BODIPY: Bifunctional Reporter for Hybrid Optical/Positron Emission Tomography Imaging. *Angew. Chem.* **2012**, *124*, 4681–4684.

© 2015 by the authors; licensee MDPI, Basel, Switzerland. This article is an open access article distributed under the terms and conditions of the Creative Commons Attribution license (http://creativecommons.org/licenses/by/4.0/).

Molecules **2015**, *20*, 12913-12943; doi:10.3390/molecules200712913

ISSN 1420-3049
www.mdpi.com/journal/molecules

Review

[68]Ga-Based Radiopharmaceuticals: Production and Application Relationship

Irina Velikyan [1,2]

[1] Section of Nuclear Medicine and PET, Department of Surgical Sciences, Uppsala University, Uppsala SE-751 85, Sweden; E-Mail: irina.velikyan@akademiska.se; Tel.: +46-0-70-483-4137; Fax: +46-0-18-611-0619

[2] PET Center, Center for Medical Imaging, Uppsala University Hospital, Uppsala SE-751 85, Sweden

Academic Editor: Svend Borup Jensen

Received: 7 June 2015 / Accepted: 6 July 2015 / Published: 16 July 2015

Abstract: The contribution of ^{68}Ga to the promotion and expansion of clinical research and routine positron emission tomography (PET) for earlier better diagnostics and individualized medicine is considerable. The potential applications of ^{68}Ga-comprising imaging agents include targeted, pre-targeted and non-targeted imaging. This review discusses the key aspects of the production of ^{68}Ga and ^{68}Ga-based radiopharmaceuticals in the light of the impact of regulatory requirements and endpoint pre-clinical and clinical applications.

Keywords: positron emission tomography; ^{68}Ga; chemistry; receptor targeting; peptide; GMP; dosimetry

---

1. Introduction

Development and availability of radiopharmaceuticals is a key driving force of nuclear medicine establishment and expansion. The role of ^{68}Ga in the growth and worldwide spreading of clinical research and routine positron emission tomography (PET) has been proven considerable especially during last two decades. Some important features influencing such progress are the generator production of ^{68}Ga, availability of commercial generators, robust labeling chemistry diversity, and potential for personalized medicine and radiotheranostics [1–3]. Small compounds, biological macromolecules as well as nano- and micro-particles have been successfully labeled with ^{68}Ga, and the resulting agents demonstrated promising imaging capability pre-clinically and clinically [1,3–6]. In particular ^{68}Ga was used for the labeling of ligands targeted to specific protein expression products such as receptors, enzymes, and antigens; small effector or hapten molecules for pre-targeted imaging; and various compounds for imaging of general biologic properties and processes such as proliferation, apoptosis, hypoxia, glycolysis, and angiogenesis. ^{68}Ga is most often utilized in radiopharmaceuticals for oncology diagnostics, however its potential has also been demonstrated for imaging of myocardial perfusion, pulmonary perfusion and ventilation as well as inflammation and infection. Feasibility of non-invasive monitoring of transplantation and survival of beta cells in diabetes mellitus is one more growing application area [7–9].

PET in combination with targeted imaging agents allows tumor-type specific non-invasive diagnosis with precise delineation of tumors and metastases and thus disease staging. Moreover, quantification of receptor expression, uptake kinetics and pre-therapeutic dosimetry may allow more efficient and effective treatment selection and planning as well as monitoring response to the therapy and early detection of recurrent disease resulting in personalized medicine and, in particular, radiotheranostics (Figure 1). The primary aims of the individualized patient management are to optimize therapeutic response and avoid futile treatments, minimize risks and toxicity as well as reduce cost and patient distress. Clinical intra-patient studies with variable amount of administered ^{68}Ga-based imaging agents demonstrated significance of individualized patient management [10–12].

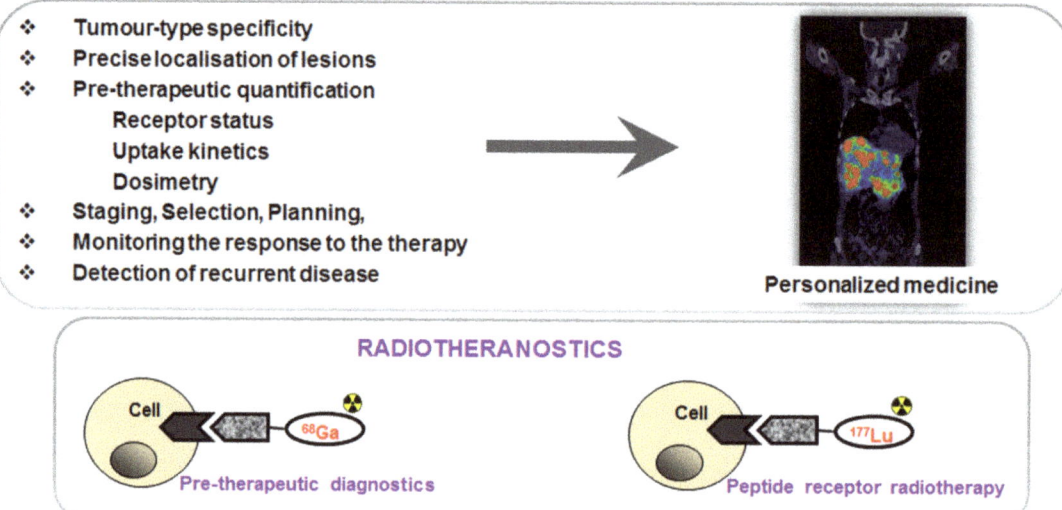

Figure 1. (**Upper panel**) Peptide receptor targeted imaging and radiotherapy provide personalized and thus more effective and efficient treatment of patients. (**Lower panel**) Drawing of the interaction of an agent, either imaging if labeled with ^{68}Ga (**left**) or radiotherapeutic if labeled with ^{177}Lu (**right**), with the cell receptor.

Such characteristics of ^{68}Ga as its availability from a generator system and amenability for kit type radiopharmaceutical preparation make this radionuclide as functional as ^{99m}Tc, but with additional advantages of higher sensitivity, resolution, quantification and dynamic scanning. Moreover, some therapeutic radionuclides resemble coordination chemistry of Ga(III) thus facilitating the radiotheranostic development wherein the pre-therapeutic imaging and radiotherapy are conducted with the same vector molecule exchanging the imaging and therapeutic radionuclides (Figure 1, lower panel).

A number of methods for ^{68}Ga-labeling have been developed allowing choice dependent on the application objectives and logistics. Production for the clinical use can be divided into three groups: manual good manufacturing practice (GMP) production; automated GMP manufacturing; and kit type preparation. This review presents such critical aspects of ^{68}Ga-radiopharmaceutical development as: generator production of ^{68}Ga and its subsequent handling; essential features of labeling chemistry in relation to the endpoint biological and clinical applications; important aspects of ^{68}Ga-radiopharmaceutical production process with respect to the regulatory issues.

2. Characteristics and Generator Production of ^{68}Ga

The advantages of ^{68}Ga have been presented in details previously and they are multiple with regard to both physical and chemical properties [1,3,4,6]. Those that are relevant to the aspects discussed in this review are summarized here. The high positron emission fraction (89%, E$_{max}$: 1899 keV, E$_{mean}$: 890 keV) and half-life of 68 min provide sufficient levels of radioactivity for high quality images while minimizing radiation dose to the patient and personnel. It requires short scanning time and allows repetitive examinations. In modern generators ^{68}Ga is obtained in ionic form compatible with subsequent highly reproducible and straightforward labeling chemistry. The only oxidation state stable at physiological pH is Ga(III) providing robust labeling chemistry with ligands that can fill the octahedral coordination sphere of Ga(III) with six coordination sites. The long shelf-life generator (t$_{½}$(^{68}Ge) = 270.95 d) is simple to use and a steady source of the radionuclide for medical centers without cyclotrons or remote from distribution site. Moreover, it is a source of the enrichment of radiopharmaceutical arsenal at centers equipped with cyclotrons. As compared to cyclotron it does not require: (i) special premises with radiation shielding constructions; (ii) consumption of energy; and (iii) highly qualified personnel for running and maintaining the equipment (Figure 2A).

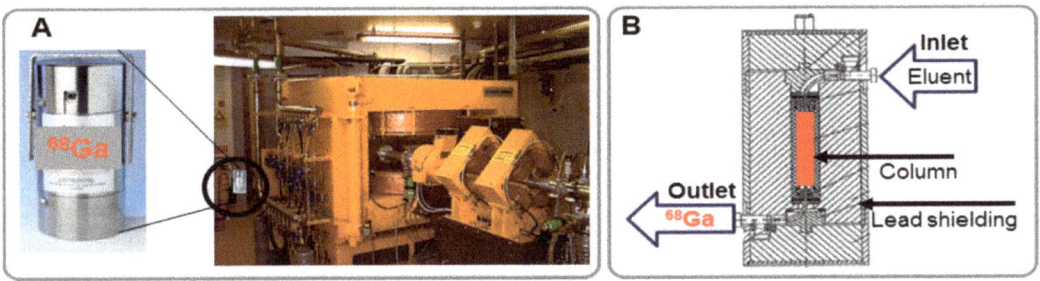

Figure 2. (**A**) Pictures of a ^{68}Ge/^{68}Ga generator and a cyclotron; (**B**) Schematic presentation of the cross section of a column-based generator.

A generator is a self-contained system housing a parent/daughter radionuclide mixture in equilibrium. Modern commercial generators consist of a small chromatographic column situated in a shielding container (Figure 2B). ^{68}Ge is produced in a high energy cyclotron from stable Ga-69 isotope (^{69}Ga(p,2n)^{68}Ge). Then, ^{68}Ge is immobilized on a column filled with inorganic, organic or mixed matrix where it spontaneously decays to ^{68}Ga (Equation (1)), which can then be extracted by an eluent. ^{68}Ga decays in its turn to stable Zn(II) (Equation (2)). Thus Ge, Ga, and Zn elements populate the generator and can be found in the eluate.

$$^{68}_{32}Ge + ^{0}_{-1}e \rightarrow ^{68}_{31}Ga + \nu \qquad (1)$$

$$^{68}_{31}Ga \rightarrow ^{68}_{30}Zn + \beta^+ + \nu; p \rightarrow n + \beta^+ + \nu \qquad (2)$$

The relation of ^{68}Ge decay and ^{68}Ga accumulation is described by secular equilibrium since the half-life of the ^{68}Ge is over 100 times longer than that of ^{68}Ga (Equation (3)).

$$\frac{t_{1/2}(^{68}_{32}Ge)}{t_{1/2}(^{68}_{31}Ga)} = 5762 \qquad (3)$$

At the equilibrium, the radionuclides have equal radioactivities achieved at 14 h post elution (Figure 3). However, already 68 min post elution 50% of the maximum achievable radioactivity is accumulated and 4 h later it is over 91%. So, the tracer production can be performed every hour or up to three productions within one working day dependent on the generator loaded radioactivity (^{68}Ge) and the age of the generator. The graph depicts a theoretical plot of ^{68}Ga generation (Figure 3, red line). In reality, ^{68}Ga recovery from the generator chromatographic column is less than quantitative and the proportion of ^{68}Ga separated during the elution process to the theoretical value expected at the secular equilibrium is defined as elution efficiency (Figure 3, black line). The precise determination of ^{68}Ga half-life time still continues reporting 67.83 min [13], 67.71 min [14], and 67.85 min [15], but for the simplicity 68 min value is used in this illustrative graph (Figure 3). It can also be mentioned that the range of the half-life values covers 62–74 min in the specification of the European Pharmacopoeia monographs [16,17].

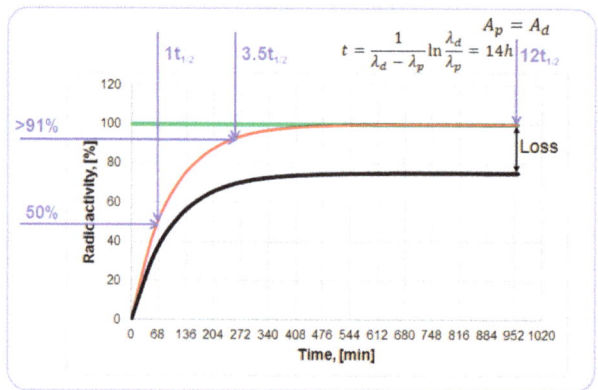

Figure 3. Graph of the secular equilibrium with ^{68}Ge decay and ^{68}Ga accumulation. The green line represents decay of ^{68}Ge described by $A_p(t) = A_p(0) * \exp(-\lambda_p t)$; the red line represents ingrowth of ^{68}Ga described by $A_d(t) = \frac{A_p(0)\lambda_d}{\lambda_d - \lambda_p}(\exp(-\lambda_p t) - \exp(-\lambda_d t))$; and the black line represents accumulation kinetics of ^{68}Ga with correction for hypothetical elution efficiency.

Commonly, modern ^{68}Ge/^{68}Ga generators demonstrate highly reproducible and robust performance, as, for example, ^{68}Ga elution yield for three generator units of various age presented in Figure 4A [18]. Elution yield has both ^{68}Ge-decay and elution efficiency components. The elution efficiency depends on the ^{68}Ge breakthrough and column matrix, and may drop in time course, however the ^{68}Ge-decay component is larger (Figure 4B).

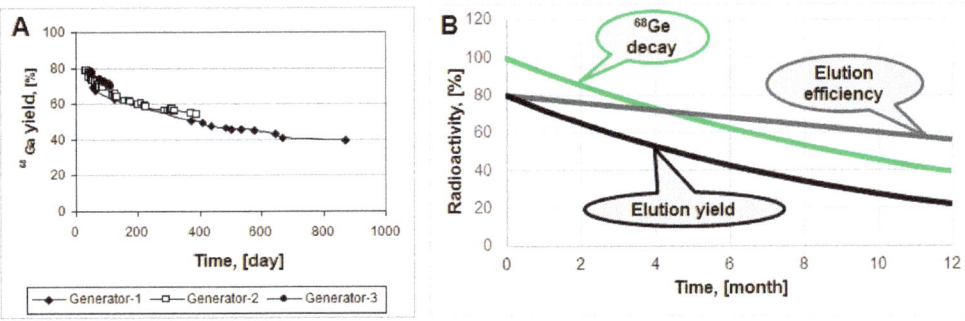

Figure 4. (**A**) ^{68}Ga elution yield for Generator-1 over 29 months, Generator-2 over 14 months and Generator-3 over three months; (**B**) Hypothetical graphs representing ^{68}Ge decay, elution efficiency and resulting non-decay corrected elution yield.

Historically, there have been two basic Ge-Ga separation methods: liquid–liquid extraction and column technology with various eluents (alkaline, acidic or complexing agents). The column technique is most widely used with various sorbents made of inorganic, organic or mixed materials (Table 1) [1,19].

Table 1. Various sorbents and respective eluents used in column based ^{68}Ge/^{68}Ga generators.

^{68}Ge/^{68}Ga Generator Column Matrix	
Inorganic (Eluent)	**Organic (Eluent)**
SnO$_2$ (1 M HCl)	N-methylglucamine (0.1 M HCl; 0.1 M NaOH; citrate; EDTA)
TiO$_2$ (0.1 M HCl)	Pyrogallol-formaldehyde (0.3 M HCl)
CeO$_2$ (0.02 M HCl)	Nanoceria-polyacrylonitrile (0.1 M HCl)
ZrO$_2$ (0.1 M HCl)	
Zr-Ti ceramic (0.5 M NaOH/KOH; 4 M HCl; acetate; citrate)	
Nano-zirconia (0.01 M HCl)	

The major parameters of a generator performance are: chemical separation specificity; radiation resistance and chemical stability of the column material; eluate sterility and apyrogenecity; ^{68}Ge breakthrough; eluent type; and elution profile. Most of the generators use acidic eluent since it provides cationic Ga(III) for the further direct chemistry. Inorganic column sorbents are used more widely as they are less sensitive to radiolysis. Development work continues in order to improve these characteristics and a number of commercial and in-house built generators have been introduced. Column matrixes that allow elution of ^{68}Ga using several different eluents dependent of the application have been developed. Organic resin with N-methylglucamine functional groups allows elution with HCl, NaOH, citrate and EDTA dependent on the subsequent synthesis and application [20]. Highly stable nanocrystalline Zr-Ti ceramic material was developed, and the respective generator elution and in-line eluate concentration/purification was automated [21]. The elution of ^{68}Ga could be performed using various eluents: basic, acidic or buffering (acetate, citrate). This generator also showed low ^{68}Ge breakthrough of <10^{-3}%, and the subsequent eluate purification not only further decreased ^{68}Ge content (<10^{-6}%) but also diminished cationic impurities. The narrow elution profile with 95% of ^{68}Ga in 2 mL volume was achieved in a generator with SnO$_2$ column sorbent [22]. A novel nanoceria-polyacrylonitrile-based generator provided high ^{68}Ga concentration eluate and low ^{68}Ge breakthrough (<10^{-5}%) [23].

The modern commercial generators rely on chromatographic separation and provide advantages such as long shelf-life of 1–2 year, stable column matrixes, cationic chemical form of ^{68}Ga(III) allowing subsequent versatile and direct labeling chemistry as well as reproducible and robust performance. There are a number of them with variation in the molarity of HCl eluent, metal cation content and ^{68}Ge breakthrough (Table 2). This diversity is a result of over six decade journey (Table 3) from the first liquid–liquid extraction generator system [24] and simple radiopharmaceuticals for clinical application such as ^{68}Ga-EDTA solution for the brain lesion imaging, ^{68}Ga-citrate for bone uptake imaging, ^{68}Ga-ferric oxide for bone marrow scanning, and ^{68}Ga-polymetaphosphates for kidney and liver scanning [25]. Further development was directed towards the generators providing cationic ^{68}Ga(III), and the first commercial one was introduced in late 1990s contributing to the blossom of the ^{68}Ga-PET together with the advent of somatostatin (SST) ligands. The first generator of pharmaceutical grade appeared on the market in 2014. More generators are on their way to the market and marketing authorization acquisition.

Table 2. Examples of commercial ^{68}Ge/^{68}Ga generators.

	Eckert & Ziegler Cyclotron Co. Ltd.	Eckert & Ziegler IGG100 and IGG101 GMP; Pharm. Grade	I.D.B. Holland B.V.	Isotope Technologies Garching
Column matrix	TiO$_2$	TiO$_2$	SnO$_2$	SiO$_2$/organic
Eluent	0.1 M HCl	0.1 M HCl	0.6 M HCl	0.05 M HCl
^{68}Ge breakthrough	<0.005%	<0.001%	~0.001%	<0.005%
Eluate volume	5 mL	5 mL	6 mL	4 mL
Chemical impurity	Ga: <1 µg/mCl; Ni < 1µg/mCl	Fe: <10 µg/GBq; Zn: <10 µg/GBq	<10 ppm (Ga, Ge, Zn, Ti, Sn, Fe, Al, Cu)	Only Zn from decay
Weight	11.7 kg	10 kg 14 kg	26 kg	16 kg

Table 3. Milestones of ^{68}Ge/^{68}Ga generator development.

Time Period	Milestone
1950–1970	First ^{68}Ge/^{68}Ga generator
	Clinical applications: ^{68}Ga-EDTA; ^{68}Ga-citrate; ^{68}Ga-colloid
1970–1980	Further development of ^{68}Ge/^{68}Ga generator: ^{68}Ga(III)
1990s	Commercial generator: ^{68}Ga(III)
2000s	Clinical use with advent of SST ligands
2011	GMP generators
2014	Marketing authorization

The ^{68}Ge/^{68}Ga generator meets criteria of an ideal generator in terms of: efficient separation of the daughter and parent elements due to their different chemical properties; physical half-life of parent allowing rapid daughter regrowth after generator elution; stable granddaughter with no radiation dose to the patient; long shelf-life; effective shielding of the generator, minimizing radiation dose to the user and expenses of hot cells; sterile and pyrogen-free output of the generator; as well as mild and versatile chemistry of the daughter ^{68}Ga amenable to automation and kit preparation. However, long shelf-life may raise concern with regard to radiolytic stability of column material, sterility of the eluate, and long-lived ^{68}Ge waste management. In addition, the volatility of GeCl$_4$ should be kept in mind and precautions to prevent the contamination of the surrounding must be taken. As the half-life of ^{68}Ge is over 100 d, it is classified as long-lived waste. The waste and storage expenses depend on radioactivity amount and container size. It would take 10 years for the loaded radioactivity of a commonly used 50 mCi generator to decay to the amount below 1 MBq (Figure 5) that would allow regular waste. In some European countries [26] according to the regulation, the eluate of the generator must contain <10 Bq/g of ^{68}Ge in order to be considered as regular waste. It has been suggested to solidify ^{68}Ge in order to decrease the concentration of ^{68}Ge in the bulky solution, so that the solution can be discarded as regular waste. The best-case scenario in terms of environment, sustainability and economy for the waste handling would be the re-cycling of costly ^{68}Ge by the manufacturers. This would provide a problem solution on the global level. It is known that the acidic environment is not favorable to the microbiological growth and it was confirmed by loading a ^{68}Ge/^{68}Ga generator column intentionally with various bacteria and fungi in exhaustive amounts and following their survival during two weeks [27]. The risk of incidental microbial contamination was found very low.

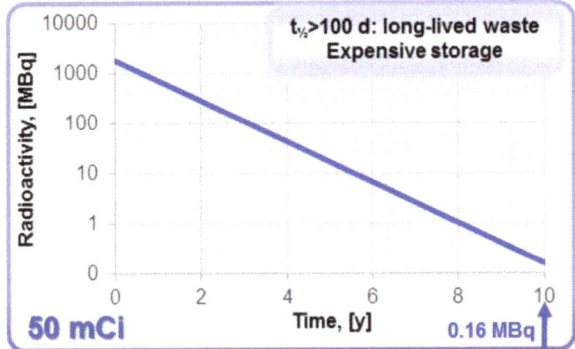

Figure 5. Decay of ^{68}Ge in a generator with loaded radioactivity of 50 mCi.

3. ^{68}Ge/^{68}Ga Generator Eluate Quality and Subsequent Labeling Chemistry

Quality and characteristics of the generator eluate including eluate volume, ^{68}Ga radioactivity concentration, HCl eluent molarity, and content of metal cationic impurities influence the efficiency of ^{68}Ga-labeling chemistry. Such aspects as pH, prevention of Ga(III) precipitation and colloid formation, radiolysis of vector molecules, competition of metal cations in the labeling chemistry are discussed in this section.

The disadvantages of the currently available generators are the large ^{68}Ga eluate volume and consequently low ^{68}Ga concentration; contamination of the eluate with long-lived parent nuclide ^{68}Ge; and also presence of cationic metal ion impurities that might compete with ^{68}Ga in the complexation reaction. The use of full generator eluate volume requires high ligand amount and long heating time resulting still in non-quantitative ^{68}Ga incorporation (Figures 6A and 7, Table 4) [18]. To overcome the drawbacks either eluate fractionation or eluate pre-concentration and pre-purification can be used (Figure 6A, Table 4). Metal cation and ^{68}Ge content can be reduced by regular elution and elution prior to the synthesis as well as eluate and product purification (Figure 6B,C).

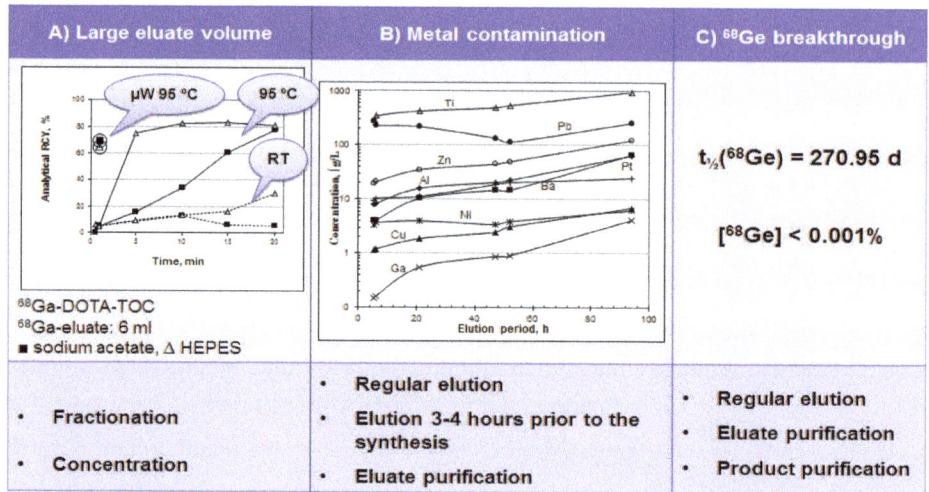

Figure 6. (**A**) Time course of ^{68}Ga complexation reaction conducted using the full original ^{68}Ga eluate (6 mL) at room temperature (**dashed line**), conventional heating in a heating block at 95 °C (**solid line**) and with microwave heating for 1 min at 90 ± 5 °C (**circled**) for two different buffer systems: ■ sodium acetate buffer, pH = 4.6, 20 nanomoles of DOTA-TOC; Δ HEPES buffer, pH = 4.2, 20 nanomoles of DOTA-TOC; (**B**) Metal ion content in 6 mL of the generator eluate as a function of the elution time period; (**C**) ^{68}Ge breakthrough with respective limit defined in European Pharmacopoeia (Ph. Eur.) monograph and methods for the reduction of the content level in the eluate and final product.

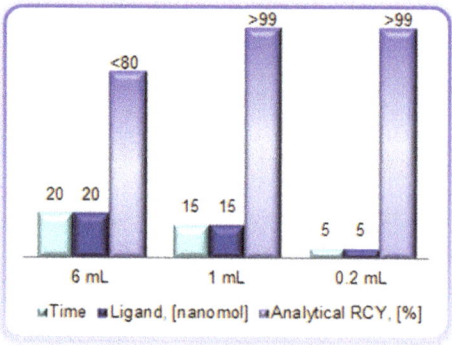

Figure 7. Reaction heating time (min), ligand amount (DOTA-TOC, (nanomole)), and analytical radiochemical yield (%) of the [^{68}Ga]Ga-DOTA-TOC synthesis using full volume of the generator eluate (6 mL), peak fraction of the generator eluate (1 mL), and pre-concentrated/pre-purified generator eluate (0.2 mL, anion exchange method).

The collection of the top fraction decreases the eluate volume and increases ^{68}Ga concentration (Figure 8A) [18,28]. Consequently, it improves the radioactivity incorporation, decreases the reaction time and required ligand amount (Figure 8B). However, it cannot remove metal cation impurities and parent ^{68}Ge.

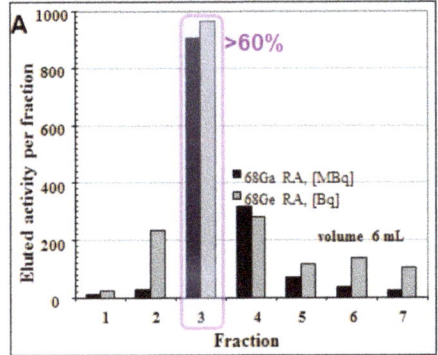
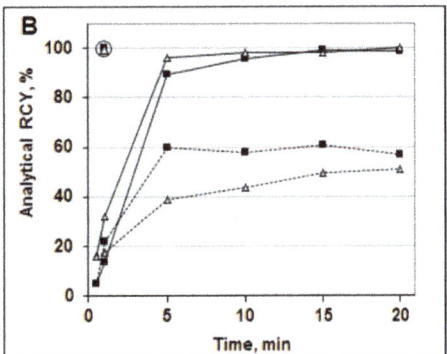

Figure 8. (**A**) Elution profile of the ^{68}Ge/^{68}Ga generator where one fraction was 1 mL (except for fraction 1 (0.3 mL) and fraction 7 (0.7 mL)), giving a total eluted volume of 6 mL. The profiles for the ^{68}Ga elution and the ^{68}Ge breakthrough are similar; the ^{68}Ge breakthrough is ~10^{-3}%. Fraction 3 (1 mL) contains over 60% of the available ^{68}Ga radioactivity; (**B**) Time course of ^{68}Ga complexation reaction conducted using 1 mL peak fraction of the generator eluate at room temperature (**dashed line**), conventional heating at 95 °C in a heating block (**solid line**), and with microwave heating at 90 ± 5 °C for 1 min (**circled**) for two different buffer systems: ■ sodium acetate buffer, pH = 4.6, 20 nmol of DOTA-TOC; ∆ HEPES buffer, pH = 4.2, 5 nmol of DOTA-TOC.

The methods for the eluate pre-concentration and pre-purification prior to the labeling synthesis are based on anion exchange chromatography, cation exchange chromatography or their combination (Table 4) [18,29–36]. Anion exchange method uses water for ^{68}Ga recovery in 200 µL [18]. Cation exchange uses acetone/HCl mixture eluent resulting in 400 µL [34]. This method was modified in order to avoid the use of acetone that might cause formation of organic impurities [37] and appearance of acetone in the formulated product. Instead of acetone, sodium chloride [29,31] and ethanol [36] eluents were introduced as well as combined method with cation exchange for the eluate purification followed by anion exchange to eliminate the acetone [32]. However, it should be mentioned that the organic impurity formation can be avoided by storing the acetone/HCl eluent without light access and in the freezer (e.g., −20 °C). Common advantage of the pre-concentration methods is the possibility to use several tandem generators or eluates collected from several generators with the same final volume and enhanced ^{68}Ga amount/concentration and generator shelf-life. Anion exchange method utilizes [^{68}GaCl$_4$]$^-$ complex formation from ~4 M HCl medium and its absorption to the anion exchange resin (Figure 9A, Table 4). While Ge(IV) forms the anionic complex at the molarity above 5 and thus does not retain on the resin and passes through. The method allows 30–90 fold eluate volume reduction to 200 µL and is independent on the generator eluate molarity. It purifies the eluate from Ge, Al, In and Ti cations, can be accomplished within 4–6 min, and is amenable to automation. The small volume and higher concentration of ^{68}Ga allows for reduced amount of the ligand and faster reaction with quantitative radioactivity incorporation and high specific radioactivity (SRA) of the tracer (Figures 7 and 9B).

Table 4. Basic methods of ^{68}Ge/^{68}Ga generator eluate utilization.

Method	Eluent	Volume	Cation Impurity Reduction	^{68}Ge Elimination
Full volume, 5–8 mL	H$_2$O/HCl	>5000 µL	Not purified	none
Fractionation, 1 mL	H$_2$O/HCl	1000 µL	Not purified	none
Eluate Concentration and Purification				
Anion exchange	H$_2$O	200 µL	One step: Al (>99%), In (>99%), Ti (90%)	Complete
Cation exchange	Acetone/HCl	400 µL	Two steps: Zn (×10^5), Ti (×10^2), Fe (×10)	10^4 fold
	NaCl/HCl	500 µL	NA	NA
	EtOH/HCl	1000 µL	Two steps: Ti (11%), Fe (×7)	400 fold
Combined cation/anion exchange	• Acetone/HCl • H$_2$O/HCl	1000 µL	NA	10^5 fold

Both fractionation and pre-concentration methods use hydrochloric acid in the eluent and thus require buffers for the correct pH adjustment necessary for the complexation. Moreover, weak buffer complexation capability is also essential in order to act as a stabilizing agent and prevent ^{68}Ga(III) precipitation and colloid formation (Figure 10A). A number of buffering systems such as HEPES, acetate, succinate, formate, tris, and glutamate were studied with HEPES, acetate and succinate buffers demonstrating better characteristics [18,38]. In particular, HEPES and acetate buffers are both biocompatible, with no toxicity issue, providing relevant pH, and functioning as stabilizing agents. However, at lower ligand concentration, HEPES is more preferable (Figure 10B). Nevertheless from the regulatory point of view acetate has an advantage since HEPES is not approved for the human use and thus purification and additional quality control (QC) analyses are required resulting in further time and resource consumption (Figure 10C).

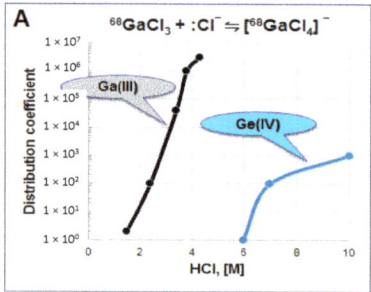
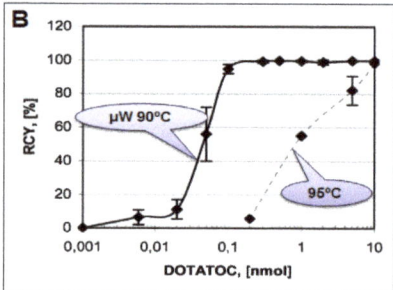

Figure 9. (**A**) Distribution coefficient D for the adsorption of Ga(III) and Ge(IV) chloride anions on an anion-exchange resin; (**B**) Influence of the DOTA-TOC amount on the decay-corrected radiochemical yield of the ^{68}Ga complexation reaction in HEPES buffer system using the full available ^{68}Ga radioactivity in 200 µL volume obtained after the pre-concentration and purification step. **Solid line**: 1 min microwave heating at 90 ± 5 °C, **dashed line**: 5 min conventional heating at 95 °C.

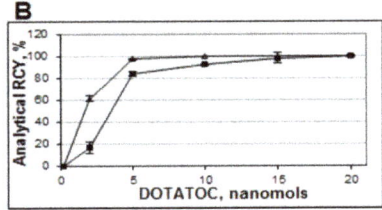

Figure 10. (A) Table showing formation of various species dependent on pH; (B) Influence of the buffering system (■ sodium acetate, Δ HEPES) on the ^{68}Ga radioactivity incorporation for different DOTA-TOC quantities (1 min microwave heating at 90 ± 5 °C). The reaction was conducted using the 1 mL peak fraction of the original generator eluate; (C) Table comparing characteristics of acetate and HEPES buffers.

The use of pre-concentration methods, especially from several generator eluates, increases ^{68}Ga concentration and thus the risk of radiolysis caused by the formation of free radicals such as hydroxyl and superoxide radicals in aqueous solutions. Thus the labeling of radiosensitive compounds, e.g., peptides and proteins comprising methionine, tryptophan and cysteine amino acid residues (Figure 11A) may require presence of radical scavengers such as ascorbic acid, gentisic acid, thiols, human serum albumin, or ethanol. For example, 10%–20% of ethanol may improve the synthesis outcome considerably (Figure 11B). Moreover, ethanol is biocompatible without toxicity or immunoreactivity issues, GMP compatible, most often does not interfere with labeling reaction, and has no biological target binding capability. In addition, it can also aid the solubility of lipophilic precursors. The radical scavengers, e.g., sodium ascorbate, can also be added post synthesis to the formulated tracer.

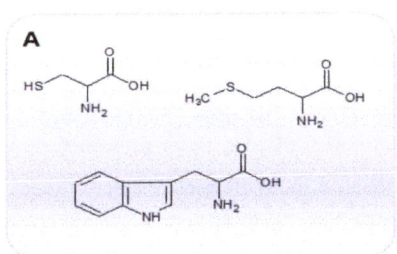
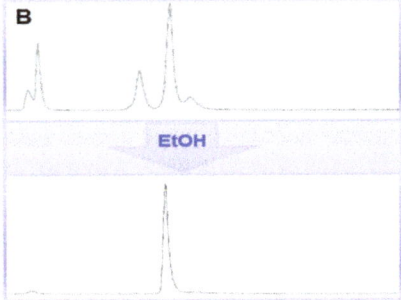

Figure 11. (A) Chemical structures of methionine, tryptophan and cysteine amino acid residues; (B) Radio-HPLC chromatograms of a crude ^{68}Ga-labeled protein comprising ~60 amino acid residues including several tryptophan and methionine amino acid residues without (**upper panel**) and with (**lower panel**) ethanol.

The generator eluate inevitably contains a number of metal cations that may interfere with the ^{68}Ga-labeling reaction (Figure 6B). Stable Zn(II) which is the product of ^{68}Ga decay (Equation (2)) is rather strong competitor in complexation with DOTA-comprising agents (Figure 12A). In generator column the accumulation of Zn(II) increases continuously (Figure 12B). At the time of secular equilibrium when the maximum amount of ^{68}Ga is accumulated, the amount of Zn(II) exceeds 10 times that of ^{68}Ga. When generator is eluted this excess is discarded and after ~2 half-lives the ratio of ^{68}Ga to Zn(II) is still over one (Figure 11B). So, regular elution and elution prior to the synthesis may allow 5.5-fold reduction of Zn(II) concentration (Figure 6B). Other solutions to overcome the reaction interference from Zn(II) could be the purification of the eluate prior to the labeling synthesis, enhanced amount of the ligand, or the use of chelators with high selectivity for Ga(III).

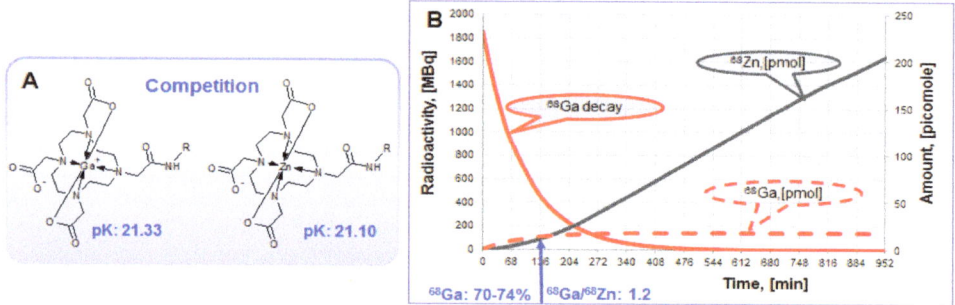

Figure 12. (**A**) Zn(II) forms thermodynamically stable complex with DOTA derivatives and interferes ^{68}Ga-labeling reaction, especially in the excessively high concentration; (**B**) Theoretical graphs (50 mCi generator) showing ^{68}Ga decay (MBq) and accumulation of radioactive ^{68}Ga and stable Zn(II) in picomoles within the time frame of secular equilibrium.

The eluate and reaction solutions contain other metal cations apart from Zn(II) that can compete with ^{68}Ga(III), requiring higher amount of the ligand and consequently decreasing specific radioactivity. The metal cations compete with Ga(III) in different manner dependent on their chemical and physical properties such as size, charge, and surface charge density. They might form faster and more stable complexes with the same chelator or like Ti(IV) and Zr(IV) form colloids and absorb ^{68}Ga(III) preventing it from the complexation. The reaction of a number of 2, 3 and 4 valent cations (Cu(II); Zn(II); Fe(II); Sn(II); Ca(II); Co(II); Ni(II); Pb(II); Al(III); Fe(III); Ga(III); In(III); Lu(III); Y(III); Yb(III); Ti(IV); Zr(IV)) relevant to the generator and labeling environment was studied with various chelators [1,39–44], especially with NOTA derivatives for which the fast labeling at room temperature was demonstrated earlier (Figure 13) [40]. Such parameters as reaction temperature, concentration, metal ion-to-ligand ratio, pH, and microwave heating mode were investigated. For DOTA derivatives, Al(III) and Ca(II) were found less critical and could be tolerated up to a concentration equal to that of the peptide bioconjugate [39,42]. The effect of the cations weakened in the following order Cu(II) > In(III) > Fe(III) > Fe(II). The radioactivity incorporation of 24% even in the presence of >1000 fold excess of Fe(III) indicated some selectivity of DOTA for Ga(III) [39]. NOTA derivatives could be radiolabeled at room temperature with over 98% yield, even in the presence of up to 10 ppm of other metal ion impurities such as Zn(II), Cu(II), Fe(III), Al(III), Sn(IV) and Ti(IV) [40,44]. Phosphinate chelators, TRAP-H, NOPO, and TRAP-Pr demonstrated efficient ^{68}Ga-labeling at a wide range of pH and as acidic as pH 1 using very low amount of the ligand and thus resulting in high SRA [41,43]. In contrast to carboxylate-based chelators (NOTA and DOTA) incorporation of ^{68}Ga(III) by the phosphinate chelators (TRAP-H, NOPO, and TRAP-Pr) was never entirely inhibited by the presence of Zn(II), not even at concentrations of 30 mM. Moreover, TRAP and NOPO were able to rapidly exchange coordinated Zn(II) with ^{68}Ga(III), indicating high selectivity of these chelators for Ga(III). Fusarinine C, a siderophore-based chelator, also demonstrated high selectivity for ^{68}Ga(III) resulting in SRA of up to 1.8 GBq/nmol [45]. Thus, the ^{68}Ga-labeling efficiency in the presence of metal cation impurities can be improved by: eluting generator regularly and prior to the synthesis; pre-purification of the generator eluate by anion or cation exchange chromatography; increasing the concentration of the ligand in order to compensate for the total metal cation concentration; and employing chelators with high selectivity for Ga(III).

^{68}Ga has been used to label small biologically active organic molecules, biological macromolecules, complexes of variable charge and lipophilicity as well as particles [1,4–6]. The majority of the imaging agents are synthesized using tagging techniques wherein a vector molecule is first conjugated to a chelator moiety for the subsequent complexation with ^{68}Ga (e.g., Figure 14A) [1]. The most frequently used chelators are derivatives of DOTA and NOTA that can stably complex ^{68}Ga(III), respectively, at elevated and room temperature (Figure 14B,C). The latter is very important in case of temperature sensitive fragile macromolecules and also it is amenable to cold type kit production in radiopharmacy practice. The advantage of DOTA is that the same vector molecule can potentially be used both for diagnosis and radiotherapy labeling it with various radionuclides such as ^{68}Ga, ^{90}Y or ^{177}Lu (Figure 1, lower panel).

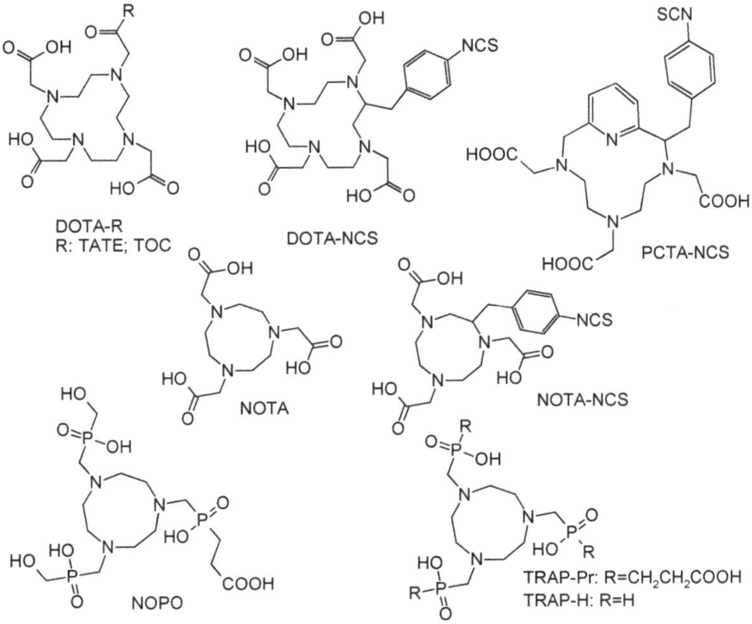

Figure 13. Chemical structures of example macrocyclic chelators and their derivatives that were studied in the competitive complexation reaction with various metal cations.

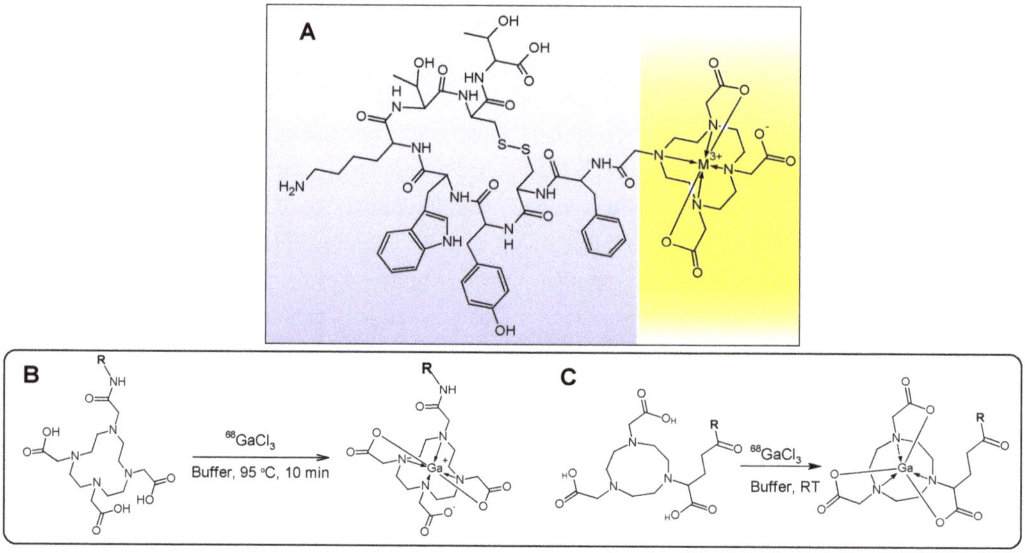

Figure 14. (**A**) Chemical structure of DOTA-TATE where the biologically active vector peptide (TATE, **purple background**) is conjugated to DOTA chelate moiety (**yellow background**) encaging the metal cation; (**B,C**) schematic presentation of ^{68}Ga-labeling, respectively, with DOTA- and NOTA-based ligands, where R stands for a macromolecule such as peptide, protein, oligonucleotide, glycoprotein, antibody or low molecular weight vector that can deliver the radionuclide to the binding site.

4. Influence of Biological and Clinical Endpoint Applications on the Chemistry Choice

Table 4 and Figure 7 summarize the three basic approaches of the generator eluate use with the respective key characteristics. Taking into consideration the poor labeling efficiency (Figures 6A and 7), the method using full generator eluate volume directly for the labeling is basically excluded. The choice between fractionation and pre-concentration methods depends on the requirements to specific radioactivity of the imaging agent that in turn is defined by target binding site limit, pharmacological side effects or ligand cost; demand on the small solution volume and high ^{68}Ga concentration; selectivity of the chelators towards ^{68}Ga(III); tracer production method, e.g., automated synthesis; or kit type preparation. If high specific radioactivity is required, then the pre-purified and pre-concentrated eluate might be preferred. While fractionation method which is simpler can be used if the amount of the ligand is not limited or the chelators have high selectivity for Ga(III). If the smallest volume and highest concentration of ^{68}Ga is required as, for example for the production of ^{68}Ga-carbon nanoparticles [46] the anion exchange concentration might be the first choice. Kit type preparation would most probably exclude pre-concentration and pre-purification methods due to the potentially high hand dose as well as multistep protocol.

SRA might be critical for endpoint biological and medical applications thus putting demand on the chemistry and such parameters as metal cation contamination, volume, and amount of the ligand. Figure 15 demonstrates an *in vitro* example where the synthesis method using anion exchange pre-concentration and pre-purification of the eluate assuring high SRA was necessary in order to obtain the saturation pattern (Figure 15B,C) and image contrast reproducibility, which could be achieved in the plateau range above SRA of 100 MBq/nmol (Figure 15D) [39]. Thus the synthesis with high SRA allowed the investigation of the influence of specific radioactivity on the binding of the tracer in frozen sections of rhesus monkey brain expressing SSTR (Figure 15A). Clear saturation allowing determination of dissociation constant, K_d and B_{max} could be observed (Figure 15B,C). The ratio of the amount of the receptor bound tracer to free tracer (Bound/Free (B/F)), which is also a signal to background ratio reflects contrast of an image. When presented as a function of SRA (Figure 15D) it provides important information with regard to the detection limit and reproducibility of a biological assay. In particular, the reduction of SRA corresponds to the decrease of B/F ratio and results in declined image contrast and poor detection. The most critical range of SRA values is around the inflection point where slight change of SRA may result in poor reproducibility of the image quantification. B/F becomes independent on SRA variation after it reaches the plateau. *In vivo* and *in vitro* experiments using tracers with optimized SRA would demonstrate high reproducibility and robustness.

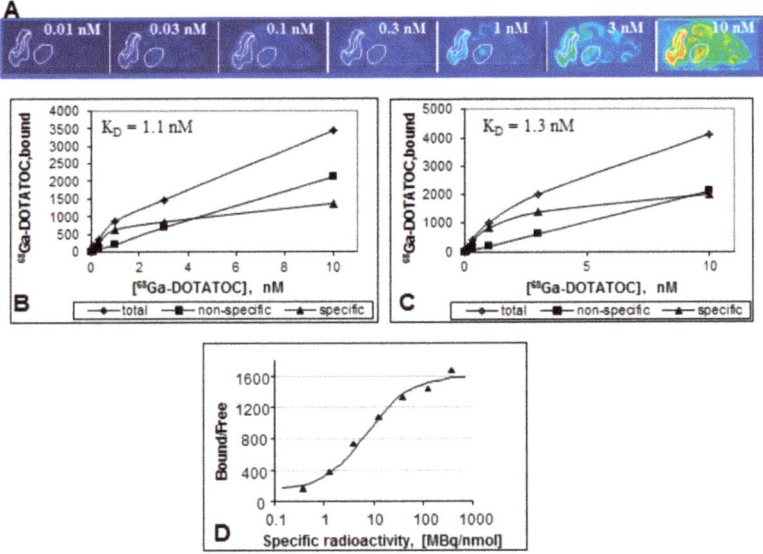

Figure 15. (**A**) Frozen section autoradiography showing [^{68}Ga]Ga-DOTA-TOC binding to SSTR of rhesus monkey thalamus and cortex. The brain sections were incubated with different concentrations of [^{68}Ga]Ga-DOTA-TOC (0.01–10 nM) for 30 min at room temperature; (**B,C**) Saturation of [^{68}Ga]Ga-DOTA-TOC binding to SSTR of Rhesus monkey thalamus (**B**) and cortex (**C**); (**D**) The ratio of the ligand bound to the receptor and the free ligand as a function of the SRA. The data were fitted to a sigmoid two-parametric model.

The choice of the synthesis method and SRA optimization are also driven by the *in vivo* agent performance in terms of the target and background uptake. It can be illustrated by a study wherein three sequential examinations with gradually increasing total amount of injected peptide were performed in the same patient on the same day (Figure 16) [10]. The anion exchange pre-concentration method was necessary to employ in order to provide a wide range of SRA values. The correlation between the lesion image contrast and SRA was not linear but followed Gaussian distribution pattern. As the peptide amount increased to 50 µg the uptake in the metastases improved, while it was decreased in liver and spleen (Figure 16, upper panel). This example also stresses the importance of individualized patient management. Thus, for all but one patient the tumor and normal tissue uptake decreased after i.v. injection of 500 µg peptide (Figure 16, upper panel). For the one patient tumor uptake increased continuously (Figure 16, lower panel). Such uptake pattern variability is most probably related to the receptor density variation and would presumably indicate different therapeutic protocols.

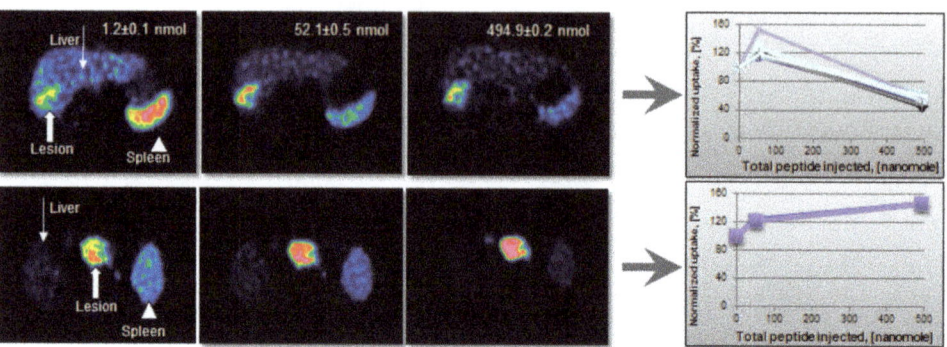

Figure 16. (Upper panel) Transaxial [^{68}Ga]-DOTA-TOC-PET images of a patient with liver metastases from a colonic carcinoid who underwent three sequential PET-CT examinations. The tracer accumulation pattern in tumor tissue (**thick arrow**) increased in the second PET examination by pre-treatment with 50 µg of unlabeled octreotide but decreased again in the third examination that was proceeded by 500 µg of octreotide. (**Lower panel**) Transaxial [^{68}Ga]-DOTA-TOC-PET images of a patient with a large endocrine pancreatic tumor who underwent three sequential PET-CT examinations. In contrast to the tumor uptake pattern in the other patients, as illustrated in the upper panel, the tumor accumulation (**thick arrow**) in this particular patient increased gradually over the three PET examinations.

The high ligand amount allowed lower SRA [10] and thus the use of simpler fractionation method that was employed in another clinical study wherein the optimized amount of the injected peptide [10] allowed high contrast and accurate comparison of two somatostatin ligand analogues ([^{68}Ga]-DOTA-TOC and [^{68}Ga]-DOTA-TATE, Figure 17) [47]. The identical imaging protocols and amount of administered peptide provided reliable comparison conditions. No statistically significant difference could be found in the tumor uptake of [^{68}Ga]Ga-DOTA-TOC and [^{68}Ga]Ga-DOTA-TATE in terms of both SUV and net uptake rate, K_i (Figure 17A, B). Thus, both tracers were found equally useful for staging and patient selection for peptide receptor radionuclide therapy (PRRT) in neuroendocrine tumors (NETs) with [^{177}Lu]Lu-DOTA-TATE. However, the marginal difference in the healthy organ distribution and excretion may render [^{68}Ga]Ga-DOTA-TATE preferable for the planning of PRRT where DOTA-TATE is used as vector. SUV did not correlate linearly with K_i and as such did not seem to reflect somatostatin receptor density accurately at higher SUV values, suggesting that K_i was the outcome measure of choice for quantification of somatostatin receptor density and assessment of treatment outcome (Figure 17B). No statistically significant difference could be observed in dosimetry estimations either (Figure 17C) [48].

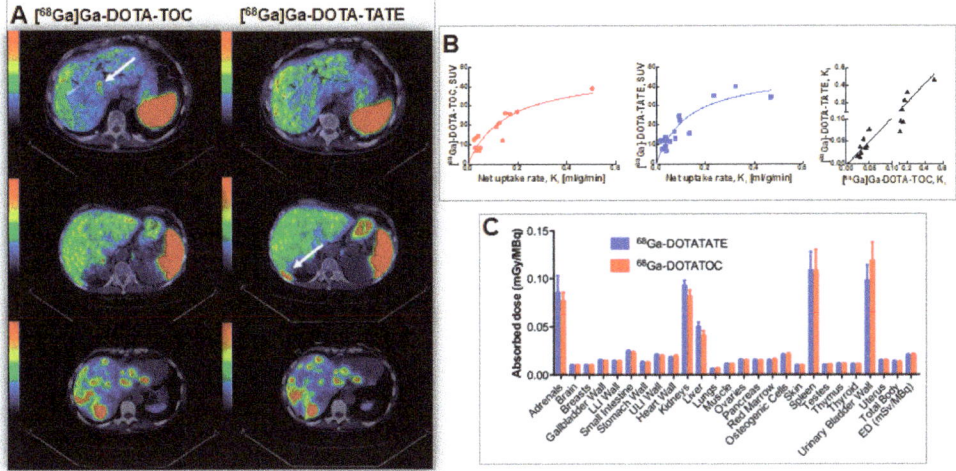

Figure 17. (**A**) Transaxial PET/CT fusion images of liver demonstrating cases of higher detection rate for: [^{68}Ga]Ga-DOTA-TOC (**A, upper panel**); [^{68}Ga]Ga-DOTA-TATE (**A, middle panel**); as well as equal detection rate (A, lower panel). Whole-body scans were conducted at 1 h p.i.. Arrows point at hepatic metastases; (**B**) SUV is presented as a function of net uptake rate K_i in tumors for [^{68}Ga]Ga-DOTA-TOC (**red**) and [^{68}Ga]-DOTA-TATE (**blue**). The solid lines are hyperbolic fits and are for visualization purposes only. Correlation between net uptake rate K_i at 1 h p.i. for [^{68}Ga]Ga-DOTA-TOC and [^{68}Ga]Ga-DOTA-TATE (**black**). The solid line is Deming regression with slope 1.06 and intercept 0.0. The axes are split in order to clarify the relationship at low uptake rates; (**C**) Absorbed doses in all organs included in OLINDA/EXM 1.1. LLI: lower large intestine; ULI: upper large intestine; ED: effective dose. Error bars indicate standard error of the mean.

Analogous results were found in another clinical study of patients affected by breast cancer using an Affibody® molecule, [^{68}Ga]Ga-ABY025 [11,12]. The detection rate and the image contrast were higher in the case of higher peptide amount. Again high ligand amounts allowed lower SRA and thus the use of simpler fractionation method. The patients received bolus intravenous administration of the low (LD) and high (HD) peptide content radiopharmaceuticals on two occasions one week apart. In some cases the lesion was not localized during the LD examination and the liver physiological uptake was rather high, while at HD the lesion was evident already at 1 h post injection and the image contrast increased with the time.

Common perception is that high affinity ligands may require tracers of high SRA. However, in the translation from cells to *in vitro* tissue and further to *in vivo* distribution, the influence of the affinity on the tracer performance may not be straightforward. The binding affinity is a parameter most often determined *in vitro* in cell cultures thus excluding *in vivo* physiological parameters. Binding affinity using frozen brain sections and biodistribution in rat were investigated for two somatostatin analogues presenting minor structural difference in C-terminal where the carboxyl group in threonine amino acid residue is exchanged to hydroxyl group (Figure 18A,D). This structural difference resulted in IC$_{50}$ value over 10-fold difference as determined in transfected cells expressing SSTR subtype 2 [49]. However, the difference could not be observed in frozen tissue section autoradiography experiment where IC$_{50}$ of octreotide against the two agents did not demonstrate a statistically significant difference (Figure 18B,E) [30]. There was no difference in *in vivo* and *ex vivo* biodistribution in rats in organs physiologically expressing SSTRs such as adrenals, pituitary gland and pancreas (Figure 18G). The pattern of peptide mass influence on the biodistribution was also similar for the two analogues (Figure 18C,F). As mentioned above, the clinical study also revealed no statistically significant differences in the uptake and detection rate dependent on the affinity difference [47]. In some cases [^{68}Ga]Ga-DOTA-TOC detected more lesions than [^{68}Ga]Ga-DOTA-TATE and vice versa (Figure 17) or the detection rate was the same. The dosimetry investigation, which is essential in order to exclude damaging effect of the radiation to the vital organs, did not reveal any statistically significant difference between the two analogues either, resulting in identical effective dose. These are illustrative examples of the complexity of the translation from the *in vitro* to *in vivo* applications. It indicated that lower values of SRA might be acceptable and thus simpler labeling techniques using fractionation method would be sufficient.

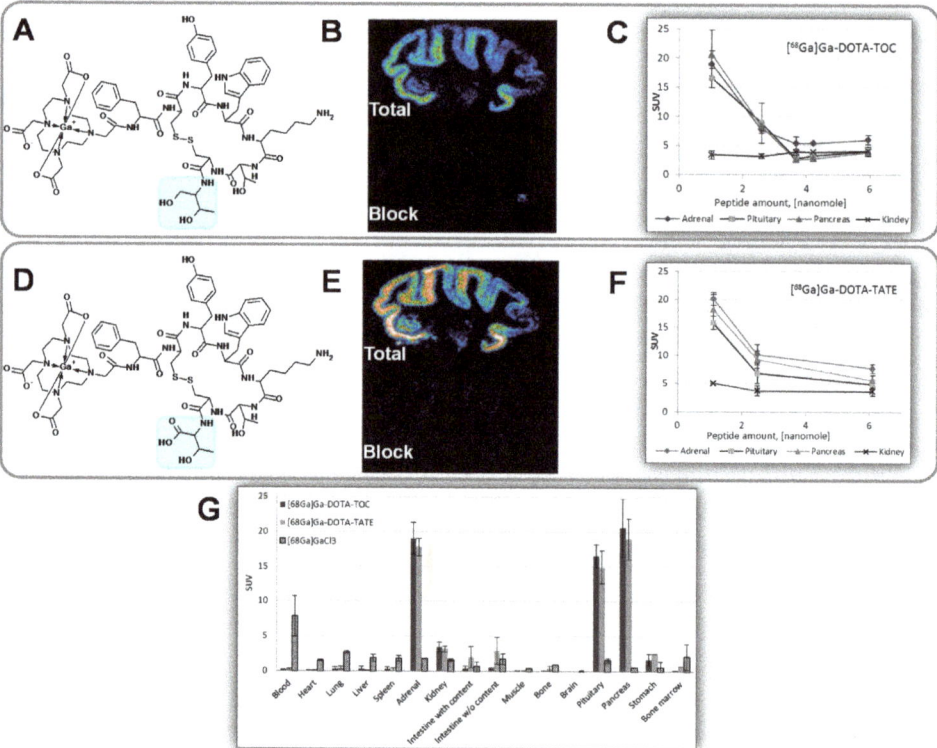

Figure 18. (**A**) Chemical structure of [^{68}Ga]Ga-DOTA-TOC (IC$_{50}$ = 2.5 ± 0.5; cell culture); (**B**) Rhesus monkey brain frozen section autoradiography (IC$_{50}$ = 23.9 ± 7.9); (**C**) *In vivo* rat organ distribution for variable injected peptide amount; (**D**) Chemical structure of [^{68}Ga]Ga-DOTA-TATE (IC$_{50}$ = 0.2 ± 0.004; cell culture); (**E**) Rhesus monkey brain frozen section autoradiography (IC$_{50}$ = 19.4 ± 8.3); (**F**) *In vivo* rat organ distribution for variable injected peptide amount; (**G**) Rat organ distribution of [^{68}Ga]Ga-DOTA-TOC and [^{68}Ga]Ga-DOTA-TATE 1 h post injection of 1 nmol of the peptide.

However, in the case of potent ligands such as Exendin-4, the amount that can be administered without induction of pharmacological effect can be very limited. Thus in the clinical examination of a patient affected by insulinoma using [^{68}Ga]Ga-DO3A-Exendin-4, the maximum amount of the administered peptide was limited to 0.5 µg/kg consequently requiring higher SRA values and respective tracer production methods involving pre-concentration and pre-purification of the generator eluate. [^{68}Ga]Ga-DO3A-Exendin-4/PET-CT demonstrated its uniqueness for the management of this disease group of patients. In particular, [^{68}Ga]Ga-DO3A-Exendin-4/PET-CT clearly localized the lesions while conventional morphological and established physiological diagnostic techniques failed to do so (Figure 19) [50]. Thus it is important to invest resources into the tracer production development.

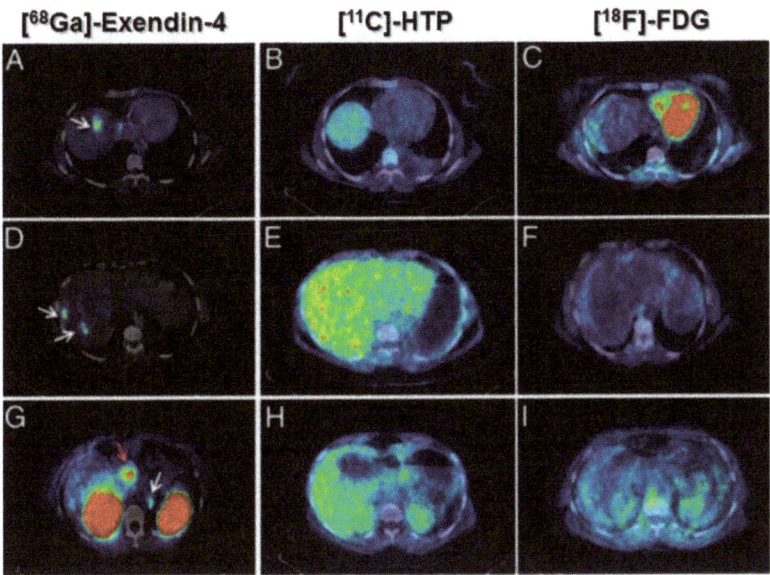

Figure 19. [^{68}Ga]Ga-DO3A-Exendin-4/PET-CT revealed several GLP-1R positive lesions (white arrows) in the liver (**A,D**) and a paraortallymph node (**G**). Beta cells in normal pancreas (red arrow) have significant expression of GLP-1R and can also be visualized by this technique (**G**). No pancreatic or hepatic lesions could be detected by PET/CT using established tumor markers such as [^{11}C]HTP (**B,E,H**) and [^{18}F]FDG (**C,F,I**).

One more aspect that influences the choice of the imaging radionuclide labeling chemistry is the possibility of subsequent radiotherapy in the light of internal radiotheranostics. In particular the choice of the chelator is essential since any structural modification of the ligand molecule may influence its biological function and it is of outmost importance to keep the *in vivo* performance and targeting properties of the imaging and radiotherapeutic analogues as similar as possible (Figure 1). It is particularly important for targeted imaging in oncology wherein the tumor-type specific precise localization of tumors and metastases becomes possible allowing for pre-therapeutic quantification of receptor status, uptake kinetics and dosimetry and thus enabling accurate therapy selection and planning as well as monitoring response to the therapy resulting in personalized medicine (Figure 1). For example, the use of DOTA derivatives for both imaging radionuclide such as ^{68}Ga(III) and therapeutic radionuclide such as ^{177}Lu(III) results in relatively minor alteration in the ligand and thus might be more preferable. These radionuclides have the same charge and fit the cavity of DOTA macrocycle forming stable complexes. However, having different coordination sphere, they form complexes of different geometry, *cis*-pseudooctahedral and monocapped square antiprismatic geometry, respectively, for Ga(III) and Lu(III) [51,52]. Even minor changes in the structure of a ligand may alter its binding parameters and biodistribution pattern and thus characterization of biological activity of each specific agent must be conducted. With regard to ^{68}Ga- and ^{177}Lu-labeled Exendin-4 analogues the biodistribution pattern and dosimetry estimations correlated despite the differences in the radionuclide-chelator complex moiety [53,54].

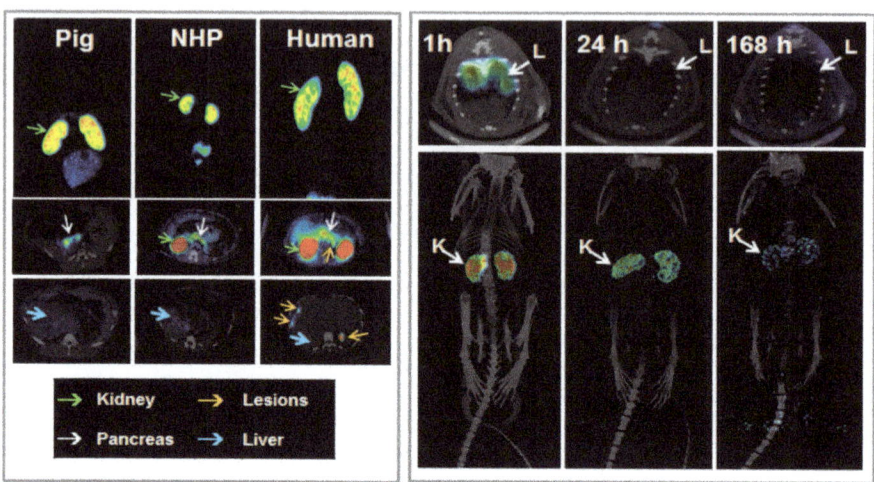

Figure 20. (**Left panel**) *In vivo* biodistribution of [^{68}Ga]Ga-DO3A-VS-Cys^{40}-Exendin-4 as analyzed by PET-CT imaging in the pig (0.025 µg/kg; 60 min), non-human primate (NHP) (0.01 µg/kg; 90 min), and human (0.17 µg/kg; 40 min, 100 min and 120 min). The pancreas (**white arrow**) was delineated within 10 min post injection in all species. The low hepatic uptake (**blue arrow**) shows the potential for outlining insulinoma tumor metastasis (**orange arrow**, human images). The MIP coronal images demonstrate the highest uptake of the tracer in the kidneys (green arrow) in all species. (**Right panel**) Representative fused SPECT-CT images of [^{177}Lu]-DO3A-VS-Cys^{40}-Exendin-4 in rats at different time points. Lungs could be outlined at 1 h p.i. and showed faster clearance in later time points (**upper panel**). MIP images of whole body scan showing dominance of kidneys as excretory organ of tracer (**lower panel**).

Dosimetry investigation plays a crucial role in the radiopharmaceutical development especially in the context of internal radiotheranostics. The dosimetry estimations conducted with ^{68}Ga-conterparts can presumably predict the applicability of the corresponding radiotherapeutic counterpart and thus save resources and expenses otherwise spent of the futile development of a radiotherapeutic pharmaceutical. The influence of the SRA on the biodistribution discussed in this section also applies here. However, SRA is a function of the half-life and thus direct comparison between the short-lived and long-lived radionuclides can be misleading. Therefore, it is the amount of the injected ligand that should be kept similar in the experiments. As mentioned above, [^{68}Ga]Ga-DO3A-Exendin-4/PET-CT demonstrated very promising results [50] indicating that internal radiotherapy using ^{177}Lu-labeled analogue might be a valuable therapeutic tool in the management of patients affected by insulinomas. However, estimation of [^{68}Ga]Ga-DO3A-Exendin-4 dosimetry (Figure 20, left panel) [54] demonstrated high kidney absorbed dose that could preclude the use of [^{177}Lu]Lu-DO3A-Exendin-4. This finding motivated the investigation of [^{177}Lu]Lu-DO3A-Exendin-4 dosimetry using rat biodistribution for the extrapolation and estimation of human dosimetry parameters (Figure 20, right panel) [53]. The amount of the injected peptide was kept similar to that of [^{68}Ga]Ga-DO3A-Exendin-4 and biodistribution pattern was comparable. The results confirmed that given the high kidney absorbed dose the amount of the acceptable administered radiation dose might be insufficient for the tumor control and thus render the treatment futile. Therefore the kidney protection and peptidase inhibition that may allow reduction of kidney absorbed dose and amplification of the tumor absorbed doses are required in order to develop [^{177}Lu]Lu-DO3A-Exendin-4 for the radiotherapy.

The *in vivo* pre-clinical and clinical examples demonstrate necessity for the optimization of the SRA of the imaging/therapeutic agents in each particular case. The optimization requires wide range of SRA values and in order to provide it the highest possible SRA must be achieved using respective labeling techniques.

5. Regulatory Aspects

Generator is involved in the GMP production process and should comply with the requirements that would assure: product quality; patient safety; traceability of the process; reliability and robustness of the performance. Quality assurance system is necessary to ensure that quality and safety of ^{68}Ga-based radiopharmaceuticals is adequate for the intended use. The qualification and validation of the performance of a chromatographic generator includes the investigation of its elution profile, elution efficiency, the extent of radionuclidic contamination of the eluate, contamination of the eluate with other metal cations and column material. To be suitable for the use in nuclear medicine, a generator must have favorable properties when these vital parameters are examined. The primary document to adhere is the European Pharmacopoeia monograph on Gallium (^{68}Ga) chloride solution and Gallium (^{68}Ga) edotreotide injection [16,17]. Other helpful documents are the EudraLex (Volume 4, GMP) annexes (Annex 1, Manufacture of sterile med products; Annex 3, Manufacture of radiopharmaceuticals; Annex 13, Investigational medicinal products); European Pharmacopoeia monographs (Radiopharmaceutical Preparations (0125) Ph. Eur.; Parenteral preparations (0520) Ph. Eur.; Bacterial endotoxins (20614) Ph. Eur.); Medical internal radiation dose format (MIRD); and International commission on radiological protection (ICRP). European Pharmacopoeia monographs on the compounding of radiopharmaceuticals and extemporaneous preparation of radiopharmaceuticals is in progress. The contribution of EANM, SNM and researchers around the world to the current advances in the regulatory aspects of PET radiopharmaceutical is considerable. Such issues as: regulatory documentation regarding small scale preparation of radiopharmaceuticals and the impact of the obligation to apply for manufacturing authorization or clinical trial; compliance with regulatory requirements for radiopharmaceutical production in clinical trials; quality of starting materials and final drug products/radiopharmaceuticals were thoroughly analyzed [55–60]. Guidelines on Good Radiopharmacy Practice (GRPP) [61]; patient examination protocols, interpretation and reporting of the patient examination results [62,63]; Investigational Medicinal Product Dossier; and Exploratory

Investigational New Drug that reduce the demand on toxicity studies and respective cost burden as well as allow easier understanding of the regulatory requirements [64–66] improve professional communication and standardization. Recognition of the microdosing concept (≤100 µg or ≤30 nanomoles for peptides/proteins) [67–70] by EMEA and FDA allows validation requirements relevant to PET radiopharmaceuticals. The GMP validation expenses could further be decreased by the reduction of toxicology studies to biodistribution and dosimetry investigation specially that the latter provides more accurate and sensitive detection of distribution throughout the organs at the same time allowing monitoring adverse effects, clinical signs, clinical chemistries, hematology, histopathology, *etc.* This approach would also reduce the number of sacrificed animals adhering to the ethical norms. The work on global standardization, growth, and dissemination conducted by International Atomic Energy Agency also play essential role in the facilitation of PET introduction into clinical routine. The advent of regulatory documentation specific to PET radiopharmaceuticals introduces more clarity and improves communication between the PET community and regulatory authorities, nevertheless it should be mentioned that currently the facilitation of the entry of novel pharmaceuticals still relies mostly on magisterial and officinal preparation in combination with compassionate use under responsibility of the prescribing physician.

The parameters of a ^{68}Ge/^{68}Ga generator that should be validated according to the specifications given in the Ph. Eur. monograph on the gallium chloride solution [16,17] are summarized in Figure 21. In addition, ^{68}Ga accumulation kinetics allows choice of the generator elution and tracer production frequency. Daily elution or elution 3–4 h prior to synthesis is recommended in order to keep the metal cation impurities at lower level. In the case of a pharmaceutical grade generator the validation might be reduced to the qualification and determination of the ^{68}Ge breakthrough and elution efficiency as well as a test synthesis with a validated tracer, if supported by the local quality assurance and regulation.

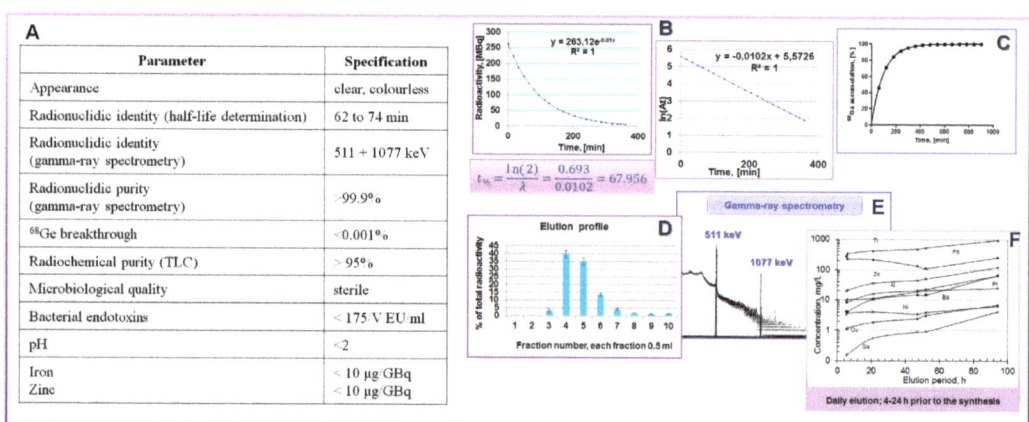

Figure 21. ^{68}Ge/^{68}Ga generator validation: (**A**) Table summarizing the validation parameters and the respective specifications; (**B**) t½ determination; (**C**) ^{68}Ga accumulation; (**D**) elution profile; (**E**) radionuclide purity; and (**F**) recommendation to elute generator prior to the synthesis.

The content of the long-lived parent ^{68}Ge in the eluate is addressed in order to assure radiation safety of the patient. The limit of 0.001% defined in the Ph. Eur. monographs was estimated assuming high and infinite accumulation of the radionuclide in sensitive organs, in particular bone marrow [71].

However, experimental evidence was necessary for the justification of the assumption. Thus, the organ distribution of ^{68}Ge(IV) in rat was conducted for the extrapolation and estimation of human dosimetry parameters in order to provide experimental evidence for the determination of ^{68}Ge(IV) limit [72]. While the dosimetry investigation of ^{68}Ge had not been performed, the metabolism, toxicity, carcinogenicity, mutagenicity, teratogenicity as well as myopathy and nephropathy of germanium in its various chemical forms had been studied previously for various administration routs [72]. In summary, germanium as a chemical element is of low risk to man without biological function or pharmacological activity, and with fast elimination without organ accumulation. Thus with regard to radioactive ^{68}Ge(IV) where the amount of the element is negligible, the safety issue is reduced to ionizing radiation and, in particular the buildup of ^{68}Ga(III) (as *in vivo* generator) at the sites of deposition of ^{68}Ge(IV). The dosimetry study showed that the maximum allowed administered radioactivity amount could be 645 MBq for female and 935 MBq for male, which was 35,000–50,000 higher than the level defined in the Ph. Eur. monograph. To put this in perspective, a fresh 50 mCi ^{68}Ge/^{68}Ga generator would allow for a breakthrough of ^{68}Ge(IV) of 35 to 50% before reaching the limit doses.

In addition, the preparation and administration of ^{68}Ge(IV) was conducted in the presence and absence of [^{68}Ga]Ga-DOTA-TOC simultaneous labeling synthesis. The presence of the tracer did not influence the distribution of the ^{68}Ge. It was also shown that ^{68}Ge(IV) was not chelated by DOTA-TOC and thus deposition of ^{68}Ge in the sites of DOTA-based imaging agents accumulation is also excluded. The content of ^{68}Ge and [^{68}Ga]Ga-DOTA-TOC was monitored by HPLC with tandem UV and radio detectors where the signal from [^{68}Ga]Ga-DOTA-TOC disappeared within 24 h while the signal from ^{68}Ge remained unchanged. The respective fractions were collected and periodically measured resulting in half-life values, respectively, for ^{68}Ga and ^{68}Ge (Figure 22).

These results imply that the ^{68}Ge(IV) limit currently recommended by monographs could be increased at least 100 times without compromising patient safety. This finding together with the availability of pharmaceutical grade generator, absence of the complexation with DOTA derivatives and knowledge that Ge(IV) does not bind to plasma proteins may facilitate the clinical introduction of kit type preparation of ^{68}Ga-based imaging agents. Kit type formulation development is ongoing at both academic and industrial establishments. Several countries are working within the frame of IAEA coordinated research project (F22050) [73].

In general terms, a radionuclide generator is defined as a medicinal product according to the current legislation [74]. However, dissimilarities between: different radionuclide generators; the use of the generator eluate; and tracer production processes are not taken into account. For example, the eluate from ^{82}Sr/^{82}Rb and ^{99}Mo/^{99m}Tc generators enters the blood stream either directly from the generator or after the labeling reaction in the product vial (kit formulation), and thus it must be assured to be sterile, isotonic and pyrogen free (medicinal product) especially considering that sodium chloride eluent is a favorable media for microbial growth. The principle difference with regard to ^{68}Ge/^{68}Ga generator used in a GMP manufacturing process is that the eluate solution is removed after the labeling reaction by product purification and the final product is sterile filtered. This can be illustrated by comparison of a typical ^{99m}Tc-based registered radiopharmaceutical preparation process with ^{68}Ga-based imaging agent manufacturing process. The preparation under pharmaceutical practice using kit formulation technique considers direct mixing of the generator eluate with the reagents in the product vial with subsequent formulation in the same vial and release without product purification, sterile filtration and quality control

(Figure 23, left column). This process results in a final radiopharmaceutical containing generator eluate components for the direct patient injection and thus the eluate quality must be assured by marketing authorization. While ^{68}Ga-radiopharmaceuticals are usually manufactured under GMP environment where generator eluate either directly or after pre-purification is added to a reaction vessel, followed by the product purification, formulation, sterile filtration and release after the quality control (Figure 23, right column). Thus the final radiopharmaceutical does not contain original generator eluate solution and is sterile filtered. Essentially the manufacturing process is similar to that of cyclotron produced radionuclide-based tracers. This implies that generators with and without marketing authorization could potentially be used in such production process. The efficiency of the worldwide dissemination of ^{68}Ga-radiopharmaceuticals with the patient benefit as priority would increase considerably if the essence of the manufacturing process would navigate the regulatory definition of generator produced ^{68}Ga. It should be specified in each particular case dependent on the production process (kit formulation or GMP manufacturing) if it is a starting material, radionuclide precursor, active pharmaceutical ingredient or active pharmaceutical ingredient starting material. The definition would influence the choice of the generator type, namely with or without marketing authorization, potentially reducing the cost and increasing accessibility.

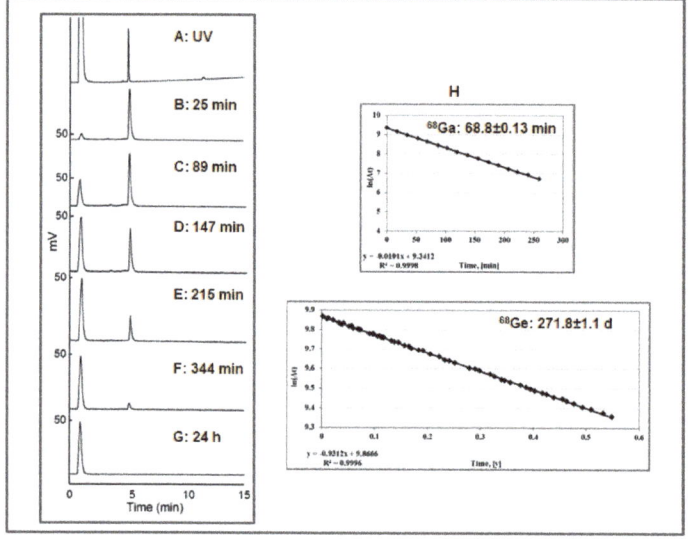

Figure 22. (**A**) UV-HPLC chromatogram of the authentic reference, [^{Nat}Ga]-DOTA-TOC (the void signal corresponds to the buffer); (**B–F**) Radio-HPLC chromatograms of [^{68}Ga]Ga-DOTA-TOC produced in the presence of [^{68}Ge]GeCl$_4$. The analysis was conducted, respectively, at 25, 89, 147, 215, and 344 min post synthesis. The signals with R$_t$ of 1.0 ± 0.02 min and 4.90 ± 0.02 min correspond, respectively, to the ionic ^{68}Ge(IV) and [^{68}Ga]Ga-DOTA-TOC; (**G**) Radio-HPLC chromatogram taken 24 h after the production of [^{68}Ga]Ga-DOTA-TOC in the presence of [^{68}Ge]GeCl$_4$. The signal with R$_t$ of 1.0 ± 0.02 min corresponds to the ionic ^{68}Ge(IV) and the signal at 4.90 ± 0.02 min corresponding to [^{68}Ga]Ga-DOTA-TOC was not detected; (**H**) Determination of the t$_{½}$ for ^{68}Ga and ^{68}Ge measuring respective collected chromatography fractions.

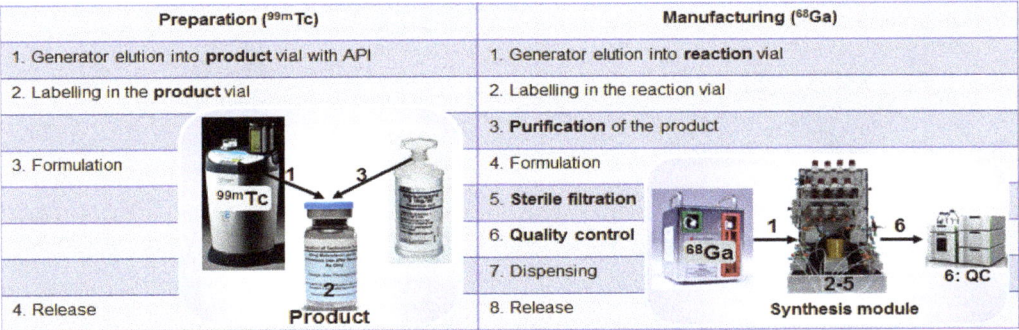

Figure 23. (**Left column**) Typical preparation steps of a ^{99m}Tc-based registered radiopharmaceutical; (**Right column**) Basic steps of an automated manufacturing of a ^{68}Ga-based radiopharmaceutical.

Radiopharmaceutical manufacturing automation provides possibility for the harmonized and standardized multicenter clinical studies that in turn may accelerate the introduction of new radiopharmaceuticals as well as their regulatory approval. Automation is important for the radioprotection, high reliability, reproducibility and robustness of the production as well as in-line traceability of the process for GMP compliance [75–77]. The increasing clinical demand of ^{68}Ga-based tracers prompted the need for the automation. There are a number of automated synthesizers either in-house built apparatus or available on the market as standard or custom made products. Stationary tubing systems require regular cleaning and cross contamination may occur in such systems. Disposable cassette systems offer improved microbiological safety with respect to sterility and endotoxin content as well as exclude chemical cross contamination. Better cGMP compliance and simplification of the process is possible since cleaning and sanitation of the tubings, containers, and purification cartridges is avoided. The stationary tubing system on the other hand provides more flexibility and lower radiopharmaceutical production price. Fractionation, anion and cation pre-concentration methods have been automated. A number of disposable cassettes for the production of tracers for the targeted imaging of SSTR, chemokine, integrin receptors, prostate specific membrane antigene as well as inflammation visualization agent, citrate, has entered the market.

6. Conclusions

Methods for the manual and automated GMP compliant production of ^{68}Ga-based agents have been developed. They are based on eluate fractionation or eluate concentration and purification approaches. The choice of the labeling method depends on the endpoint pre-clinical and clinical application with respective requirements to the imaging agent characteristics. The market of generators and automated synthesis systems is expanding. The automated production improves the practicality of harmonized and standardized multicenter clinical studies facilitating the introduction of new radiopharmaceuticals. The development of kit type preparation is also feasible, although there are yet no registered ^{68}Ga-radiopharmaceuticals on the market at present. Considerable advances have been made in PET radiopharmaceutical regulation and legislation still there is a number of particular questions and aspects to be addressed. Currently, understanding and support from national authorities prioritizing the benefit of patients is of outmost importance for the introduction of new radiopharmaceuticals into clinical practice.

Conflicts of Interest

The author declares no conflict of interest.

References and Notes

1. Velikyan, I. Prospective of ^{68}Ga-radiopharmaceutical development. *Theranostics* **2014**, *4*, 47–80.
2. Velikyan, I. Molecular imaging and radiotherapy: Theranostics for personalized patient management. *Theranostics* **2012**, *2*, 424–426.
3. Velikyan, I. Radionuclides for Imaging and Therapy in Oncology; In *Cancer Theranostics*; Chen, X., Wong, S., Eds.; Elsevier: Amsterdam, The Netherlands, 2014; pp. 285–325.
4. Velikyan, I. Positron emitting [^{68}Ga]Ga-based imaging agents: Chemistry and diversity. *Med. Chem.* **2011**, *7*, 338–372.
5. Velikyan, I. The diversity of ^{68}Ga-based imaging agents. *Recent Results Cancer Res.* **2013**, *194*, 101–131.
6. Velikyan, I. Continued rapid growth in Ga applications: Update 2013 to june 2014. *J. Label. Compd. Radiopharm.* **2015**, 99–121.
7. Blom, E.; Langstrom, B.; Velikyan, I. ^{68}Ga-labeling of biotin analogues and their characterization. *Bioconjugate Chem.* **2009**, *20*, 1146–1151.
8. Eriksson, O.; Carlsson, F.; Blom, E.; Sundin, A.; Langstrom, B.; Korsgren, O.; Velikyan, I. Preclinical evaluation of a ^{68}Ga-labeled biotin analogue for applications in islet transplantation. *Nucl. Med. Biol.* **2012**, *39*, 415–421.
9. Selvaraju, R.K.; Velikyan, I.; Johansson, L.; Wu, Z.; Todorov, I.; Shively, J.; Kandeel, F.; Korsgren, O.; Eriksson, O. *In vivo* imaging of the glucagonlike peptide 1 receptor in the pancreas with ^{68}Ga-labeled do3a-exendin-4. *J. Nucl. Med.* **2013**, *54*, 1458–1463.
10. Velikyan, I.; Sundin, A.; Eriksson, B.; Lundqvist, H.; Sorensen, J.; Bergstrom, M.; Langstrom, B. *In vivo* binding of [^{68}Ga]-dotatoc to somatostatin receptors in neuroendocrine tumours—Impact of peptide mass. *Nucl. Med. Biol.* **2010**, *37*, 265–275.
11. Sorensen, J.; Velikyan, I.; Wennborg, A.; Feldwisch, J.; Tolmachev, V.; Sandberg, D.; Nilsson, G.; Olofsson, H.; Sandstrom, M.; Lubberink, M.; *et al.* Measuring her2-expression in metastatic breast cancer using ^{68}Ga-aby025 pet/ct. *Eur. J. Nucl. Med. Mol. Imaging* **2014**, *41*, S226.
12. Velikyan, I.; Wennborg, A.; Feldwisch, J.; Orlova, A.; Tolmachev, V.; Lubberink, M.; Sandstrom, M.; Lindman, H.; Carlsson, J.; Sorensen, J. Gmp compliant preparation of a ^{68}Gallium-labeled affibody analogue for breast cancer patient examination: First-in-man. *Eur. J. Nucl. Med. Mol. Imaging* **2014**, *41*, S228–S229.
13. Bé, M.M.; Schönfeld, E. Table de Radionucléide. 2012. Available online: http://www.nucleide.org/DDEP_WG/DDEPdata.htm (accessed on 25 June 2015).
14. McCutchan, E.A. Nuclear Data Sheets for A = 68. *Nucl. Data Sheets* **2012**, *113*, 1735–1870.
15. García-Toraño, E.; Peyrés Medina, V.; Romero, E.; Roteta, M. Measurement of the half-life of ^{68}Ga. *Appl. Radiat. Isot.* **2014**, *87*, 122–125.
16. European Pharmacopeia 7.7 (01/2013:2482 Gallium (68Ga) Edotreotide injection). *Eur Pharm.* **2011**, *23*, 310–313.

17. European Pharmacopeia. Gallium (68 Ga) chloride solution for radiolabelling. European Directorate for the Quality of Medicines. *Eur Pharm.* **2013**, *2464*, 1060–1061.
18. Velikyan, I.; Beyer, G.J.; Langstrom, B. Microwave-supported preparation of ^{68}Ga-bioconjugates with high specific radioactivity. *Bioconjugate Chem.* **2004**, *15*, 554–560.
19. Rosch, F. Past, present and future of ^{68}Ge/^{68}Ga generators. *Appl. Radiat. Isot.* **2013**, *76*, 24–30.
20. Nakayama, M.; Haratake, M.; Koiso, T.; Ishibashi, O.; Harada, K.; Nakayama, H.; Sugii, A.; Yahara, S.; Arano, Y. Separation of Ga-68 from Ge-68 using a macroporous organic polymer containing n-methylglucamine groups. *Anal. Chim. Acta* **2002**, *453*, 135–141.
21. Le, V.S. ^{68}Ga generator integrated system: Elution-purification-concentration integration. *Recent Results Cancer Res.* **2013**, *194*, 43–75.
22. Saha Das, S.; Chattopadhyay, S.; Alam, M.; Madhusmita; Barua, L.; Das, M.K. Preparation and evaluation of sno2-based ^{68}Ge/^{68}Ga generator made from ^{68}Ge produced through natzn(α,xn) reaction. *Appl. Radiat. Isot.* **2013**, *79*, 42–47.
23. Chakravarty, R.; Chakraborty, S.; Ram, R.; Dash, A.; Pillai, M.R.A. Long-term evaluation of "barc ^{68}Ge/^{68}Ga generator" based on the nanoceria-polyacrylonitrile composite sorbent. *Cancer Biother. Radiopharm.* **2013**, *28*, 631–637.
24. Gleason, G.I. A positron cow. *Int. J. Appl. Radiat. Isot.* **1960**, *8*, 90–94.
25. Yano, Y. Radiopharmaceuticals from Generator-Produced Radionuclides; In *Preparation and Control of ^{68}Ga Radiopharmaceuticals*; International Atomic Energy Agency: Vienna, Austria, 1971; pp. 117–125.
26. De Blois, E.; Chan, H.S.; Roy, K.; Krenning, E.P.; Breeman, W.A.P. Reduction of ^{68}Ge activity containing liquid waste from ^{68}Ga pet chemistry in nuclear medicine and radiopharmacy by solidification. *J. Radioanal. Nucl. Chem.* **2011**, *288*, 303–306.
27. Petrik, M.; Schuessele, A.; Perkhofer, S.; Lass-Florl, C.; Becker, D.; Decristoforo, C. Microbial challenge tests on nonradioactive tio2-based ^{68}Ge/^{68}Ga generator columns. *Nucl. Med. Commun.* **2012**, *33*, 819–823.
28. Breeman, W.A.; de Jong, M.; de Blois, E.; Bernard, B.F.; Konijnenberg, M.; Krenning, E.P. Radiolabelling dota-peptides with ^{68}Ga. *Eur. J. Nucl. Med. Mol. Imaging* **2005**, *32*, 478–485.
29. Schultz, M.K.; Mueller, D.; Baum, R.P.; Leonard Watkins, G.; Breeman, W.A. A new automated nacl based robust method for routine production of gallium-68 labeled peptides. *Appl. Radiat. Isot.* **2013**, *76*, 46–54.
30. Velikyan, I.; Xu, H.; Nair, M.; Hall, H. Robust labeling and comparative preclinical characterization of dota-toc and dota-tate. *Nucl. Med. Biol.* **2012**, *39*, 628–659.
31. Mueller, D.; Klette, I.; Baum, R.P.; Gottschaldt, M.; Schultz, M.K.; Breeman, W.A.P. Simplified nacl based ^{68}Ga concentration and labeling procedure for rapid synthesis of ^{68}Ga radiopharmaceuticals in high radiochemical purity. *Bioconjugate Chem.* **2012**, *23*, 1712–1717.
32. Loktionova, N.S.; Belozub, A.N.; Filosofov, D.V.; Zhernosekov, K.P.; Wagner, T.; Turler, A.; Rosch, F. Improved column-based radiochemical processing of the generator produced ^{68}Ga. *Appl. Radiat. Isot.* **2011**, *69*, 942–946.
33. Gebhardt, P.; Opfermann, T.; Saluz, H.P. Computer controlled Ga-68 milking and concentration system. *Appl. Radiat. Isot.* **2010**, *68*, 1057–1059.

34. Zhernosekov, K.P.; Filosofov, D.V.; Baum, R.P.; Aschoff, P.; Bihl, H.; Razbash, A.A.; Jahn, M.; Jennewein, M.; Rosch, F. Processing of generator-produced ^{68}Ga for medical application. *J. Nucl. Med.* **2007**, *48*, 1741–1748.
35. Meyer, G.J.; Macke, H.; Schuhmacher, J.; Knapp, W.H.; Hofmann, M. ^{68}Ga-labelled dota-derivatised peptide ligands. *Eur. J. Nucl. Med. Mol. Imaging* **2004**, *31*, 1097–1104.
36. Eppard, E.; Wuttke, M.; Nicodemus, P.L.; Rosch, F. Ethanol-based post-processing of generator-derived ^{68}Ga toward kit-type preparation of ^{68}Ga-radiopharmaceuticals. *J. Nucl. Med.* **2014**, *55*, 1023–1028.
37. Petrik, M.; Ocak, M.; Rupprich, M.; Decristoforo, C. Impurity in ^{68}Ga-peptide preparation using processed generator eluate. *J. Nucl. Med.* **2010**, *51*, 495.
38. Bauwens, M.; Chekol, R.; Vanbilloen, H.; Bormans, G.; Verbruggen, A. Optimal buffer choice of the radiosynthesis of Ga-68-dotatoc for clinical application. *Nucl. Med. Commun.* **2010**, *31*, 753–758.
39. Velikyan, I.; Beyer, G.J.; Bergstrom-Pettermann, E.; Johansen, P.; Bergstrom, M.; Langstrom, B. The importance of high specific radioactivity in the performance of ^{68}Ga-labeled peptide. *Nucl. Med. Biol.* **2008**, *35*, 529–536.
40. Velikyan, I.; Maecke, H.; Langstrom, B. Convenient preparation of ^{68}Ga-based pet-radiopharmaceuticals at room temperature. *Bioconjugate Chem.* **2008**, *19*, 569–573.
41. Notni, J.; Simecek, J.; Hermann, P.; Wester, H.J. Trap, a powerful and versatile framework for Gallium-68 radiopharmaceuticals. *Chem. Eur. J.* **2011**, *17*, 14718–14722.
42. Oehlke, E.; Le, V.S.; Lengkeek, N.; Pellegrini, P.; Jackson, T.; Greguric, I.; Weiner, R. Influence of metal ions on the ^{68}Ga-labeling of dotatate. *Appl. Radiat. Isot.* **2013**, *82*, 232–238.
43. Simecek, J.; Hermann, P.; Wester, H.J.; Notni, J. How is ^{68}Ga labeling of macrocyclic chelators influenced by metal ion contaminants in ^{68}Ge/^{68}Ga generator eluates? *ChemMedChem* **2013**, *8*, 95–103.
44. Chakravarty, R.; Chakraborty, S.; Dash, A.; Pillai, M.R.A. Detailed evaluation on the effect of metal ion impurities on complexation of generator eluted ^{68}Ga with different bifunctional chelators. *Nucl. Med. Biol.* **2013**, *40*, 197–205.
45. Zhai, C.; Summer, D.; Rangger, C.; Haas, H.; Haubner, R.; Decristoforo, C. Fusarinine c, a novel siderophore-based bifunctional chelator for radiolabeling with Gallium-68. *J. Label. Compd. Radiopharm.* **2015**, *58*, 209–214.
46. Borges, J.B.; Velikyan, I.; Långström, B.; Sörensen, J.; Ulin, J.; Maripuu, E.; Sandström, M.; Widström, C.; Hedenstierna, G. Ventilation distribution studies comparing technegas and gallgas using ^{68}GaCl$_3$ as the label. *J. Nucl. Med.* **2011**, *52*, 206–209.
47. Velikyan, I.; Sundin, A.; Sörensen, J.; Lubberink, M.; Sandström, M.; Garske-Román, U.; Lundqvist, H.; Granberg, D.; Eriksson, B. Quantitative and qualitative intrapatient comparison of ^{68}Ga-dotatoc and ^{68}Ga-dotatate: Net uptake rate for accurate quantification. *J. Nucl. Med.* **2014**, *55*, 204–210.
48. Sandström, M.; Velikyan, I.; Garske-Román, U.; Sörensen, J.; Eriksson, B.; Granberg, D.; Lundqvist, H.; Sundin, A.; Lubberink, M. Comparative biodistribution and radiation dosimetry of ^{68}Ga-dotatoc and ^{68}Ga-dotatate in patients with neuroendocrine tumors. *J. Nucl. Med.* **2013**, *54*, 1755–1759.

49. Reubi, J.C.; Schär, J.C.; Waser, B.; Wenger, S.; Heppeler, A.; Schmitt, J.S.; Mäcke, H.R. Affinity profiles for human somatostatin receptor subtypes sst1-sst5 of somatostatin radiotracers selected for scintigraphic and radiotherapeutic use. *Eur. J. Nucl. Med. Mol. Imaging* **2000**, *27*, 273–282.
50. Eriksson, O.; Velikyan, I.; Selvaraju, R.K.; Kandeel, F.; Johansson, L.; Antoni, G.; Eriksson, B.; Sörensen, J.; Korsgren, O. Detection of metastatic insulinoma by positron emission tomography with [^{68}Ga]exendin-4-a case report. *J. Clin. Endocrinol. Metab.* **2014**, *99*, 1519–1524.
51. Aime, S.; Barge, A.; Botta, M.; Fasano, M.; Danilo Ayala, J.; Bombieri, G. Crystal structure and solution dynamics of the lutetium(iii) chelate of dota. *Inorg. Chim. Acta* **1996**, *246*, 423–429.
52. Heppeler, A.; Froidevaux, S.; Macke, H.R.; Jermann, E.; Behe, M.; Powell, P.; Hennig, M. Radiometal-labelled macrocyclic chelator-derivatised somatostatin analogue with superb tumour-targeting properties and potential for receptor-mediated internal radiotherapy. *Chem. Eur. J.* **1999**, *5*, 1974–1981.
53. Velikyan, I.; Bulenga, T.N.; Selvaraju, K.R.; Lubberink, M.; Espes, D.; Rosenstrom, U.; Eriksson, O. Dosimetry of [177Lu]-DO3A-VS-Cys40-Exendin-4—impact on the feasibility of insulinoma internal radiotherapy. *Am. J. Nucl. Med. Mol. Imaging* **2015**, *5*, 109–126.
54. Selvaraju, R.; Bulenga, T.N.; Espes, D.; Lubberink, M.; Sörensen, J.; Eriksson, B.; Estrada, S.; Velikyan, I.; Eriksson, O. Dosimetry of [^{68}Ga]Ga-DO3A-VS-Cys40-Exendin-4 in rodents, pigs, non-human primates and human-repeated scanning in human is possible. *Am. J. Nucl. Med. Mol. Imaging* **2015**, *5*, 259–269.
55. Decristoforo, C.; Penuelas, I.; Elsinga, P.; Ballinger, J.; Winhorst, A.D.; Verbruggen, A.; Verzijlbergen, F.; Chiti, A. Radiopharmaceuticals are special, but is this recognized? The possible impact of the new clinical trials regulation on the preparation of radiopharmaceuticals. *Eur. J. Nucl. Med. Mol. Imaging* **2014**, *41*, 2005–2007.
56. Decristoforo, C.; Peñuelas, I. Towards a harmonized radiopharmaceutical regulatory framework in europe? *Q. J. Nucl. Med. Mol. Imaging* **2009**, *53*, 394–401.
57. Verbruggen, A.; Coenen, H.H.; Deverre, J.R.; Guilloteau, D.; Langstrom, B.; Salvadori, P.A.; Halldin, C. Guideline to regulations for radiopharmaceuticals in early phase clinical trials in the EU. *Eur. J. Nucl. Med. Mol. Imaging* **2008**, *35*, 2144–2151.
58. Norenberg, J.P.; Petry, N.A.; Schwarz, S. Operation of a radiopharmacy for a clinical trial. *Semin. Nucl. Med.* **2010**, *40*, 347–356.
59. Lange, R.; Ter Heine, R.; Decristoforo, C.; Peñuelas, I.; Elsinga, P.H.; van Der Westerlaken, M.M.L.; Hendrikse, N.H. Untangling the web of european regulations for the preparation of unlicensed radiopharmaceuticals: A concise overview and practical guidance for a risk-based approach. *Nucl. Med. Commun.* **2015**, *36*, 414–422.
60. Aerts, J.; Ballinger, J.R.; Behe, M.; Decristoforo, C.; Elsinga, P.H.; Faivre-Chauvet, A.; Mindt, T.L.; Peitl, P.K.; Todde, S.C.; Koziorowski, J. Guidance on current good radiopharmacy practice for the small-scale preparation of radiopharmaceuticals using automated modules: A european perspective. *J. Label. Compd. Radiopharm.* **2014**, *57*, 615–620.
61. Elsinga, P.; Todde, S.; Penuelas, I.; Meyer, G.; Farstad, B.; Faivre-Chauvet, A.; Mikolajczak, R.; Westera, G.; Gmeiner-Stopar, T.; Decristoforo, C. Guidance on current good radiopharmacy practice (cgrpp) for the small-scale preparation of radiopharmaceuticals. *Eur. J. Nucl. Med. Mol. Imaging* **2010**, *37*, 1049–1062.

62. Virgolini, I.; Ambrosini, V.; Bomanji, J.B.; Baum, R.P.; Fanti, S.; Gabriel, M.; Papathanasiou, N.D.; Pepe, G.; Oyen, W.; de Cristoforo, C.; *et al.* Procedure guidelines for pet/ct tumour imaging with ^{68}Ga-dota-conjugated peptides: ^{68}Ga-dota-toc, ^{68}Ga-dota-noc, ^{68}Ga-dota-tate. *Eur. J. Nucl. Med. Mol. Imaging* **2010**, *37*, 2004–2010.
63. Janson, E.T.; Sorbye, H.; Welin, S.; Federspiel, B.; Gronbaek, H.; Hellman, P.; Ladekarl, M.; Langer, S.W.; Mortensen, J.; Schalin-Jantti, C.; *et al.* Nordic guidelines 2014 for diagnosis and treatment of gastroenteropancreatic neuroendocrine neoplasms. *Acta Oncol.* **2014**, *53*, 1284–1297.
64. Marchetti, S.; Schellens, J.H.M. The impact of fda and emea guidelines on drug development in relation to phase 0 trials. *Br. J. Cancer* **2007**, *97*, 577–581.
65. Mills, G. The exploratory ind. *J. Nucl. Med.* **2008**, *49*, 45N–47N.
66. Todde, S.; Windhorst, A.D.; Behe, M.; Bormans, G.; Decristoforo, C.; Faivre-Chauvet, A.; Ferrari, V.; Gee, A.D.; Gulyas, B.; Halldin, C.; *et al.* Eanm guideline for the preparation of an investigational medicinal product dossier (impd). *Eur. J. Nucl. Med. Mol. Imaging* **2014**, *41*, 2175–2185.
67. Lappin, G.; Kuhnz, W.; Jochemsen, R.; Kneer, J.; Chaudhary, A.; Oosterhuis, B.; Drijfhout, W.J.; Rowland, M.; Garner, R.C. Use of microdosing to predict pharmacokinetics at the therapeutic dose: Experience with 5 drugs. *Clin. Pharmacol. Ther.* **2006**, *80*, 203–215.
68. Garner, R.C.; Lappin, G. The phase 0 microdosing concept. *Br. J. Clin. Pharmacol.* **2006**, *61*, 367–370.
69. Bergstrom, M.; Grahnen, A.; Langstrom, B. Positron emission tomography microdosing: A new concept with application in tracer and early clinical drug development. *Eur. J. Clin. Pharmacol.* **2003**, *59*, 357–366.
70. *Guidance for Industry, Investigators, and Reviewers*; Exploratory ind Studies. U.S. Department of Health and Human Services: Washington, DC, USA, 2006.
71. Decristoforo, C.; Pickett, R.D.; Verbruggen, A. Feasibility and availability of ^{68}Ga-labelled peptides. *Eur. J. Nucl. Med. Mol. Imaging* **2012**, *39*, S31–S40.
72. Velikyan, I.; Antoni, G.; Sorensen, J.; Estrada, S. Organ biodistribution of germanium-68 in rat in the presence and absence of [^{68}Ga]ga-dota-toc for the extrapolation to the human organ and whole-body radiation dosimetry. *Am. J. Nucl. Med. Mol. Imaging* **2013**, *3*, 154–165.
73. International Atomic Energy Agency. Development of Ga-68 Based Pet-Radiopharmaceuticals for Management of Cancer and Other Chronic Diseases. Available online: http://cra.iaea.org/cra/explore-crps/all-active-by-programme.html (accessed on 25 June 2015).
74. Directive 2001/83/ec of the European Parliament and of the Council of 6 November 2001 on the Community Code Relating to Medicinal Products for Human Use. Available online: http://ec.europa.eu/health/files/eudralex/vol-1/dir_2001_83_consol_2012/dir_2001_83_cons_2012_en.pdf (Updated 2012) (accessed on 25 June 2015).
75. Decristoforo, C. Gallium-68—A new opportunity for pet available from a long shelflife generator—automation and applications. *Curr. Radiopharm.* **2012**, *5*, 212–220.
76. Boschi, S.; Lodi, F.; Malizia, C.; Cicoria, G.; Marengo, M. Automation synthesis modules review. *Appl. Radiat. Isot.* **2013**, *76*, 38–45.

77. Boschi, S.; Malizia, C.; Lodi, F. Overview and perspectives on automation strategies in ^{68}Ga radiopharmaceutical preparations. *Recent Results Cancer Res.* **2013**, *194*, 17–31.

© 2015 by the authors; licensee MDPI, Basel, Switzerland. This article is an open access article distributed under the terms and conditions of the Creative Commons Attribution license (http://creativecommons.org/licenses/by/4.0/).

MDPI
St. Alban-Anlage 66
4052 Basel
Switzerland
www.mdpi.com

Molecules Editorial Office
E-mail: molecules@mdpi.com
www.mdpi.com/journal/molecules

Disclaimer/Publisher's Note: The statements, opinions and data contained in all publications are solely those of the individual author(s) and contributor(s) and not of MDPI and/or the editor(s). MDPI and/or the editor(s) disclaim responsibility for any injury to people or property resulting from any ideas, methods, instructions or products referred to in the content.

www.ingramcontent.com/pod-product-compliance
Lightning Source LLC
LaVergne TN
LVHW070418100526
838202LV00014B/1484